T0214497

Lecture Notes in Computer Science   11473

More information about this series at http://www.springer.com/series/7409

Jianhui Li · Xiaofeng Meng ·
Ying Zhang · Wenjuan Cui ·
Zhihui Du (Eds.)

Big Scientific Data Management

First International Conference, BigSDM 2018
Beijing, China, November 30 – December 1, 2018
Revised Selected Papers

 Springer

Editors
Jianhui Li
CNIC, CAS
Beijing, China

Xiaofeng Meng
Renmin University of China
Beijing, China

Ying Zhang
Centrum Wiskunde and Informatica
Amsterdam, The Netherlands

Wenjuan Cui
CNIC, CAS
Beijing, China

Zhihui Du ⓘ
Tsinghua University
Beijing, China

ISSN 0302-9743   ISSN 1611-3349 (electronic)
Lecture Notes in Computer Science
ISBN 978-3-030-28060-4   ISBN 978-3-030-28061-1 (eBook)
https://doi.org/10.1007/978-3-030-28061-1

LNCS Sublibrary: SL3 – Information Systems and Applications, incl. Internet/Web, and HCI

This Springer imprint is published by the registered company Springer Nature Switzerland AG
The registered company address is: Gewerbestrasse 11, 6330 Cham, Switzerland

Preface

It is our great pleasure to present the proceedings of the First International Conference on Big Scientific Data Management (BigSDM 2018). BigSDM 2018 was jointly organized by the Chinese Academy of Sciences, Peking University, Tsinghua University, Renmin University of China, CWI, and MonetDB Solutions. BigSDM 2018 provided an open forum for international experts from natural and computer sciences to explore the life-cycle management of big scientific data, to discuss major methodologies and key technologies in big data-driven discovery making, to exchange experience and knowledge about big scientific data across various fields through specific cases, and to identify the challenges and opportunities in scientific discovery acceleration.

BigSDM 2018 was held in Beijing, China, during November 30 – December 1, 2018. The theme of BigSDM 2018 was data-driven scientific discovery, focusing on the best practices, standards, architectures, and data infrastructures for fostering cross-disciplinary discovery and innovation. The topics involved application cases in the big scientific data management, paradigms for enhancing scientific discovery through big data, data management challenges posed by big scientific data, machine learning methods to facilitate scientific discovery, science platforms and storage systems for large scale scientific applications, data cleansing and quality assurance of science data, and data policies.

We received 86 submissions, each of which was assigned to at least 3 Program Committee (PC) members to review. The peer review process was double-blind. Each paper was judged by its relevance to the conference, technical merits, innovations, and presentation. After the thoughtful discussions by the PC, 24 full research papers (an acceptance rate of 27.91%) and 7 short papers were selected. The conference program included keynote presentations by Prof. Barry C. Barish (the winner of 2017 Nobel Prize in Physics), Prof. Huadong Guo (Academician of Chinese Academy of Sciences), and Prof. Alexander S. Szalay (Bloomberg distinguished professor of the Johns Hopkins University). The program of BigSDM 2018 also included several talks by the principal researchers from major scientific experiments, such as the Five-hundred-meter Aperture Spherical radio Telescope (FAST), and more than 15 topic-specific talks by famous experts in big scientific management, to share their cutting edge technology and views about the big scientific data management.

We are grateful to the general chairs, Prof. Xiaofeng Meng (Renmin University of China), Prof. Martin Kersten (CWI, University of Amsterdam, MonetDB Solutions), and Prof. Xuebin Chi (Computer Network Information Center, Chinese Academy of Sciences), as well as to all the members of the Organization Committee, for their efforts in the organization of the conference. We are thankful to all the PC members who contributed their time and expertise to the paper reviewing process. In particular, we would like to thank the local Organization Chairs, Prof. Zhihong Shen (Computer Network Information Center, Chinese Academy of Sciences) and Hui Zhang

(Computer Network Information Center, Chinese Academy of Sciences), as well as many volunteers, for their great efforts in the arrangement of the conference venue, accommodation, and all the other relating issues. Without their support, the conference would not have been successful. We would also like to thank the authors who submitted their papers to the conference, and all the presenters who shared their knowledge and experiences at the conference.

February 2019   Jianhui Li
                                                                                                       Wenjuan Cui

Organization

General Chairs

Xiaofeng Meng   Renmin University of China, China
Martin Kersten   University of Amsterdam, The Netherlands
Xuebin Chi   Computer Network Information Center, CAS, China

Program Committee Chairs

Jianhui Li   Computer Network Information Center, CAS, China
Ying Zhang   MonetDB Solutions, The Netherlands
Bill Howe   University of Washington, USA

Session Chairs

Data Policy and Infrastructure

Mingqi Chen   Chinese Academy of Sciences, China

Big Science

Jianyan Wei   National Astronomical Observatories, CAS, China

Machine Learning for Discovery

Ying Zhang   MonetDB Solutions, The Netherlands

Big Scientific Data Management

Juncai Ma   Institute of Microbiology, CAS, China

Local Organization Chairs

Zhihong Shen   Computer Network Information Center, CAS, China
Hui Zhang   Computer Network Information Center, CAS, China

Publicity Chairs

Xin Chen   Computer Network Information Center, CAS, China
Yang Zhang   Computer Network Information Center, CAS, China

Publication Chairs

Lei Zou   Peking University, China
Wenjuan Cui   Computer Network Information Center, CAS, China

Program Committee

Zhihui Du	Tsinghua University, China
Jianfeng Zhan	Institute of Computing Technology, CAS, China
Yong Qi	Xi'an Jiaotong University, China
Ziming Zou	National Space Science Center, CAS, China
Guoqing Li	Institute of Remote Sensing and Digital Earth, CAS, China
Chenzhou Cui	National Astronomical Observatories, CAS, China
Juncai Ma	Institute of Microbiology, CAS, China
Jungang Xu	University of Chinese Academy of Sciences, China
Ying Liu	University of Chinese Academy of Sciences, China
Wei Wang	Institute of Software, CAS, China
Lei Liu	Fudan University, China
Yangyong Zhu	Fudan University, China
Yaodong Cheng	Institute of High Energy Physics, CAS, China
Linhuan Wu	Institute of Microbiology, CAS, China
Hu Zhan	National Astronomical Observatories, CAS, China
Zhifei Zhang	Capital Medical University, China
Haibo Li	Institute of High Energy Physics, CAS, China
Li Zha	Institute of Computing Technology, CAS, China
Lizhen Cui	Shandong University, China
Yue Liu	Shanghai University, China
Stefan Manegold	Leiden University, CWI, The Netherlands
Manolis Marazakis	FORTH, Greece
Mourad Khayati	University of Fribourg, Switzerland
Nan Tang	Qatar Computing Research Institute, Qatar
Georgios Goumas	National Technical University of Athens, Greece
Vasileios Karakostas	National Technical University of Athens, Greece
Romulo Pereira Gonçalves	Netherlands eScience Center, The Netherlands
Rob van Nieuwpoort	Netherlands eScience Center, The Netherlands
Ben van Werkhoven	Netherlands eScience Center, The Netherlands
Ioannis Athanasiadis	Wageningen University & Research, The Netherlands
Bart Scheers	Data Spex, The Netherlands
Edwin A. Valentijn	University of Groningen, The Netherlands
Andrea Borga	Oliscience, The Netherlands
Hannes Mühleisen	Centrum Wiskunde & Informatica, The Netherlands
Raül Sirvent Pardell	Barcelona Supercomputing Center, Spain
Daniele Lezzi	Barcelona Supercomputing Center, Spain
Marta Patiño	Universidad Politécnica de Madrid, Spain
Tony Hey	Science and Technology Facilities Council, UK
Holger Pirk	Imperial College London, UK
Thomas Heinis	Imperial College London, UK
John Goodacre	University of Manchester and ARM Ltd., UK
Julian Chesterfield	OnApp, UK

Sponsors

Chinese Academy of Sciences (CAS)

National Natural Science Foundation of China (NSFC)

Ministry of Science and Technology of People's Republic of China

Contents

Gravitational Waves: Detectors to Detections

Barry C. Barish[1,2(✉)]

[1] Caltech, Pasadena, USA
barish@caltech.edu
[2] UC Riverside, Riverside, USA
barry.barish@ucr.edu

We announced the observation of gravitational waves from the Laser Interferometer Gravitational-wave Observatory (LIGO) on 11-February-2016 [1], one hundred years after Einstein first proposed the existence of gravitational waves [2, 3]. Whether gravitational waves existed or not was controversial among theorists for the first 50 years, and then the actual observation came after another fifty years to develop a detector sensitive enough to observe the tiny distortions in spacetime from gravitational waves. The instrument was developed through the LIGO Laboratory, a Caltech/MIT collaboration, and the scientific exploitation through the LIGO Scientific Collaboration (LSC), having more than a thousand scientists, from around the world, who co-author the gravitational wave observational papers. In addition, many others have made and are making important contributions to our research.

In this lecture, I describe the LIGO project and, the improvements that led to detection of merging black holes in Advanced LIGO (Fig. 1). I also describe some key features of the interferometers, some implications of the discoveries, and finally, I comment on the evolution of LIGO planned for the coming years.

The basic scheme used in LIGO employs a special high power stabilized single-line NdYAG laser that enters the interferometer and is split into two beams transported in perpendicular directions. The LIGO vacuum pipe is 1.2 m diameter and is at high vacuum (10^{-9} tor). The 'test' masses serve as very high-quality interferometer mirrors that are suspended, in order to keep them isolated from the earth. The mirrors are made of fused Silica and are hung in a four-stage pendulum for Advanced LIGO.

In the simplest version of a suspended mass interferometer, the equal length arms are adjusted such that the reflected light from mirrors at the far ends arrive back at the same time, and inverting one, the two beams cancel each other and no light is recorded in the photodetector. This is the normal state of the interferometer working at the 'dark port.' Many effects make the beams not completely cancel and the actual optical configuration is more sophisticated.

When a gravitational wave crosses the interferometer, it stretches one arm and compresses the other, alternating at the frequency of the gravitational wave. Consequently, the light from the two arms returns at slightly different times (or phase) and the two beams no longer completely cancel. The process reverses itself, stretching the other arm and squeezing the initial arm at the frequency of the gravitational wave. The resulting frequency and time-dependence of signal is recorded from a photo-sensor. This signal is the waveform from the passage of a gravitational waves and it is directly compared with the predictions of general relativity. The experimental challenge is to

© Springer Nature Switzerland AG 2019
J. Li et al. (Eds.): BigSDM 2018, LNCS 11473, pp. 1–5, 2019.
https://doi.org/10.1007/978-3-030-28061-1_1

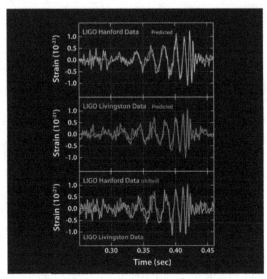

Image Credit: Caltech/MIT/LIGO Lab

Fig. 1. The gravitational wave discovery figure showing the waveforms as they were observed by the LIGO Scientific Collaboration (LSC) within minutes after the event was recorded in Advanced LIGO. Each of the three figures show the detected "strain" signals in units of 10^{-21} vs time. The top trace is the observed waveform detected in the Hanford, Washington interferometer, the middle trace is the observed waveform in Livingston, Louisiana The two signals are almost identical, but are shifted by 6.9 ms and have been superposed in the bottom trace.

make the interferometer sensitive enough to record the incredibly tiny distortions of spacetime that come from a gravitational wave over the various background noise sources.

The spacetime distortions from the passage of an astrophysical source are expected to be of the order of $h = \Delta L/L \sim 10^{-21}$, a difference in length that is a tiny fraction of the size of a proton. In LIGO, we have made the length of the interferometer arms as long as is practical (4-km), and this results in a difference in length of about 10^{-18} m. We achieve this precision by using very sophisticated instrumentation to reduce the seismic and thermal noise sources, and by effectively making the statistics very high by having many photons traversing the interferometer arms.

We built the initial version of LIGO between 1994 and 2000, at which time we began to search for gravitational waves with sensitivities beyond that of any previous instruments. We improved the interferometer sensitivity in steps over the next ten years, alternating improving the detector and searching for gravitational waves. We failed to observe gravitational waves during this period, and embarked on a major upgrade to Advanced LIGO in about 2010.

The Advanced LIGO laser is a multi-stage Nd:YAG laser. Our goal for Advanced LIGO is to raise the power from ~ 18 W used in Initial LIGO laser in steps to 180 W for the Advanced LIGO laser, improving the high frequency sensitivity, accordingly. The pre-stabilized laser system consists of the laser and a control system to stabilize the laser in frequency, beam direction, and intensity. For the results presented here, due to stability issues, heating and scattered-light effects, the laser power has only been increased modestly over Initial LIGO. We will bring the power up systematically in steps for future data runs, carefully eliminating issues from scattered light, heating mirrors, etc.

The key improvement in Advanced LIGO that enabled the detection of the black hole merger event was the implementation of an active seismic isolation and quadruple suspension system (Fig. 2).

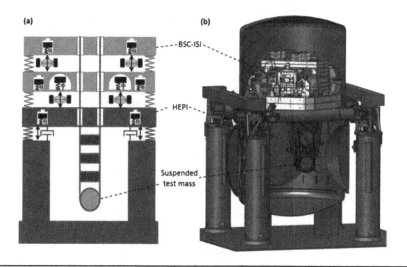

Fig. 2. Advanced LIGO multi-stage suspension system for the test masses with active-passive seismic

The multiple suspension system moved all active components off the final test masses, resulting in better isolation. Initial LIGO used 25-cm, 11-kg, fused-silica test masses, while for Advanced LIGO the test masses are 34 cm in diameter to reduce thermal noise contributions and are 40 kg, which reduces the radiation pressure noise to a level comparable to the suspension thermal noise. The test mass is suspended by fused silica fibers, rather than the steel wires used in initial LIGO. The complete suspension system has four pendulum stages, increasing the seismic isolation and providing multiple points for actuation.

The active seismic isolation senses motion and is combined with the passive seismic isolation using servo techniques to improve the low frequency sensitivity by a factor of x100. Since the rate for gravitational events scales with the volume, this improvement increases the rate for events by 10^6. This improvement enabled Advanced LIGO to make a detection of a black hole merger in days, while Initial LIGO failed to detect gravitational waves in years of data taking.

The observation of the first Black Hole merger by Advanced LIGO [4–7] was made 14 September 2015. Figure 1 shows the data, and Fig. 3 reveals the key features of the observed compact binary merger. These are a result of the analysis of the observed event shown in Fig. 1. At the top of Fig. 3, the three phases of the coalescence (inspiral, merger, and ringdown) are indicated above the wave forms. As the objects inspiral together, more and more gravitational waves are emitted and the frequency and amplitude of the signal increases (the characteristic chirp signal). This is following by the final merger, and then, the merged single object rings down. The bottom pane shows on the left scale that the objects are highly relativistic and are moving at more than 0.5 the speed of light by the time of the final coalescence. On the right side, the scale is units of Schwarzschild radii and indicate that the objects are very compact, only a few hundred kilometers apart when they enter our frequency band.

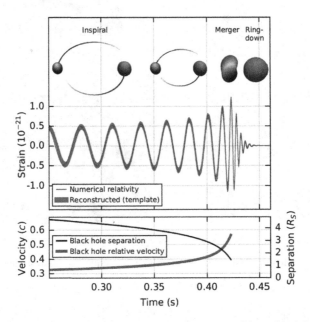

Fig. 3. The physics interpretation of the observed event as a binary black hole merger

By comparing and fitting our waveform to General Relativity, we have concluded that we have observed the merger of two heavy compact objects (black holes), each ~ 30 times the mass of the sun, and going around each other separated by only a few hundred kilometers and moving at relativistic velocities.

On the top of the figure, we see the three different phases of the merger: the inspiral phase of the binary black holes system, then their merger of the two compact objects, and finally, there is a ringdown. The characteristic increasing frequency and amplitude with time (the so-called chirp signal) can be seen in the inspiral phase. The largest amplitude is during the final merger, and finally, there is a characteristic ringdown frequency. The two Black Holes inspiral and merge together due to the emission of gravitational radiation coming from the acceleration.

The bottom of Fig. 3 is even more revealing. The right-hand axis is basically in units of about a hundred kilometers. That means the separation between the two merging objects at the beginning was only ~ 400 km, and at the end ~ 100 km. Therefore, these ~ 30 Solar Mass objects are confined in a volume only about twice the size of Stockholm, yet the final Black Hole has 60 times the mass of our sun, or 10 million times as massive as the earth. From the axis on the left, we see that when we first observed this event, the two objects were moving at about three-tenths the speed of light and that increases to over half the speed of light by the time of the final merges!

Since announcing our observation of GW150914, we have detected a total of ten black hole mergers, and one neutron star merger. These observations establish the new field of gravitational wave astronomy, as well as providing new important tests of general relativity. As we look to the future, we expect to detect more types of phenomena giving gravitational wave signals, and eventually open up a new cosmology with gravitational waves.

References

1. Abbott, B.P., et al.: Observation of gravitational waves from a binary black hole merger. Phys. Rev. Lett. **116**, 061102 (2016)
2. Einstein, A.: Approximative integration of the field equations of gravitation. Sitzungsber. Preuss. Akad. Wiss. Berlin (Math. Phys.) **1916**, 688–696 (1916)
3. Einstein, A.: Über Gravitationswellen. Sitzungsber. Preuss. Akad. Wiss. Berlin (Math. Phys.) **1918**, 154–167 (1918)
4. Abbott, B.P., et al.: GW150914: the advanced LIGO detectors in the era of first discoveries. Phys. Rev. Lett. **116**, 131103 (2016)
5. Abbott, B.P., et al.: GW150914: first results from the search for binary black hole coalescence with advanced LIGO. Phys. Rev D **93**, 122003 (2016)
6. Abbott, B.P., et al.: Tests of general relativity with GW150914. Phys. Rev. Lett. **116**, 221101 (2016)
7. Abbot, B.P., et al.: Astrophysical implications of the binary black hole merger GW150914. Astrophys. J. Lett. **833**, 1 (2016)

Big Data Challenges of FAST

Youling Yue[1,2] and Di Li[1,2,3(✉)]

[1] National Astronomical Observatories, Chinese Academy of Sciences,
Beijing 100101, People's Republic of China
dili@nao.cas.cn
[2] CAS Key Laboratory of FAST, NAOC,
Beijing 100101, People's Republic of China
[3] University of Chinese Academy of Sciences,
Beijing 100049, People's Republic of China

Abstract. We present the big-data challenges posed by the science operation of the Five-hundred-meter Aperture Spherical radio Telescope (FAST). Unlike the common usage of the word 'big-data', which tend to emphasize both quantity and diversity, the main characteristics of FAST data stream is its single-source data rate at more than 6 GB/s and the resulting data volume at about 20 PB per year. We describe here the main culprit of such a high data rate and large volume, namely pulsar search, and our solution.

Keywords: Big data · FAST · Pulsar

1 Introduction

The Five-hundred-meter Aperture Spherical radio Telescope (FAST) (Nan et al. 2011) is a Chinese mega-science project, originated from a concept for the Square Kilometer Array (SKA). Once the SKA concept moves toward small-dish-large-number, FAST becomes a stand-alone project, which was formally funded in 2007. The construction begun in 2011 and finished in 2016. The ensuing commissioning phase will come to an end in September, 2019.

With an unprecedented gain of 2000 m^2/K, superseding any antenna or antenna array ever built, FAST poises to increase volume of knowledge, i.e. the known sources and pixels of sky images, of pulsar and HI, by one order of magnitude. To accomplish these goals in an efficient manner, a Commensal Radio Astronomy FAST Survey (CRAFTS; Li et al. 2018) has been designed and tested. Utilizing the FAST L-band Array of 19-beams (FLAN) in drift-scan mode, CRAFTS aims to obtain data streams of pulsar search, HI imaging, HI galaxies, and transients (Fast Radio Burst–FRB, in particular), simultaneously. Our own innovative technologies have made such commensal observation feasible, which was not the case in past at, e.g. GBT, Arecibo, Parkes, etc. CRAFTS requires acquiring and processing data in four conceptually independent backends. The fast-varying nature of pulsar signals demand a sampling rate

J. Li et al. (Eds.): BigSDM 2018, LNCS 11473, pp. 6–9, 2019.
https://doi.org/10.1007/978-3-030-28061-1_2

of 10000 Hz or more, which makes pulsar search the main component of FAST's big-data volume. FAST is thus facing a big data challenge comparable to that of Large Synoptic Survey Telescope (LSST), which also expects data over 100 PB.

We quantify the data requirement of CRAFTS, pulsar search in particular, in Sect. 2, describe our current data solution in Sect. 3, and discuss our future plan and challenges in the final section.

2   CRAFTS Pulsar Data Challenge

The golden standard of pulsar surveys is the Parkes multi-beam pulsar survey (PMPS) (Manchester et al. 2001) started in August 1997. PMPS discovered over 800 Pulsars. It utilizes 250 μs 1 bit sampling, 96 frequency channels, 288 MHz bandwidth, 13 beams, amounting to a 0.64 MB/s data rate. The total data volume is about 4 TB, which is trivial for today's desktop.

In the two decades following the PMPS, the computing power available has grown exponentially. So does the appetite of astronomers, even outpace the industrial trend, just like Otto in "A Fish out of Walter". Assuming an increase in pulsar search data volume by 100 fold in 10 years, we should expect 6.4 GB/s data rate and 40 PB total volume for FAST pulsar survey. Forecasting technology requirements on decades time scale is necessarily unreliable, especially when Moore's law seems to be hitting the quantum limit in recent years. These numbers, however, are amazingly close to the reality of FAST.

FLAN has 19 beams, with each has 2 polarization of 500 MHz bandwidth, generating a total of 38 analog signals. With 1 GHz 8-bit sampling, this leads to 38 GB/s total raw data rate. Store raw data is not an economically feasible solution, not to mention the processing pressure. The solution is shrink the data volume with acceptable loss.

3   CRAFTS Data Processing

There are two main challenges of CRAFTS pulsar survey. One is to store, transfer, and process the data. The other is to sift through billions of candidate signals to find about 1000 new pulsars. AI techniques are now common in such tasks.

Pulsar data can be simplified as a 2D matrix or a image, usually 8-bit. Pulsar signal is repeating with characteristic patterns, but also with much variation. An example is given in Fig. 1.

The processing of pulsar data is a multi-dimensional problem. Each dimension is computation of $O(n)$ or $O(n * \log(n))$. Since most pulsar signal is week, we need to use FFT to find the period. For pulsar in binaries, the pulsar period is not exact the same, it is modulated by the orbital period. This add 4 dimensions of freedom. Searching will be computing intensive. Each dimension add the computing complexity by a factor of 100 to 1000. This increase the computing of PFLOPS at least by a factor of 10^8, i.e. 10^{23} FLOPS (0.1 YFLOPS).

First step of solving storage and transfer problem is data compression. With FLAN, the on-board processing provided by ROACH2, facilitates 4k channels,

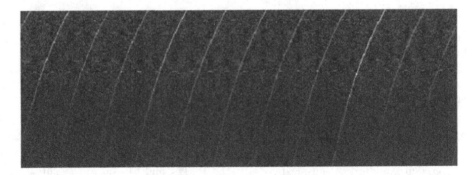

Fig. 1. A sample image of pulsar data. Each bright line a pulsar signal. This is a strong pulsar. Most pulsar signal is weaker than noise, thus need to be folded according to the period to be shown.

full polarization, 8 bit sampling, and time resolution of 8 μs. The fastest pulsar has period around 1 ms. Time resolution of 8 μs is overkill, 0.05–0.1 ms is enough and leaves some room for submillisecond pulsars. By compressing to time resolution ∼50 μs, we cut the data rate to 1/6 of the raw, thus 6.3 GB/s or 544 TB/day (24 h), which can be transferred through 100 GbE. This means around 200 PB/yr first stage data, which is still huge. It must be further compressed after the first round processing. Being irreversible, the second stage of compression has to be done with caution and necessary compromise. The current consideration is to reduce the channel number from 4k to 1k and data sampling 8 bit to 2 bit, a further compression factor of 16 can be achieved, corresponding to ∼12 PB per year. Considering maintenance time and other science data, FAST will store about 7–10 PB pulsar data per year.

4 Discussion and Conclusion

For single-dish telescopes, especially FAST, their data volumes are manageable by modern technology standards. The bigger challenge lies in optimizing the solution under budget constraints.

In data processing, pulsar search has several unsolved problems. Searching for pulsar in binaries, e.g., requires over ZFLOPS computational power, close to YFLOPS. If exponential growth, even one slower than Moore's law, still holds, pulsar search problem for FAST can be solved after 20 years or more. By then though, much larger pulsar survey will take place, e.g. phased array receiver with 100 beams or more for FAST, SKA surveys, and etc. Emerging techniques such as quantum computing, AI, new algorithms will also see wider application.

Acknowledgement. This work is supported by National Key R&D Program of China grant No. 2017YFA0402600, the National Natural Science Foundation of China grant No. 11725313 and CAS "Light of West China" Program.

References

CASPER. https://casper.berkeley.edu

Li, D., et al.: FAST in space: considerations for a multibeam, multipurpose survey using china's 500-m aperture spherical radio telescope (FAST). IEEE Microwave Mag. **19**(3), 112–119 (2018)

LSST. https://www.lsst.org/about/dm/technology

Manchester, R.N., et al.: The Parkes multi-beam pulsar survey - I. Observing and data analysis systems, discovery and timing of 100 pulsars. Mon. Not. R. Astron. Soc. **328**(1), 17–35 (2001)

Nan, R., et al.: The five-hundred aperture spherical radio telescope (FAST) project. Int. J. Mod. Phys. **20**, 989 (2011)

ROACH2. https://casper.berkeley.edu/wiki/ROACH-2_Revision_2

SKA. https://www.skatelescope.org

CSSDC Big Data Processing and Applications in Space Science Missions

Yunlong Li, Jia Zhong, Fuli Ma, and Ziming Zou[✉]

National Space Science Center,
Zhongguancun, Haidian District, Beijing 100190, China
{liyunlong,zhongjia,mafuli007,mzou}@nssc.ac.cn

Abstract. Chinese space science starts to accelerate its pace with the support of Strategic Priority Program on Space Science, planning to launch more space science projects in the next five years. It brings great requirements in timeliness and accuracy of data processing, raising new challenges in heterogeneous data management and analysis. Chinese Space Science Data Center (CSSDC) is constantly exploring ways to face the potential challenges, keeping up with the rapid pace of new space science projects. This paper introduces CSSDC's preliminary explorations of the big data processing and application in space science, including the dynamic and scalable scientific data processing system that supports multi-satellites and multi-tasks, a hierarchical storage strategy for large-scale space science data, and the application of the advanced data mining methods to the space science researches with collaboration of domain experts.

Keywords: Data processing · Space Science Missions · CSSDC

1 Introduction

In 2010 the China's State Council approved the "Innovation 2020" plan to support a series of programs to promote scientific and technological innovation and development. The Strategic Priority Program on space science (SPP) supported by the plan was to launch various independent space science missions and international cooperations so as to deepen the understanding of the universe as well as to make new discoveries and breakthroughs in space science. The first stage of SPP (SPP I, from 2011–2017) has successfully launched the Dark Matter Particle Explorer (DAMPE), the Quantum Experiments at Space Scale (QUESS), the Shijian-10 recoverable satellite and the Hard X-ray Modulation Telescope (HXMT). All of the four missions are achieving promising results. The second stage of SPP (SPP II) was officially approved in 2017 addressing its scientific questions such as the origin and evolution of the universe and life, search for extraterrestrial life, and the impact of the Sun and the solar system on Earth and human development. SPP II will launch four satellites, the Einstein Probe (EP),

ⓒ Springer Nature Switzerland AG 2019
J. Li et al. (Eds.): BigSDM 2018, LNCS 11473, pp. 10–15, 2019.
https://doi.org/10.1007/978-3-030-28061-1_3

the Advanced Space-borne Solar Observatory (ASO-S), the Solar wind Magneto-sphere Ionosphere Link Explorer (SMILE) and Gravitational wave high-energy Electromagnetic Counterpart All-sky Monitor (GECAM).

As the data center of SPP, the Chinese Space Science Data Center (CSSDC) is responsible for the data lifecycle management of the SPP missions, including data processing, data management, archiving and sharing. Before the initiation of SPP I, CSSDC kept around 1.5 TB of data from early space science missions. Since 2011 the data volume and the number of data files in CSSDC shot up by two orders of magnitude. The data generated by the SPP I project has accu-mulated to about 170 TB by 2018, and the number of the data files are more than one million. The new space science missions supported by the SPP II are expected to produce more than 20 PB data, signifying that there will be another two orders of magnitude increases in the amount of data in the next five years.

The exponential growth of data brings impact on our data processing and management system, the existing data analysis methods based on Consultative Committee for Space Data System (CCSDS) standard format are difficult to meet the requirements of correctness and timeliness of data processing systems for the new space science missions, the CCSDS packet counts are frequently reset thus overlapped due to the large amount of data. CSSDC proposes a fast processing method to transform the problem of large data processing into the index of less data and the corresponding source package processing to improve the efficiency of data processing, and a data-driven and business-driven collab-orative processing framework is designed using scientific workflow technology which supports diversified data processing flow of space science satellites and parallel scheduling of various payload data processing tasks. This method has the advantages of expandable processing efficiency and less memory usage. It is applied to the ground system and meets the data processing requirements of space science missions.

Space science missions usually have a variety of instruments, generating many types of data with varying levels of importance and different frequency of access. CSSDC applies a three-tier hierarchical storage to keep the data. The system automatically assigns various kinds of storage resources to keep nearly 2000 kinds of data products generated by DAMPE, QUESS, SJ-10 and HXMT with effective and correct access. Thousands of data files distribution or transfer simultane-ously brings huge challenge in I/O access, which had given a lesson in SJ-10 mission implementation. Data management and distribution in data product granularity instead of file granularity is a good choice, which decreases the I/O access to a large extent.

The experiences from the space missions, data management and data anal-ysis enable CSSDC to be aware of that the massive heterogeneous data are posing great challenges to the scientific data analysis - the growing mountains of data piled up in the big data era always hide deeply the critical paths to new knowledge. In this context, CSSDC starts to search for cooperation with domain experts to apply the advanced data mining methods to the space science researches, expecting to get insightful views on the scientific data, which might be the first step of a science or technology innovation.

2   Difficulties of Data Processing

There are several difficulties in the data processing of space science satellites. One is that the generation speed of detection data is fast, although single detection data is several hundred KB at most, the detection frequency is high. Taking the DAMPE telemetry source package as an example, the fastest source package generation is in 1 s and the slowest is in 32 s; the satellite data source package is counted by 16 bits of continuous binary number, and the circulation is every 16384 counts. Source packet count overflow happens. It is hard to make source packet rearrangement and splicing by traditional counting.

In addition, there are varieties of detection data and the amount of data processing and calculation is large. For example, the HXMT has 18 main detectors, 3 particle detectors, 3 on-orbit calibration detectors, 6 top anti-coincidence shielding detectors and 12 side anti-coincidence shielding detectors. It has 13 virtual channels and 36 kinds of data source packages. The average amount of raw data downloaded by HXMT is 10 GB. Every time data is downloaded, virtual channel separation and source packet extraction are needed in time. The amount of computation in the parsing process is enormous. The number of parsed data units is also very large, which has certain pressure on data storage, management, retrieval and release.

Another difficulty is the diversity of data processing algorithms. Each satellite's data product has a different processing algorithm. Take the SJ-10 satellite, which has the maximum detectors, as an example, there are 46 kinds of scientific data products and 47 kinds of derived quick-look data products. It means that the data processing system needs to implement as least the same number of different algorithms. From the scheduling mode of algorithm module, there are three kinds of data processing methods in space science: post-receiving trigger, timing trigger and manual trigger, which meet the requirements of monorail original data processing, multi-track joint processing and historical data processing respectively.

3   CSSDC Data System

To tackle with the problems above, CSSDC applies the unified data processing system with the architecture of general processing platform + satellite dedicated processing plug-ins [1]. The data system is mainly composed by four parts. The data processing package engages in multi-orbit satellites data parallel processing, quick-look and visualization. The data management package engages in multi-satellites all level data products and archive data permanent reservation and management. The data distribution package provides near real time data distribution to satellites mission teams for multiple missions. A common service platform provides common service for the three packages, including data exchange, unified scheduler, PBS and unified database access service. Once the data processing flow runs, the working flow engine invokes processing algorithm module and processing parameters automatically according to the process definition. The system subsequently implements mass data co-processing of multi-satellites and data product outputting of multi-level standards by process driven means. The general processing platform supports rapid, reasonable and efficient distribution to processing nodes of satellite data processing tasks. The satellite dedicated processing plug-ins, which meet the different satellite data processing requirements, provide fast access to data products, satisfying all kinds of scientific users' needs for continuing the follow-up scientific data analysis (Fig. 1).

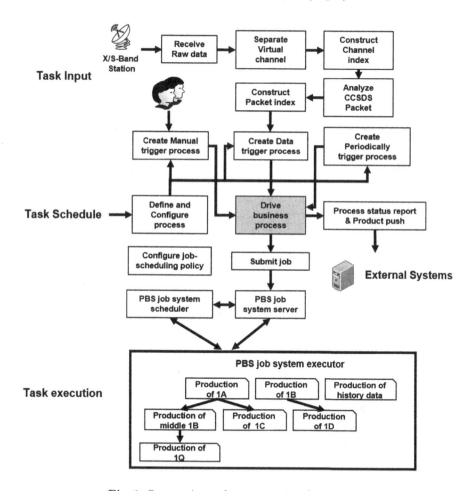

Fig. 1. Space science data processing framework.

The framework has several advantages such as it is data and business driven approaches for the space science missions. Various pipelines are triggered corresponding to their different scenarios, the system has more convenience for executing the processing in parallel. The system applies the scientific work flow technology which makes it easy to configure the data pipelines for each individual user and add more pipelines and processing tasks. With one unified scheduler to schedule all the tasks, the data processing for different missions are parallelized (Fig. 2).

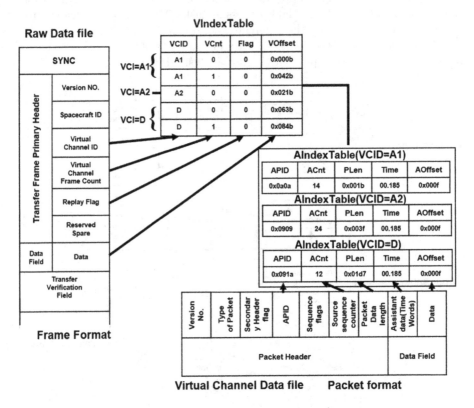

Fig. 2. Joint index structure for space science satellite data.

The data system proposes a joint index structure includes all the needed information for preprocessing to simplify the processing for many cases such as differentiation of the X band transmitter's work modes, or finding the packets gaps or the inconsistency of the time format. The joint index structure enhances the efficiency of the processing, avoiding of data swapping and repeat data parsing.

4   Anomaly Detection

The data processing system adopts data quality examines. They check the process to ensure that the data products are valid, try to identify the errors in the data products files and eliminate them ultimately. The main method is to check the correctness of the key value and the file format. CSSDC has applied several machine learning techniques to help detecting and locating data anomalies. A typical case is that the non-linear dimensionality reduction algorithm auto-encoder successfully detect the anomalies in the HXMT's engineering data. It helps us merge different thresholds of different channels into one threshold. In this way, it becomes clear at what time the anomaly occurred (Fig. 3).

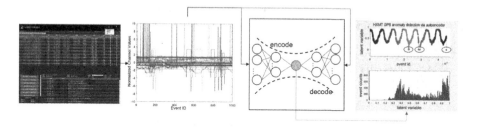

Fig. 3. Anomaly detection of satellite engineering data via auto-encoder.

5   Discussion

The data processing and management platform has successfully supported four space science missions. Service more than 500 scientists every day. And also the following critical and characteristic problems has been successfully addressed. The SPP brings opportunities of new discoveries in space science with great big data challenges. The data system needs to support scalability to capture and process space science data which are expected to be more than 20 PB in next five years. More than 1 TB downlinked data needs to be processed within very limited time, especially for the astronomy mission due to the scientific objectives short lifetime. CSSDC needs to apply data product-driven approaches across the entire data lifecycle from the onboard computing to data analysis and avoid of transfer delay. The system also needs to increased data science services for on-demand, interactive visualization and analytics, target specific analytic, such as for space physic, astronomy, gravitational wave, solar flare, etc. Last but not the least the data system needs to take full advantage of the very limited computing resources, with very effective scheduler and processing algorithm. More works needs to be done with limited resources.

Reference

1. Sun, X.-J., et al.: A rapid data processing method for space science satellites. Comput. Eng. Sci. **40**(08), 1351–1357 (2018)

Information Services of Big Remote Sensing Data

Guojin He[1,2(✉)] , Guizhou Wang[1,2] , Tengfei Long[1,2] , Huichan Liu[1,2] ,
Weili Jiao[1,2] , Wei Jiang[1,3] , Ranyu Yin[1,3] , Zhaoming Zhang[1,2] ,
Wanchun Leng[1,3] , Yan Peng[1,2] , Xiaomei Zhang[1,2] , and Bo Cheng[1,2]

[1] Institute of Remote Sensing and Digital Earth,
Chinese Academy of Sciences, Beijing 100094, China
hegj@radi.ac.cn
[2] Key Laboratory of Earth Observation Hainan Province, Hainan 572029, China
[3] University of Chinese Academy of Sciences, Beijing 100049, China

Abstract. With the rapid development of earth-observation technologies, different kinds of remote-sensing data are available, including SAR, multi-spectral optical data and hyper-spectral optical data. Remote sensing is already in an era of big data, which results in great challenges of information services. On the one hand, the potential value behind remote-sensing data has not yet been effectively mined for the thematic applications; on the other hand, for some emergency applications such as quick response of natural disasters (earthquake, forest fires, etc.), remote-sensing data and its derived information should be instantly processed and provided. However, the traditional data service mode of remote sensing is unable to meet the needs of such kind of applications. This paper proposes a concept of information services for big remote sensing data based on the needs of remote-sensing applications. The challenges of a big remote sensing data service are first discussed, and then the concept, framework, and key technologies of information service for big remote sensing data are also addressed.

Keywords: Remote sensing big data · Information services ·
Active service model · Data mining and knowledge discovery

This work was supported by the Strategic Priority Research Program of the Chinese Academy of Sciences (XDA19090300), the National Natural Science Foundation of China (61731022 and 61701495), the Open Research Fund of Key Laboratory of Digital Earth Science, Institute of Remote Sensing and Digital Earth, Chinese Academy of Sciences (No. 2016LDE006), the National Key Research and Development Programs of China (Grant No. 2016YFA0600302 and 2016YFB0502502), the Hainan Provincial Department of Science and Technology under the grant No. ZDKJ2016021, ZDKJ2016015-1 and ZDKJ2017009. The authors thank the anonymous reviewers for their valuable suggestions and comments.

J. Li et al. (Eds.): BigSDM 2018, LNCS 11473, pp. 16–31, 2019.
https://doi.org/10.1007/978-3-030-28061-1_4

1   Introduction

Thanks to the aerospace-, airborne-, and ground-based earth-observation technologies, the distribution and location of the status, phenomena, and process of the earth's surface can objectively and rapidly be retrieved. The earth-observation data has also been used for the survey of land and resources, land use dynamic monitoring, investigation and evaluation of the eco-environment, emergency response for disasters, post-disaster reconstruction, and security and military purposes [10,26,37]. The thirteenth annual plenary meeting of the Group on Earth Observation (GEO-XIII) held in St. Petersburg, Russian Federation from 7th–10th of November, 2016 put forward the "2017–2019 Work Programme". A global movement is underway in terms of earth-observation applications for the sustainable development goals (SDGs), including biodiversity and ecosystems sustainability, disaster resilience, energy and mineral resource management, food security and sustainable agriculture, infrastructure and transportation management, public health surveillance, sustainable urban development, and water resources management [5].

A lot of remote-sensing data are obtained from multi-sensors with multi-resolution, multi-temporal and multi-spectral [31,34,35]. This is not only because of the development of satellite and sensor technologies, but also because of the increasing demands for earth-observation applications. It was reported that the number of launched earth-observation satellites would be 2,427 by 2014, of which the highest spatial and spectral resolution of optical satellite images would reach 10 cm and 5 nm, respectively [36]. In addition, the small satellite constellations, such as Planet Labs and Skybox, have also been rapidly developing in recent years. It was stated that the daily volume of satellite remote-sensing data acquired globally was up to 1 TB [12]. Remote sensing is in an era of big data. For the past 30 years, China has made considerable progress in developing Chinese earth-observation systems, which mainly consist of meteorology satellites, resource satellites, and GF satellites. Moreover, China also has a small satellite constellation for environment and disaster monitoring. Furthermore, the small commercial satellite constellations, such as BJ and JL, are also members of the Chinese earth-observation system. These kinds of earth-observation systems have the ability to provide around-the-clock, full-weather, and global observation [9].

As far as the satellite data services are concerned, the satellite ground system traditionally works in a fixed pattern of "receiving, processing, archiving, and distributing [14,17]". The earth-observation data and products are downloaded and browsed online, and the thematic target information should be extracted by offline data processing. Therefore, this traditional satellite service pattern cannot meet the needs for rapid response applications, especially in real-time emergency monitoring purposes, such as flood monitoring [21].

In this paper, the challenges of big remote sensing data (BRSD) services are first analyzed [32,33], and then a framework for instant and active information services of BRSD are proposed, followed by a discussion of the key technologies for the proposed framework. Instead of satellite data and products being ordered after offline processing, the customized earth-observation products and the thematic information could be sent to users instantly.

2   Challenges of Big Remote Sensing Data Services

Satellite remote-sensing data has helped scientific decision making in emergency issues. For example, Chinese government used satellite images for directing disaster relief actions when the earthquake happened in Wenchuan County, Sichuan Province, China, in 2008. Instead of the raw data, the users prefer ready-to-use products and even the information derived from the raw data. How can these demands be met? What are the challenges?

(1) Inefficient Services of Remote-Sensing Data

Traditionally, a remote-sensing data service runs under the mode of "receiving, processing, archiving and distribution [14,17]". That is, data are regularly preprocessed in the catalog after being received by ground stations. Metadata are disseminated on the Internet for users to search, and data products are produced by data-processing systems according to the user's orders. It normally takes several days for the users to receive the products. In the case of emergencies, such as an earthquake or forest fire, the urgent acquisition mode of remote sensing data is usually activated; however, there will still be a delay of several hours in the traditional remote-sensing data service chain. The information extraction from the data for emergency events needs more time than the procedure of traditional remote sensing data service mode. Thus, inefficient service of remote-sensing data is the first challenge faced.

(2) How to Mine Information Intelligently from big remote sensing data

An era of BRSD is coming because of the unprecedented ability of earth observation data acquisition. However, the efficiency of data processing and information mining is less than that of data receiving, which leads to a situation of "big data but lack information or knowledge [12,19]". One of the reasons is that the process of traditional information extraction is vastly different than the cognitive process of the human brain. Thus, new methods based on artificial intelligence, such as deep learning, migrating learning, and human visual perception should be developed to solve this kind of challenge. These techniques can be used to create a collaborative cognitive computing model that might improve the efficiency of mining information from remote-sensed data. Currently, Google Earth Engine is one cloud-based platform that allows users to analyze petabytes of satellite data for time-series information mining. With Google Earth Engine, Pekel and colleagues processed three million Landsat satellite images taken from between 1984 and 2015; with these images, they extracted and analyzed the dynamics of land surface water bodies by using expert system, visual analysis, and evidence reasoning methods [27].

(3) **How to Meet the Users' Demands in the Era of big remote sensing data**

Traditional distribution service of remote-sensing data that follows "receiving, processing, archiving, distribution" is a passive service mode [14]. The products are produced by ground stations and data centers per the user's orders. Instead of the thematic information or target knowledge, the data service mainly supplies the raw data or the standard products, and the users must process the data by themselves, which again results in a gap between the user's demands and the data provided. The acquisition capacity of multi-source remote-sensing data has been sharply increasing, and a new service mode of remote-sensing data should be constructed, one which has the capability of data aggregation, intelligent computing, and active pushing service. The new service mode of remote sensing data can not only provide real-time distribution of satellite data, but also supply instant data computing and thematic information extracting.

3   Framework for Information Services of Big Remote Sensing Data

The framework for BRSD information services mainly includes three steps (shown in Fig. 1): The first is to instantly harvest the satellite's downlink data from multiple receiving satellite stations and the historical data from multi-source archiving databases. The second is to compute the data intelligently and discover the knowledge from BRSD. The third is to actively push the information to users.

The satellite downlink data, one of the most essential components of information services, may be distributed worldwide to different satellite ground stations. Thus, the first step is to integrate the downlink data from multiple satellite ground stations. Mainly, remote-sensing satellites can be divided into land satellites, marine satellites, and meteorological satellites. These different satellites normally correspond to different ground-receiving station network systems, and the different remote-sensing data sources are characterized by massive volumes, heterogeneous data types, and high-dimensional structures. It breaks down the shortage of mining information from single sensor data by integrating the satellite downlink data and multi-source historic remote-sensing data, so it has the chance to excavate the spatial relations, invariant features, and dynamic changes of the features hidden in the BRSD.

For BRSD information services, one of the greatest challenges is instant geopositioning. Fast, high-precision geometric positioning is the basis of integration, advanced processing, and multi-source satellite remote-sensing images. Accurate on-orbit geometric calibration of space-borne optical sensors is a key for the high-precision geo-referencing of satellite images, and many satellite data, such as IKONOS, ALOS, Pleiades, QuickBird, and WorldView, have been equipped with mature calibration procedures and methods. For instance, the ALOS calibration group utilizes geometric calibration fields that are distributed in Japan, Italy,

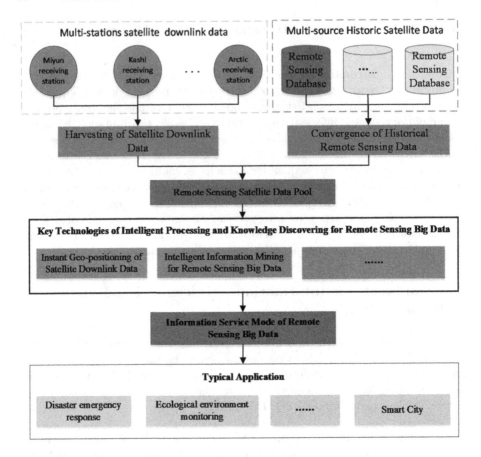

Fig. 1. The framework for information service of BRSD.

Switzerland, South Africa, and other areas. Without any ground control points (GCPs), the positioning accuracy (90% circular error) of an ALOS satellite image is 9.4 m [15].

Because of the accuracy limitations of a satellite altitude orbit control system (AOCS), improving the positioning accuracy of the space-borne optical sensor has entered a bottleneck stage, and higher positioning accuracy without GCPs is not expected until the breakthrough of AOCS technology. Therefore, to realize accurate geometric positioning of space-borne optical images at this stage, a certain number of ground control points or geometric registrations with reference images are commonly required. However, high-precision ground control points and high-resolution reference images are usually difficult to obtain; thus, the instant geometric positioning accuracy of satellite downlink data is limited. First, accurate on-orbit geometric calibration and rigorous sensor models are used to ensure the positioning accuracy of the satellite's downlink data. Second, to achieve accurate geometric positioning of a satellite's downlink data, the imaging geometric models of multi-source satellites needs to be optimized with the help of ground geo-reference or multi-satellite co-observations. On the one hand, the general geometric imaging model (RPC model) can be solved or optimized using sparse GCPs, which can be automatically collected from Bing Map, Google Map, and other online image maps with the help of online matching technology used on multi-source remote sensing images [23,24,30]. On the other hand, in the absence of ground control data, the geometrical constraints between different images in the same area can be used to compensate the errors of the orientation parameters of space-borne sensors, thus achieving fast and accurate geometric positioning without GCPs. In addition, because of the large volume of the satellite downlink data, making the computation-intensive procedures, such as automated image matching and orthorectification, near real-time by parallel computing is also an essential issue.

Recently, an efficient and accurate geo-positioning approach based on the geometric constrains of a multi-view image sequence and low-resolution reference image (as shown in Fig. 2) was applied to eight scenes of GaoFen-1 (GF-1) PAN images in absence of GCPs [22]. The spatial resolution of GF-1 PAN images is 2 m while that of the low-resolution reference image (Landsat-8) is 15 m. By performing multi-view matching and guided block adjustment, the geometric accuracy of the GF-1 images reached 3.27 m (root mean square error), and the process took about 90 seconds on an ordinary personal computer.

The second challenge for information services is how to mine information intelligently from BRSD. The remote-sensing datasets are from multiple sources, and have different passing times, and have different spatial and spectral resolutions; this means a new mechanism of multi-source remote-sensing data fusion and information coupling should be taken into account first. Note that when a satellite orbits the earth, it normally takes a fixed time interval to collect data from a region, with a specific spatial and spectral resolution. Therefore, an image from a satellite can hardly hold the high spatial, spectral, and temporal resolution simultaneously in terms of the bottleneck of the onboard sensor technologies [38]. A temporal-spatial data fusion model that might integrate the advantages from the remote-sensing images with different temporal and spatial resolutions can break the limitations of a single sensor and can thus alleviate the issue between temporal resolution and spatial resolution. Zhao et al. strengthened the resolution of MODIS images by using compressed sensing technology and forecasted images by a STARFM method [39]. Then, Song and Huang forecasted enhanced images by EXTARFM and also found results [28]. Funded by the Strategic Priority Research Program of the Chinese Academy of Sciences, CASEarth DataBank system will be constructed, which is a intelligent service system of big remote sensing data with satellite data based RTU(Ready To Use)products processing function, data engine,computing engine and data visualization engine [6–8,11,13].

With the development of artificial intelligence, cognitive computing, and other popular technologies, there is an opportunity to use these technologies to process BRSD. Deep learning can train many remote-sensing image

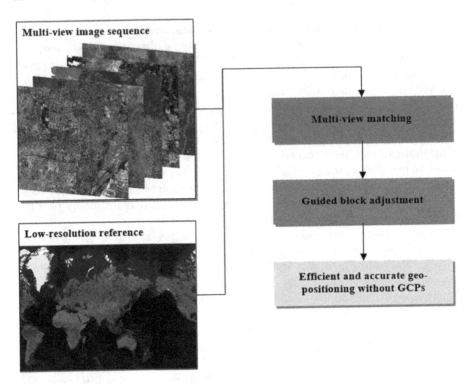

Fig. 2. Efficient and accurate geo-positioning based on the geometric constrains of multi-view image sequence and low-resolution reference image.

samples, construct a strategy that can learn multiple features automatically, and conduct fast extractions of specific targets of multi-source historical archives of remote-sensing data. Jean et al. used the convolutional neural network to study the characteristics of Google Earth multi-source, high-resolution satellite remote-sensing images to accurately predict poverty in Africa [16]. Luus et al. proposed a multi-scale input depth learning strategy for supervised classification of land use from multi-spectral satellite images which resulted in classification accuracy significantly higher than that of other classification methods such as unsupervised feature learning (UFL) and the vector of locally aggregated tensors (VLAT) method [25]. Cheng et al. achieved target detection of high resolution remote-sensing imagery by constructing a rotation invariant convolution neural network, which can improve detection accuracy compared to traditional methods [3]. Looking at the similarities in geographical spatial distribution and the sequence evolution of time in remote-sensing images, transfer learning can move the priori-training geological rules into the target model so that the existing model can quickly process remote-sensing geoscience information automatically. Cao et al. proposed supervised transfer learning by using a small number of samples to extract super resolution remote-sensing image vehicle information, which can be achieved with a high-accuracy vehicle detection rate [2]. Zhou et al. took the historical data as the label sample to carry on the classification and realized the fast target sample classification through the transfer learning algorithm [41]. With the rapid increase of remote-sensing downlink data from onboard satellites, real-time services for specific applications, such as quick responses to natural disasters, will become possible. Therefore, it is necessary to construct a collaborative cognitive model by integrating deep learning and transfer learning, which can combine the examples, features, parameters, and knowledge to maximize the information value of historical archive data and satellite downlink data. A built-up area extraction model has been constructed using a deep convolutional neural network (as shown in Fig. 3). The landsat-5 image and reference built-up area map (as shown in Fig. 4) were used to create 20,000 training samples and 10,000 test samples. These samples were used to train the model.

With the accumulated magnanimous time-series remote-sensing data, humans can understand the dynamic changes that happen on the earth's surface over a long period of time. Because changes in the earth's surface are closely related to global change regarding land-use/land-cover change and, the change detection research with time-series remote-sensing data is significant [1,18,40]. The traditional information mining methods mainly rely on the models or rules generated from historical remote-sensing data to map the changes in the earth's surface. In an information service mode, the ever-growing amount of data from new sensors, particularly downlink data will help to improve the models generated from historical archives. On the one hand, satellite downlink data can be used as a new data sample for the revision of the original model or rules; on the other hand, the satellite downlink data can also provide current surface status information. This means the previous situation might have been recorded in the historical archived data. Therefore, it is significant to develop a time-series information mining method that can combine historical data with satellite downlink data. An incremental learning method provides a new way of using real-time information extraction with big remote sensing data [4]. It is a dynamic construction technique because the incremental learning method acquires new knowledge from the continuous input samples without forgetting existing knowledge. An incremental learning method uses the latest data to extend the existing model knowledge so that it can effectively obtain information and knowledge from the combination of historical cognitive models and newly added data. With respect to real-time information extraction from big remote sensing data, the incremental learning method has the advantage of acquiring the downlink data continuously, progressively updating the original historical data model in an incremental way. In addition, because the incremental learning method does not require a relearning of all the data, it reduces the time and space requirements for the processing system; thus, it is more favorable for big remote sensing data processing. Based on the incremental learning method, the time-series information extraction model that combines the satellite downlink data with the original historical archive data can be studied. The involvement of satellite downlink data can promote the optimization of the model, improve

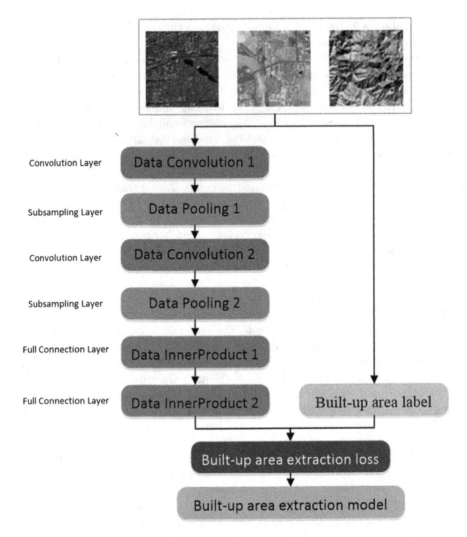

Fig. 3. Built-up area extraction model based on deep convolutional neural network.

the accuracy of surface information extraction, and quickly produce the results (for example, providing real-time flood condition information).

By using the sequential pattern mining method, the hidden and meaningful land-cover change patterns of Wuhan city in Hubei province, China were discovered (as shown in Figs. 5 and 6) [20]. This method regards each classification image as a temporal profile existing at every pixel location within the time-series and searches for a recurrent pixel's change patterns in space over time to represent land-cover change dynamics. This study provides a new way to detect the land-cover change, and experimental results showed that this approach can quantify the percentage area affected by land-cover changes, map the

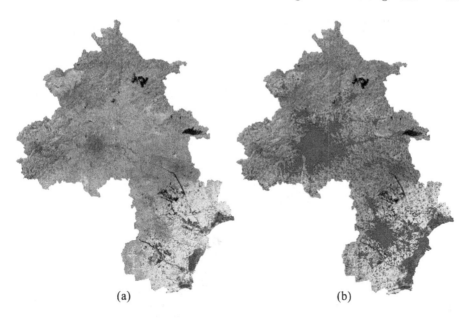

(a)   (b)

Fig. 4. (a) The landsat-5 image of Beijing and Tianjin cities, China (Combined RGB-543); (b) The reference built-up area map of Beijing and Tianjin cities, China (the built-up area is shown in red) (Color figure online)

spatial distribution of these land cover changes, and reveal interesting information regarding the trajectories of change. In the future, with the combination of all the available satellite downlink data, it would put real-time monitoring changes in the Earth's surface within reach.

In recent years, a visual attention model has been used for target recognition from high-resolution, remote-sensing images. An improved selective visual attention mechanism was applied for the rare earth mineral area detection in Dingnan County of Jiangxi province [29]. In this case, ALOS 2.5 m panchromatic and 10 m multi-spectral images were collected for a study on the extraction of mining-area information, as shown in Fig. 7. By using a visual attention model, the extracting area was obtained, as shown in Fig. 6. The results show that the visual attention model can extract useful information from remote-sensing images, such as the exact number, the target area, and edge area with a high accuracy, which indicates that this method has considerable value and can be widely used in big data processing, especially for high-resolution, remote-sensing images.

How to deliver the information and knowledge actively is another key technical issue. First, it is necessary to take advantage of Internet services, not only the traditional Web client, but also the mobile Internet client. The information from BRSD should be actively pushed to the most ordinary users. Second, it is very important to explore the behaviors and demands of the particular users based on big data analysis technology and to provide a personalized, custom push service to the users. Finally, the crowdsourcing service approach can be

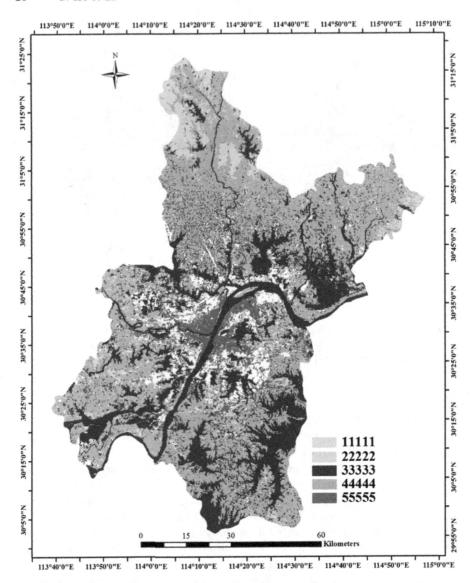

Fig. 5. Spatial distribution of the land-cover stable patterns of Wuhan city in Hubei province, China.

used for remote-sensing training sample collection and precision verification. In this way, more and more users can participate in information BRSD services. Taking instant fire detection services for example, the information of the suspected fire's location can be quickly extracted from the downlink data. To verify the accuracy of information extraction, a message can be pushed to certain users, such as the users near the suspected fire spot, so that they can confirm if there

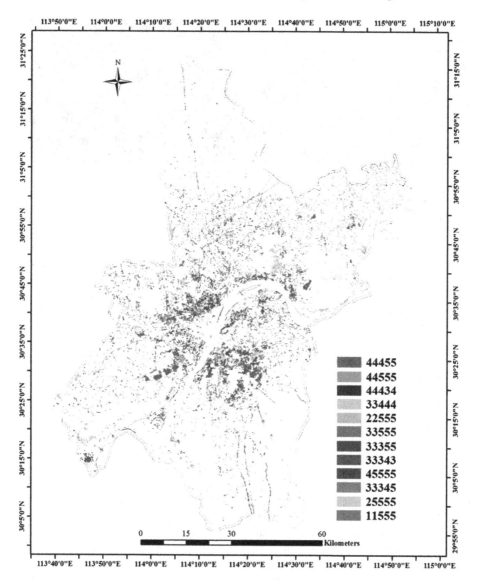

Fig. 6. Spatial distribution of the human-induced patterns of Wuhan city in Hubei province, China.

is a fire. This crowdsourcing service can improve the accuracy and efficiency of information services.

4   Discussions

Currently, more and more remote-sensing information is needed by many emergency applications such as earthquake quick response, forest fires monitoring,

Fig. 7. (a) The image is the raw remote sensing image of ALOS (Combined RGB-341); (b) The image is the mining area information map.

agricultural yield estimation and so on. The efficiency and ability of BRSD service is becoming more and more important for these applications. From the above, the concept of information service is defined, and the framework and key technologies for information service are constructed and discussed. We not only pay attention to the application of new technologies, but also focus on the integration of traditional and new methods.

The instant geo-positioning of satellite downlink data is the basis of multi-source satellite remote sensing image integration and subsequent processing and application. Generally, fast and high-precision georeferencing of satellite down-link data requires three aspects of essential technology: accurate imaging geometric mode and on-orbit geometric calibration of space-borne optical sensors, efficiently collecting ground control points (GCPs) from offline or online reference images, and robustly optimizing the imaging geometric model with GCPs of limited quantity (number) or limited quality (geometric accuracy) or even without GCPs. How to mine information intelligently from BRSD is another challenge for information services. For this purpose, we need to develop the intelligent information extraction model on the basis of artificial intelligence, cognitive computing and other popular technologies, which is the most important part in the framework of the information service. It is very important and necessary to build up a collaborative cognitive model which has the ability to integrate deep learning, transfer learning and incremental learning techniques in order to combine the examples, features, parameters, and knowledge to maximize the information value of historical archive data and satellite downlink data.

BRSD information services extend data sharing to informational knowledge sharing, and covert the traditional passive mode to the active services pushing mode. Through this information service framework, remote-sensing data and its derived information can be instantly acquired, processed and provided.

5   Conclusions

An era of remote sensing big data is coming, which has brought us a new situation for information services. On the one hand, the potential value behind remote-sensing data has not been effectively mined for thematic applications. On the other hand, for some emergency applications such as quickly responding to natural disasters, the BRSD and its derived information should be instantly processed and provided. However, the traditional data service mode of remote sensing is unable to meet the needs of these kinds of applications. The challenges of BRSD services were discussed first. Then, the concept, framework, and key technologies of information services for BRSD were addressed.

References

1. Boriah, S.: Time series change detection: algorithms for land cover change. Ph.D. thesis (2010)
2. Cao, L., Wang, C., Li, J.: Vehicle detection from highway satellite images via transfer learning. Inf. Sci. **366**, 177–187 (2016). https://doi.org/10.1016/j.ins.2016.01.004
3. Cheng, G., Zhou, P., Han, J.: Learning rotation-invariant convolutional neural networks for object detection in VHR optical remote sensing images. IEEE Trans. Geosci. Remote Sens. **54**(12), 7405–7415 (2016)
4. Elwell, R., Polikar, R.: Incremental learning of concept drift in nonstationary environments. IEEE Trans. Neural Netw. **22**(10), 1517–1531 (2011). https://doi.org/10.1109/TNN.2011.2160459
5. GEO: 2017–2019 Work Programme (2016)
6. Guo, H.: Big data drives the development of earth science. Big Earth Data **1**(1–2), 1–3 (2017). https://doi.org/10.1080/20964471.2017.1405925
7. Guo, H.: Big earth data: a new frontier in earth and information sciences. Big Earth Data **1**(1–2), 4–20 (2017). https://doi.org/10.1080/20964471.2017.1403062
8. Guo, H.: A project on big earth data science engineering. Bull. Chin. Acad. Sci. **33**(08), 818–824 (2018). https://doi.org/10.16418/j.issn.1000-3045.2018.08.008
9. Guo, H., Wang, L., Liang, D.: Big earth data from space: a new engine for earth science. Sci. Bull. **61**(7), 505–513 (2016). https://doi.org/10.1007/s11434-016-1041-y
10. Hansen, M.C., Potapov, P.V., Moore, R., Hancher, M., Turubanova, S.A., Tyukavina, A.: High-resolution global maps of 21st-centuary forest cover change. Science **342**(6160), 850–853 (2013). https://doi.org/10.1126/science.1244693
11. He, G., et al.: Opening and sharing of big earth observation data: challenges and countermeasures. Bull. Chin. Acad. Sci. **33**(08), 783–790 (2018). https://doi.org/10.16418/j.issn.1000-3045.2018.08.003. http://www.bulletin.cas.cn/publish%20article/2018/8/20180805.htm
12. He, G., et al.: Processing of earth observation big data: challenges and countermeasures. Chin. Sci. Bull. **60**(5–6), 470 (2015). https://doi.org/10.1360/N972014-00907
13. He, G., et al.: Generation of ready to use (RTU) products over China based on Landsat series data. Big Earth Data **2**(1), 56–64 (2018). https://doi.org/10.1080/20964471.2018.1433370. https://www.tandfonline.com/doi/full/10.1080/209644 71.2018.1433370

14. Irons, J.R., Dwyer, J.L., Barsi, J.A.: The next landsat satellite: the landsat data continuity mission. Remote Sens. Environ. **122**, 11–21 (2012). https://doi.org/10. 1016/j.rse.2011.08.026

15. JAXA - Japan Aerospace Exploration Agency: Calibration Result of JAXA standard products (As of 6 September 2011) (2011)

16. Jean, N., Burke, M., Xie, M., Davis, W.M., Lobell, D.B., Ermon, S.: Combining satellite imagery and machine learning to predict poverty. Science **353**(6301), 790–4 (2016). https://doi.org/10.1126/science.aaf7894. http://www.ncbi.nlm.nih.gov/pubmed/27540167

17. Justice, C.O., et al.: An overview of MODIS Land data processing and product status. Remote Sens. Environ. **83**(1–2), 3–15 (2002). https://doi.org/10.1016/S0034-4257(02)00084-6

18. Kuenzer, C., Dech, S., Wagner, W.: Remote sensing time series revealing land surface dynamics: status quo and the pathway ahead. In: Kuenzer, C., Dech, S., Wagner, W. (eds.) Remote Sensing Time Series. RSDIP, vol. 22, pp. 1–24. Springer, Cham (2015). https://doi.org/10.1007/978-3-319-15967-6_1

19. Li, D.R., Tong, Q.X., Li, R.X., Gong, J.Y., Zhang, L.P.: Current issues in high-resolution earth observation technology. Sci. China Earth Sci. **55**(7), 1043–1051 (2012). https://doi.org/10.1007/s11430-012-4445-9

20. Liu, H., He, G., Jiao, W., Wang, G., Peng, Y., Cheng, B.: Sequential pattern mining of land cover dynamics based on time-series remote sensing images. Multimedia Tools Appl. **76**(21), 22919–22942 (2017)

21. Liu, J., Ma, C., Chen, F., Zhang, J., Qu, Q.: Design and implementation of active-based instant remote sensing data service. Remote Sens. Inf. **31**(3), 61–67 (2016)

22. Long, T.: Efficient positioning technology for instant satellite image service. Ph.D. thesis, Beijing (2016)

23. Long, T., Jiao, W., He, G.: RPC RPC estimation via 1-norm-regularized least squares (L1LS). IEEE Trans. Geosci. Remote Sens. **53**(8), 4554–4567 (2015). https://doi.org/10.1109/TGRS.2015.2401602

24. Long, T., Jiao, W., He, G., Zhang, Z.: A fast and reliable matching method for automated georeferencing of remotely-sensed imagery. Remote Sens. **8**(1), 56 (2016). https://doi.org/10.3390/rs8010056

25. Luus, F.P.S., Salmon, B.P., van den Bergh, F., Maharaj, B.T.J.: Multiview deep learning for land-use classification. IEEE Geosci. Remote Sens. Lett. **12**(12), 2448–2452 (2015)

26. Mialhe, F., Gunnell, Y., Ignacio, J.A.F., Delbart, N., Ogania, J.L., Henry, S.: Monitoring land-use change by combining participatory land-use maps with standard remote sensing techniques: showcase from a remote forest catchment on Mindanao, Philippines. Int. J. Appl. Earth Obs. Geoinf. **36**, 69–82 (2015). https://doi.org/10. 1016/j.jag.2014.11.007

27. Pekel, J.F., Cottam, A., Gorelick, N., Belward, A.S.: High-resolution mapping of global surface water and its long-term changes. Nature **540**(7633), 418–422 (2016). https://doi.org/10.1038/nature20584

28. Song, H., Huang, B.: Spatiotemporal satellite image fusion through one-pair image learning. IEEE Trans. Geosci. Remote Sens. **51**(4), 1883–1896 (2013). https://doi. org/10.1109/TGRS.2012.2213095

29. Song, X., He, G., Zhang, Z., Long, T., Peng, Y., Wang, Z.: Visual attention model based mining area recognition on massive high-resolution remote sensing images. Cluster Comput. **18**(2), 541–548 (2015). https://doi.org/10.1007/s10586-015-0438-8

30. Tengfei, L., Weili, J., Guojin, H.: Nested regression based optimal selection (NRBOS) of rational polynomial coefficients. Photogram. Eng. Remote Sens. **80**(3), 261–269 (2014). https://doi.org/10.14358/PERS.80.3.261

31. Wang, L., Lu, K., Liu, P., Ranjan, R., Chen, L.: IK-SVD: Dictionary learning for spatial big data via incremental atom update. Comput. Sci. Eng. **16**(4), 41–52 (2014). https://doi.org/10.1109/MCSE.2014.52

32. Wang, L., Ma, Y., Yan, J., Chang, V., Zomaya, A.Y.: pipsCloud: high performance cloud computing for remote sensing big data management and processing. Future Gener. Comput. Syst. **78**, 353–368 (2018). https://doi.org/10.1016/j.future.2016.06.009

33. Wang, L., Song, W., Liu, P.: Link the remote sensing big data to the image features via wavelet transformation. Cluster Comput. **19**(2), 793–810 (2016). https://doi.org/10.1007/s10586-016-0569-6

34. Wang, L., et al.: G-Hadoop_Mapreduce across distributed data centers for data-intensive computing.pdf. Future Gener. Comput. Syst. **29**(3), 739–750 (2013). https://doi.org/10.1016/j.future.2012.09.001

35. Wang, L., Zhang, J., Liu, P., Choo, K.K.R., Huang, F.: Spectral-spatial multi-feature-based deep learning for hyperspectral remote sensing image classication. Soft. Comput. **21**(1), 213–221 (2017). https://doi.org/10.1007/s00500-016-2246-3

36. Yuan, M.: Development of remote sensing satellites and their business models. Satell. Appl. **3**, 15–19 (2015)

37. Zhang, Z., He, G., Wang, M., Wang, Z., Long, T., Peng, Y.: Detecting decadal land cover changes in mining regions based on satellite remotely sensed imagery: a case study of the stone mining area in Luoyuan County. SE China. Photogram. Eng. Remote Sens. **81**(9), 745–751 (2015). https://doi.org/10.14358/PERS.81.9.745

38. Zhao, S.: The technology and application of multi-source remote sensing image fusion. Nanjing University Press (2008)

39. Zhao, Y., Huang, B., Wang, C.: Multi-temporal MODIS and Landsat reflectance fusion method based on super-resolution reconstruction. Yaogan Xuebao-J. Remote Sens. **17**(3), 590–608 (2013)

40. Zhao, Z., et al.: Review of remotely sensed time series data for change detection. Yaogan Xuebao/J. Remote Sens. **20**(5), 1110–1125 (2016). https://doi.org/10.11834/jrs.20166170

41. Zhou, Y., Lian, J., Han, M.: Remote sensing image transfer classification based on weighted extreme learning machine. IEEE Geosci. Remote Sens. Lett. **13**(10), 1405–1409 (2016). https://doi.org/10.1109/LGRS.2016.2568263

Data Management in Time-Domain Astronomy: Requirements and Challenges

Chen Yang[1], Xiaofeng Meng[1(✉)], Zhihui Du[2], Zhiqiang Duan[1], and Yongjie Du[1]

[1] School of Information, Renmin University, Beijing, China
xfmeng@ruc.edu.cn
[2] Department of Computer Science and Technology, Tsinghua University, Beijing, China

Abstract. In time-domain astronomy, we need to use the relational database to manage star catalog data. With the development of sky survey technology, the size of star catalog data is larger, and the speed of data generation is faster. So, in this paper, we make a systematic and comprehensive introduction to process the data in time-domain astronomy, and valuable research questions are detailed. Then, we list candidate systems usually used in astronomy and point out the advantages and disadvantages of these systems. In addition, we present the key techniques needed to deal with astronomical data. Finally, we summarize the challenges faced by the design of our database prototype.

Keywords: Distributed database · Time-domain astronomy · Catalog data

1 Introduction

The origin of information explosion is astronomy, which is first facing challenges of big data [13]. In the new century, with the development of astronomical observation technique, astronomy has already entered a informative big data era and astronomical data is rapid growth in terabytes (TB) or even petabytes (PB). When Sloan Digital Sky Survey project started in 2000, data being collected by telescope in New Mexico in few weeks is greater than all historical data. In 2010, information files contained 1.4×2^{42} bytes. But, Large Synoptic Survey Telescope (LSST) can be used in Chile in 2019, which can get the same information within 5 days. Now, a number of countries are running large-scale sky survey project. Except SDSS, these projects include PanSTARRS (Panoramic Survey Telescope and Rapid Response System), WISE (Widefield Infrared Survey Explorer), 2MASS (Two Micron All Sky Survey), Gaia of the European Space Agency (ESA), UKIDSS (UKIRT Infrared Deep Sky Survey), NVSS (NRAO VLA Sky Survey), FIRST (Faint Images of the Radio Sky at Twenty-cm), 2df (Two-degree-Field Galaxy Redshift Survey), LAMOST (Large Sky Area Multi-Object Fiber Spectroscopic Telescope), GWAC (Ground Wide Angle Camera)

© Springer Nature Switzerland AG 2019
J. Li et al. (Eds.): BigSDM 2018, LNCS 11473, pp. 32–43, 2019.
https://doi.org/10.1007/978-3-030-28061-1_5

in China and so on. These sky surveys are generating a large number of astronomical data.

Astronomical data also has four V characteristics of big data: Volume, Velocity, Variety and Veracity [8]. For example, LSST covers all sky and store in the database one cycle per 7 days. GWAC covers 5000 degree^2 and stores in real-time per 15 s. Astronomy moves on to a big data era. Therefore, it is important to study big data which is generated by astronomy [9]. However, the existing database systems cannot support the demand of astronomical data, especially for the real-time and GWAC's scalability. We need to design our database system for the discovery of transient celestial phenomena in short-timescale.

The rest of the paper is organized as follows. Section 2 surveys the background. Then, there are the problem definition and basic knowledge in Sect. 3. And, in Sect. 6, we list candidate systems used in astronomy and point out the advantages and disadvantages of these systems. Then, in Sect. 4, we give the functional analysis of our database prototype. In addition, we point out the challenges to the design of our database prototype in Sect. 5. Finally, Sect. 7 concludes this paper.

2   Background

GWAC is built in China, which consists of 36 wide angle telescopes with 18 cm aperture. Each telescope equips 4k × 4k charge coupled device (CCD) detector. Cameras cover 5000 degree^2. Temporal sampling is 15 s. Cameras detect objects of fixed sky area for 8 h each observation night. GWAC has special adventure in the time-domain astronomical observation, according to size of observation field and sampling frequency of observation time. It is great challenges for data management and processing in giant data and high temporal sampling.

Table 1. Dada volume generated by GWAC

Cameras	One day (8 h)		One year (260 days)		Ten years	
	Records	Size	Records	Size	Records	Size
1	3.37×10^8	61.88 GB	8.77×10^{10}	15.71 TB	8.77×10^{11}	157.1 TB
36	1.21×10^{10}	2.17 TB	3.16×10^{12}	565.62 TB	3.16×10^{13}	5.52 PB

As shown in Table 1, GWAC works 8 h a night, 260 days a year on average. The index of star catalog data in GWAC includes: each star catalog in one image having 1.756×10^5 records. So, camera array can generate 6.3×10^6 records in 15 s, contain 1920 × 36 = 69120 images and occupy about 2.17 TB storage space in each night. The requirements of database management systems (DBMS) are: (1) rapid big data storage capacity that all star catalogs are ingested within 15 s and 2.17 TB star catalog data in each observation night should be stored before next observation night. (2) high-speed data acquisition can be analyzed in real-time and rapid contextual computing capacity when facing mass of incessancy and high-density star catalogs. This means relevance star catalog data generated by one CCD within 15 s and reference star catalog to form light curves. (3) In 10-year design cycle, GWAC will generate about 5.53 PB size of star catalogs. Therefore, DBMS for storing star catalog data must have the great management ability for massive data.

For GWAC, the most immediate way to data management and design of processing system is the database (only for data storage) and peripheral program (rapid operation and the result obtaining). Yang et al. [18] researched by cross-match for key technique developed a sky partitioning algorithm based on longitude and latitude of space and increased the speed of cross match compute rapidly. Based on the advantage of parallel compute by graphics processor, Zhao et al. [21] used graphics processor accelerate method to speed up the image subtraction processing. Zhao et al. [20] developed point source extracting program SEXtractor in the field of astronomy which is developed by graphic processor accelerate. Wang et al. [17] developed a cross-match accelerating algorithm based on graphic processor. The advantages of this plan are the straightforward thoughts and many mature techniques. The disadvantages are that database is exchanging with peripheral programs and bring useless time loss of I/O. Combination of program results in lack of optimizations as a whole.

Jim Gray directed the development of Skyserver and purposed Zone algorithm [4,5]. It means that we can use SQL of DBMS to realize hyperspace index replacement classical Hierarchical Triangular Mesh (HTM). This method can reduce data interaction and increase speed. This is the principle of large-scale scientific computation and database architecture design: designing philosophy to bring computation to data rather than putting data to computation [16]. This paper inspired by this idea, purposes design idea that combines data processing of GWAC and data management to a database platform.

Massive astronomical data is great challenge for data storage and management. Therefore, rapid processing of massive astronomical data is very important. GWAC astronomical database can provide inquiry service and form a light curve. The database contains two scientific targets:

- **Rapid big data storage capacity**. All star catalogs generated by cameras can be stored within 15 s, and 2.17 TB star catalog data in each observation night should be stored before next observation night.
- **High-speed data collection**. Data can be analyzed in real-time, including efficient detection and dynamic recognition astronomical object.

The main scientific target of GWAC is to search optical transient sources in real time and locate in observation sky and formulate star catalog index. GWAC works sky survey every 15 s. Facing incessancy observation intensive stellar field and massive star catalog in short timescale, data processing system must have relevance computing ability to rapidly recognize celestial objects and data processing algorithm, meaning relevance star catalog data generated by each CCD in 15 s and reference star catalog to generate light curve. So, the goal is that we need to develop a database which can integrate the algorithm of the point source identification and has high expansibility.

3   Preliminaries

3.1   Problem Definition

Point Source Extraction. The camera array (Ω) is consist of 36 wide angle telescopes. Each wide angle telescope $T_i (i = 1, 2, ..., 36)$ equips 4k × 4k CCD detector. Point source extraction is to transform optical image into figure signal by CCD detector which forms star catalog data.

Each row of data in star catalog is used to record one star. Property information [12] for each star shows on Table 2. Each image in star catalog has about $1.756 × 10^5$ records.

Cross-Match. It contrasts and matches the object catalog with template catalog. As shown in Fig. 1, if object catalog could match template catalog, the pipeline will enter timing sequence photometry channel to process and manage light curve. If the star cannot be matched in template catalog, it is transient source (candidate). Cross-match is the key algorithm in GWAC searching transient source and generation light curve. Cross-match issue must depend on effective partition strategy. We will divide the sky into each horizontal strip in the pixel coordinate and each source has a strip belonging to itself. At first, cross-match can compare strip property and decrease times of comparing. Strip can be integrated inside database as the basic unit of data processing.

Light Curve. It is the change of the brightness of the object relative to the time. It is the function of time, which usually shows a particular frequency interval.

3.2   Camera Array Processing Flow

There is a flow graph for GWAC data processing in Fig. 2. According to basic preprocessing of original image, it will extract point source and astrometric calibration of star catalog. Then, it completes tasks about relative discharge calibration, real-time dynamic identification of astronomical object and light curve mining based on cross-match observation data and reference star catalog.

Overall, objectives of our database prototype are: (1) capacity of rapid storage data, object catalogs generated by cameras should be ingested within 15 s. 2.17TB star catalog data in each observation night should be stored before next observation night. (2) high-speed data acquisition can be analyzed in real time and has rapid contextual computing capacity when facing mass of incessancy and high-density star catalog. This means relevance star catalog data generated by one CCD in 15 s and reference star catalog to generate light curve.

Table 2. Star catalog attributes of each source

Name	Type	Description
ID	long int	Every inserted source measurement gets a unique id. generated by the source extraction procedure
imageid	int	The reference ID to the image from which the source was extracted
zone	small int	The zone ID in which a source declination resides, calculated by the source extraction procedure
ra	double	Right ascension of a source (J2000 degrees), calculated by the source extraction procedure
dec	double	Declination of a source (J2000 degrees) as above
mag	double	The magnitude of a source
mag_error	double	The error of magnitude
pixel_x	double	The instrumental position of s source on CCD along x
pixel_y	double	The instrumental position of s source on CCD along y
ra_err	double	The 1-sigma error on ra (degrees)
dec_err	double	The 1-sigma error on declination (degrees)
x	double	Cartesian coordinates representation of RA and declination, calculated by the source extractor procedure
y	double	Cartesian coordinates representation of RA and declination, as above
z	double	Cartesian coordinates representation of RA and declination, as above
flux	double	The flux measurements of a source, calculated from the mag value
flux_err	double	The flux error of a source
calmag	double	Calibrated mag
flag	int	The source extraction uses a flag for a source to tell for instance if an object has been truncated at the edge of the image
background	double	The source extraction estimates the background of the image
threshold	double	The threshold indicates the level from which the source extraction should start treating pixels as if they were part of objects
ellipticity	double	Ellipticity is how stretched the object is
class_star	double	The source extractions classification of the objects

4   Functional Analysis and Requirements

The key steps in processing GWAC's data are shown in Fig. 3. Within 15 s, we need to ingest 1.756×10^5 records into the database, complete the cross-match, generate the light curve and fulfill the task of data mining. The core functional analysis is as follows:

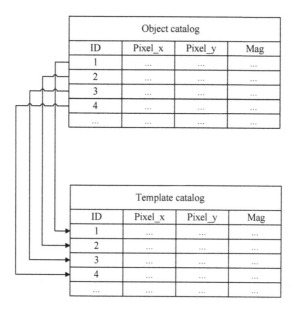

Fig. 1. Example of cross-match

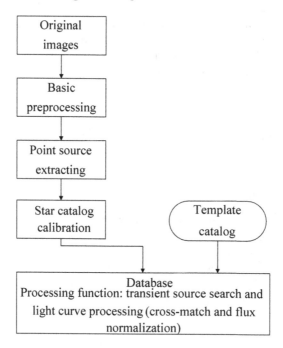

Fig. 2. GWAC data processing

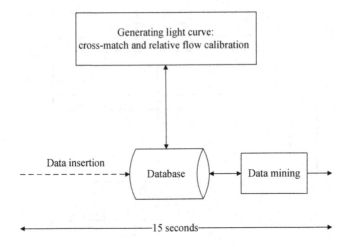

Fig. 3. The goal of core processing

4.1   Real-Time Storage

The system could rapidly store object star catalog generated by cameras in 15 s. 2.17 TB star catalog can be stored as increment data and make sure that data could be stored in real time. Also, system merges increment data daily during data storage low cycle (non observation night and other time) to increase data storage capacity and decrease storage delay.

4.2   Cross-Match

The system should establish efficient index mechanism, optimize database join operation, increase star catalog relevance and efficiency of cross-match.

4.3   High Scalability

As the observation data of each wide angle telescope is increasing, it is hard to use one server storage and analysis different telescopes data. We need to design high reliable distributed clusters architecture. There are different subset clusters for every wide angle telescope storage star catalog data to ensure consistency of star catalog data. Also it realize whole system processing capacity linear growth and high throughput rate and low delay.

4.4   Data Mining

In database, we need to use technique of data mining to find meaningful astronomical phenomena. The process of data mining can divide into online mining and offline mining (shown in Fig. 4).

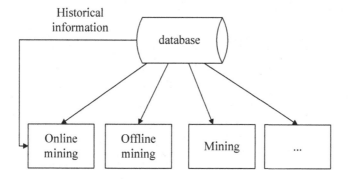

Fig. 4. Data mining

Online Mining. For data flow which observation time is shorter than one night, real-time monitoring and inner analysis windows should be used for dynamic recognition of astronomical targets.

Offline Mining. For long-timescale data which observation time is longer than one night, using full scale historical data predict waveform of light curve and judging change cycle of curve to analyze fluctuation features of star body is better.

5   Major Challenges

Based on the functional analysis in Sect. 4, the main challenge issues for our database prototype can be concluded as below.

5.1   Customized Operators

For characteristics of astronomical data, we can customize operators in our database prototype. There are the customized operators as below:

- Increment storage operator "DeltaInsert" ensures data real-time storage.
- Range join operator "RangeJoin" ensures rapid cross-match.
- Data mining operator library ensures association analysis, classify, cluster, outliers detector and other functions.

In addition, our database prototype is designed with batch processing and query plan adapter of streaming processing, by uniform query system interface. it cannot only get real-time data flow query result, but also it can get query result of full-history form of offline historical data.

5.2   Large-Scale Data Management

With the increment of the amount of data generated by GWAC, our database prototype needs to deal with the problem of data management in PB level. A large-scale data management engine needs to be designed to ensure data consistency and integrity, and easy to scale out.

Table 3. Different system characteristics

Features	SciDB	OceanBase	MonetDB	MongoDB	Spark
Storage engine	Array data model	Memory affair storage	Binary association table	Assembly storage	RDD
Advantage	Shared-nothing design, SciDB-R	Support ACID Affair, Fault tolerant automatically and load balancing, Division storage	Automatic indexing, index not taking extra storage space	Powerful query language, support dynamic query and fully index	Easy to scale out clusters and fast iterative computations
Disadvantage	Not support complex search condition	Low open source version	Low speed of insert operation	Not support transaction operation	Low data storage efficiency

5.3   Scalable Query Processing

In a large-scale cluster environment, our database prototype needs to ensure a low query response time. We should use the design philosophy of massively parallel processing (MPP) to implement scalable query processing.

5.4   Long-Term Data Storage

Since the life cycle of GWAC is 10 years, it is essential to provide the hardware and the storage strategy to save all historical data. Original image data can verify correctness of related analysis and provide original data for image analysis in depth.

6   Candidate Systems

For characteristic and functional requirements of astronomical data, there are some candidate systems preparing to be compared and their characteristics are summarized in Table 3.

6.1   SciDB

SciDB is a new science database for scientific data. Application areas include astronomy, particle physics, fusion, remote sensing, oceanography, and biology [15]. Scientific data often does not fit easily into a relational model of data. Searching in a high dimensional space is natural and fast in a multi-dimensional array data model, but often slow and awkward in a relational data model.

Array DBMS is a natural fit for science data. So, SciDB uses array data model as the storage engine. Another characteristic of SciDB is sparse or dense array.

Some arrays have a value for every cell in an array structure. It have ability to process skewed data. Moreover, SciDB uses the shared-nothing distributed storage framework. This is easy to scale out the cluster. In addition, SciDB has built an interface for R language that lets R scripts access data residing in SciDB. However, SciDB can not support complex search condition. And, the effect of data storage in real-time is not good.

6.2 OceanBase

OceanBase [2] is a high-performance distributed database. It can realize cross row and cross table affairs based on hundred billions records and hundreds TB data.

Because OceanBase is a relational DBMS, it uses memory affair storage as the storage engine, and can support ACID affair. And, in the distributed environment, OceanBase provides automatic fault tolerance, load balancing and division storage. However, the open source version of OceanBase is low, and the stability of the system is not strong.

6.3 MonetDB

MonetDB is also a relational DBMS for high-performance applications in data mining, scientific databases, XML Query, text and multimedia retrieval, that is developed at the CWI database architectures research group since 1993 [11].

MonetDB is designed to exploit the large main memory of modern computer systems effectively and efficiently during query processing, while the database is persistently stored on disk. The core architecture of MonetDB has proved to provide efficient support not only for the relational data model and SQL, but also for the non-relational data model, e.g., XML and XQuery [7,10]. In addition, MonetDB supports column-store by using binary association table as storage engine. Moreover, It can build the index automatically, the index can not take extra storage space. Yet, the efficiency of insertion operation is low, especially for incremental insertion.

6.4 MongoDB

MongoDB [1] is an agile database that allows schemas to change quickly as applications evolve, while still providing the functionality that developers expect from traditional databases.

Figure 5 shows the basic difference between the schema-free document database structure and the relational database [6]. While the tables in a relational database have a fixed format and fixed column order, a MongoDB collection can contain entities of different types in any order. The element dbRef allows the creation of an explicit reference to another document in the same database or in another database on another server.

MongoDB uses assembly storage as storage engine to support dynamic query and fully index, and has a powerful query language. However, it can not support transaction operation well.

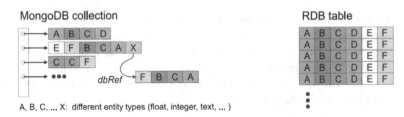

A, B, C, ... X: different entity types (float, integer, text, ...)

Fig. 5. Basic data structure between mongoDB and a relational database

6.5 Spark

Apache Spark [3] is an open-source cluster computing framework for big data processing. It has the distributed data frame, and goes far beyond batch applications to support a variety of compute-intensive tasks including interactive queries, streaming, machine learning and graph processing [14].

Spark use RDD [19] as storage engine to ensure the correctness and fault tolerance of the query processing. It is easy to scale out clusters, and has fast iterative computing power. However, because spark needs to rely on other storage frameworks, its data storage efficiency is low.

7 Summary

We investigate and survey requirements of databases in astronomy and introduce the background knowledge, points out core problems and the main challenges. However, none of these candidates is suitable for large time-domain surveys and that a new system should be developed to meet the challenges.

Acknowledgement. This research was partially supported by the grants from the National Key Research and Development Program of China (No. 2016YFB1000602, 2016YFB1000603); the Natural Science Foundation of China (No. 91646203, 61532016, 61532010, 61379050, 61762082); the Fundamental Research Funds for the Central Universities, the Research Funds of Renmin University (No. 11XNL010); and the Science and Technology Opening up Cooperation project of Henan Province (172106000077).

References

1. Mongodb. http://www.mongodb.org/
2. Oceanbase. https://github.com/alibaba/oceanbase/tree/master/oceanbase_0.4
3. Spark. http://spark-project.org/
4. Zone. https://arxiv.org/ftp/cs/papers/0408/0408031.pdf
5. Zone project. http://research.microsoft.com/apps/pubs/default.aspx?id=64524
6. Ameri, P., Lutz, R., Latzko, T., Meyer, J.: Management of meteorological mass data with mongodb. In: Einviroinfo (2014)
7. Boncz, P., Grust, T., Van Keulen, M., Manegold, S., Rittinger, J., Teubner, J.: MonetDB/XQuery: a fast XQuery processor powered by a relational engine. In: SIGMOD, pp. 479–490 (2006)

8. Bryant, R.E., Katz, R.H., Lazowska, E.D.: Bigdata computing: creating revolutionary breakthroughs in commerce, science, and society (2008)
9. Cui, C., et al.: Astronomy research in big-data era. Chin. Sci. Bull. **60**(z1), 445–449 (2015). (in Chinese)
10. Idreos, S., Groffen, F.E., Nes, N.J., Manegold, S., Mullender, K.S., Kersten, M.L.: MonetDB: two decades of research in column-oriented database architectures. IEEE Data Eng. Bull. **35**(1), 40–45 (2012)
11. Manegold, S., Kersten, M.L., Boncz, P.: Database architecture evolution: mammals flourished long before dinosaurs became extinct. Proc. VLDB Endow. **2**(2), 1648–1653 (2009)
12. Wan, M.: Column store for GWAC: a high cadence high density large-scale astronomical light curve pipeline and distributed shared-nothing database. Publ. Astron. Soc. Pac. **128**(969), 114501 (2016)
13. Naimi, A.I., Westreich, D.J.: Big data: a revolution that will transform how we live, work, and think. Information **17**(1), 181–183 (2014)
14. Shanahan, J.G., Dai, L.: Large scale distributed data science using apache spark. In: ACM SIGKDD International Conference on Knowledge Discovery and Data Mining (2015)
15. Stonebraker, M., et al.: Requirements for science data bases and SciDB. In: CIDR (2009)
16. Szalay, A.S., Blakeley, J.A., Szalay, A.S., Blakeley, J.A.: Gray's laws: database-centric computing in science (2009)
17. Wang, S., Zhao, Y., Luo, Q., Wu, C., Xv, Y.: Accelerating in-memory cross match of astronomical catalogs. In: IEEE International Conference on Escience, pp. 326–333 (2013)
18. Yang, X., et al.: A fast cross-identification algorithm for searching optical transient sources. Astron. Res. Technol. **10**(3), 273–282 (2013)
19. Zaharia, M., et al.: Resilient distributed datasets: a fault-tolerant abstraction for in-memory cluster computing. In: Usenix Conference on Networked Systems Design and Implementation, pp. 141–146 (2012)
20. Zhao, B., Luo, Q., Wu, C.: Parallelizing astronomical source extraction on the GPU. In: IEEE International Conference on Escience, pp. 88–97 (2013)
21. Zhao, Y., Luo, Q., Wang, S., Wu, C.: Accelerating astronomical image subtraction on heterogeneous processors. In: IEEE International Conference on Escience, pp. 70–77 (2013)

AstroServ: A Distributed Database for Serving Large-Scale Full Life-Cycle Astronomical Data

Chen Yang[1], Xiaofeng Meng[1(✉)], Zhihui Du[2], JiaMing Qiu[2], Kenan Liang[3], Yongjie Du[1], Zhiqiang Duan[1], Xiaobin Ma[2], and Zhijian Fang[3]

[1] School of Information, Renmin University, Beijing, China
xfmeng@ruc.edu.cn
[2] Department of Computer Science and Technology, Tsinghua University, Beijing, China
[3] School of Computer Science and Technology, Shandong University, Jinan, China

Abstract. In time-domain astronomy, STLF (Short-Timescale and Large Field-of-view) sky survey is the latest way of sky observation. Compared to traditional sky survey who can only find astronomical phenomena, STLF sky survey can even reveal how short astronomical phenomena evolve. The difference does not only lead the new survey data but also the new analysis style. It requires that database behind STLF sky survey should support continuous analysis on data streaming, real-time analysis on short-term data and complex analysis on long-term historical data. In addition, both insertion and query latencies have strict requirements to ensure that scientific phenomena can be discovered. However, the existing databases cannot support our scenario. In this paper, we propose AstroServ, a distributed system for analysis and management of large-scale and full life-cycle astronomical data. AstroServ's core components include three data service layers and a query engine. Each data service layer serves for a specific time period of data and query engine can provide the uniform analysis interface on different data. In addition, we also provide many applications including interactive analysis interface and data mining tool to help scientists efficiently use data. The experimental results show that AstroServ can meet the strict performance requirements and the good recognition accuracy.

Keywords: Distributed database · Astronomical data · Full life-cycle

1 Introduction

In recent years, many large optical instruments in time-domain astronomy have brought unprecedented observation capabilities to us. As shown in Fig. 1, these instruments have made great progress in three factors, including field-of-view (FoV), spatial resolution and temporal resolution, which means that the telescope can search larger area and darker objects with higher frequency, respectively. Due to the cost limitation, three factors cannot currently be met at the

© Springer Nature Switzerland AG 2019
J. Li et al. (Eds.): BigSDM 2018, LNCS 11473, pp. 44–55, 2019.
https://doi.org/10.1007/978-3-030-28061-1_6

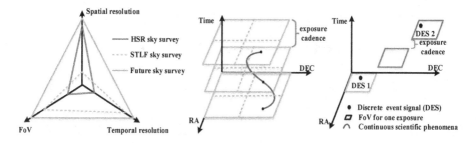

Fig. 1. The left is a radar graph, which compares characteristics of different sky surveys. STLF sky survey (middle) can generate continuous time-series data of the same star in one night. HSR sky survey (right) can only generate discrete data points in one night. RA and DEC are right ascension and declination, respectively.

same time, so that two ways are attempted to design observation instruments, including HSR (High Spatial Resolution) sky survey and STLF sky survey.

The study about HSR sky survey has been around for a long time, such as PTF [17], Skymapper [8], SDSS [11], Pan-STARRS [13], and LSST [9]. They finish a survey within 3–5 days and scan different regions (about 3–9.6 square degrees) at every exposure. Thus, HSR sky survey cannot catch short and continuous scientific phenomena or only catch discrete scientific events of different observed regions.

STLF sky survey as a new observation approach cannot only simultaneously observe lots of spacial objects, but also a huge amount of data is collected at a high exposure frequency. For example, GWAC [19] finishes a survey within 15 s and scans the same region (about 5000 square degrees) at every exposure. Through STLF sky survey, human being can understand how short astronomical phenomena evolve. Although the new observation ability can help scientists reveal more natural laws, the database systems behind STLF sky survey will face unprecedented challenges.

The new survey data generated by STLF sky survey provides scientists a completely new way to achieve scientific discovery, being very different from HSR sky survey. The databases behind HSR sky survey, such as SkyServer [18] and Qserv [20], only support the management of long-term historical data. However, continuous time-series data has higher value and supports more analysis methods on them, compared with discrete data points. Thus, scientists often expect to launch an analytical query on streaming data, short-term data and long-term historical data to confirm a scientific phenomenon and issue an alert as soon as possible. It causes the analysis requirement not limited to long-term historical data, but extend to full life-cycle of data. Obviously, it is more helpful for scientists to reveal scientific laws on site since many astronomical physical phenomena are transient and hard to reproduce, such as microlensing.

It is a challenging work to design a database behind STLF sky survey to support continuous analysis on streaming data, real-time analysis on short-term

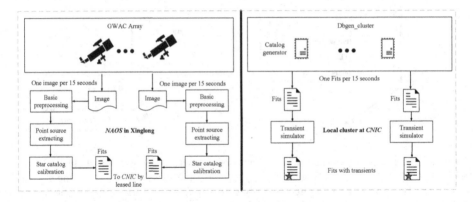

Fig. 2. The left is the real data generation pipeline and the right is data generator to simulate GWAC's working mode.

data and complexity analysis on long-term historical data. We face stringent data challenges as follows.

- **High density.** Hundreds of thousands of objects will be extracted after an exposure, causing the database to require high throughput. For example, GWAC can simultaneously observe about 3.5 million objects.
- **High frequency.** Many data will be generated in very short time, causing the database to require low request latency. For example, GWAC will produce data every 15 s so that The ingest latency must be less than 15 s and a real-time query must be also finished less than 15 s.
- **Tremendous amount.** As time goes on, a lot of data will be accumulated, causing the database to require efficient storage capacity. For example, GWAC can collect 6.7 billion high-dimensional data points per night. Finally, produce about 2.24PB size of data over 10 years. Thus, the storage resource should be used as little as possible from an economic point of view.

The existing astronomical databases are designed for HSR sky survey, so they do not have a good support for continuous analysis and real-time analysis. Other high-performance databases, such as MonetDB [19] and Hbase [2], cannot support the scientific analysis. Thus, in this paper we follow GWAC's data feature to design a database AstroServ serving full life-cycle astronomical data of STLF sky survey. AstroServ mainly includes five major contributions as follows.

- **Detect service on streaming.** It receives the newly arriving data, normalizes data with a scientific pipeline and recognizes abnormal star from huge amounts of objects. This module can follow the status of observation instruments in real time and find early phase of scientific phenomena.
- **Real-time data service.** It receives normalized data and identified abnormal star information and manage one night of data. This module can efficiently ingest them into in-memory store to provide real-time query service by a set of compact data store and index structures.

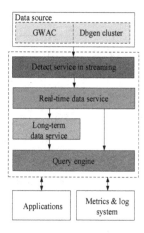

Fig. 3. Major components in AstroServ

- **Long-term data service.** It ingests one night of data every time and finally manage all historical data. This module can append data with a high through-put and provide off-line query service by compact data scheme and efficient metadata management.
- **Query engine.** We design a framework to build query engine to ensure that it can query data from both real-time data service and long-term data service with a unified way.
- **Applications.** Combined with core service of our database, we present two applications including interactive analysis interface and off-line integrated detector to help user fast tracking and data mining.

The rest of the paper is organized as follows. Section 2 introduces the GWAC project and data sources. Section 3 presents our work. Section 4 describes our experimental results. Section 5 is related work. Section 6 summaries our work and presents directions for future work.

2 Background

2.1 GWAC Project

The Ground-based Wide-Angle Camera array (GWAC) with 20 cameras[1], which was built in China, was a part of the SVOM space mission. Each camera can observe about 175,600 objects. Thanks to the low exposure cycle and large FoV, the survey cycle of GWAC was equal to the exposure cycle 15 s. Thus, it is ideal for searching for optical transients of various types by continuously imaging the same region.

GWAC can produce 3.5 million rows of data per 15 s and 6.7 billion rows of data are produced in one night (about 8 h). 2.24PB size of data will be produced

---

[1] GWAC will eventually expand to 36 cameras.

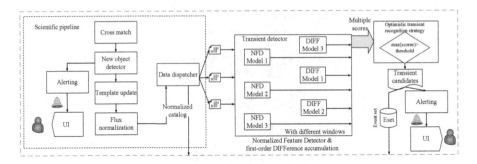

Fig. 4. Scientific pipeline and detector (the rest) in detector service on streaming

over 10 years. It suggests that the worst performance of our database must meet (1) the throughput is more than 234,133 rows per second in detect service on streaming, (2) the throughput is also more than 234,133 rows per second and the query latency on 6.7 billion rows of data is less than 15 s, and (3) long-term data service can also ingest 6.7 billion rows of data into database within 12 h and can manage 2.24PB data and provide the tolerable query latency.

2.2 Data Sources

GWAC as a camera array only produces the image and the image will be transformed into a relational table named as the catalog file through point source extracting, etc, as shown in Fig. 2. The catalog file sent from remote place (e.g., Xinglong) to local servers (e.g., Beijing) will be used as AstroServ's final input. To easily design our database, we first develop a distributed data generator on local cluster where each machine simulates a GWAC's camera and synchronously generates the catalog file. In addition, we also design a transient simulator to randomly attach the scientific phenomenon to stars (e.g., attaching continuous microlensing to some star) to test our detect service.

3   AstroServ's Architecture

As shown in Fig. 3, AstroServ mainly includes six major components, where detect service in streaming, real-time data service, long-term data service and query engine belong to the core components. The core components manage different phases of data and provide the query service. Finally, achieve the full life-cycle management.

3.1   Detect Service on Streaming

This component follows the "master-slave" mode and each slave node serves the data generated by a camera, including two functions: (1) scientific pipeline and (2) detector [10]. Scientific pipeline can assign the IDs to stars, find the new

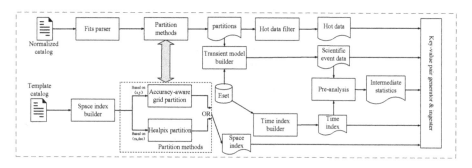

Fig. 5. Architecture of real-time data service

stars and normalizes the catalog data to help detector improve the recognition precision. Then, several key steps have to be done. First, the catalog file where stars have no IDs is sent to cross-match module [16], stars will be given IDs by searching the nearest star in the template table which includes all of identified stars. In cross-match module, we use the pixel coordinates to improve the performance [21]. If the star in catalog does not appear in the template, it suggests a new star and AstroServ will send an alerting signal. When the new stars appear, they will be added into the template table. We design an encode strategy to define the star ID and maintain the template table automatically. In addition, flux normalization was carried out through comparison to standard stars in the same fields to correct the magnitude error due to external interference, such as the cloud. After that, the normalized catalog could be just used for transient detection and storage (Fig. 4).

The detector is used for recognizing the scientific phenomena. After each exposure, it receives the normalized catalogs and returns the set of abnormal star IDs, called as Eset (Event set). The detector uses the integration framework based on sliding window, including six models to detect different types of scientific phenomena. Six models derive from two methods being NFD (Normalized Feature Detector) and DIFF (first-order DIFFerence accumulation) [16] with different window sizes. Each model will give a score representing the probability of an abnormal star to next module. An optimistic transient recognition strategy will be used to decide Eset according to the maximum score of every stars. Finally, send the alerting signals.

3.2 Real-Time Data Service

This component focuses on efficient storage and query of short-term data (e.g., one night of data). For example, one night of data in GWAC has high value, so that the low-latency access to these data can help user analyze scientific phenomena on site. Thus, both the ingest and query latency must be less than the exposure interval. As shown in Fig. 5, we use a set of optimization methods to improve the performance. We build scientific phenomenon model based on key-value schema to organize data to speed up the access to time-series data.

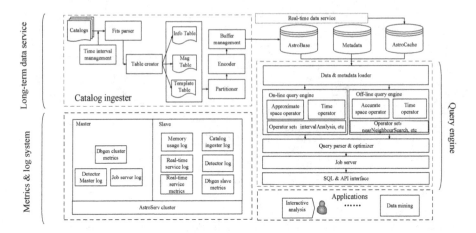

Fig. 6. Long-term data service, query engine, metrics & log system and applications

In addition, we use partition methods to partition both catalog and template table and physically cluster star data of the same partition to improve the ingest due to less partition keys. Noting that we implement two partition methods: (1) accuracy-aware grid partition presented by us and (2) Healpix partition [12]. Accuracy-aware grid partition supports the approximate query to improve the query performance. For improving the access to scientific phenomenon data, we build a time index based on time-evolving feature. It has the insert-friendly structure and supports the distributed scan. For speeding up the analysis to scientific phenomenon data, we first pre-analyze the abnormal stars inside every partition and generate intermediate statistics to avoid the access to original data. In addition, we also store hot data separately to improve their access. All of generated data are finally transformed into key-value pairs and inserted into a distributed key-value in-memory store. Although we insert multiple types of data but they are clustered into different partitions so that it does not reduce the ingest performance. Using the partition as the insert unit can also reduce the unnecessary data structure overhead.

3.3   Long-Term Data Service

This component is actually a catalog ingester which ingests one night of catalogs into a disturbed on-disk store. The major challenge is to ensure that the insert latency is less than 8 h. Thus, we overlap the processing between real-time data service and long-term data service. As shown in Fig. 6, long-term data service (i.e., time interval management) can dynamically monitor the generation of normalized catalog and ingest catalogs when real-time data service is idle. We totally use three logical tables to keep data: (1) original time-series information except magnitude attribute in Info table, (2) magnitude attribute in Mag table and (3) non-time-series information such as locations in Template table. This organization form can improve the performance of the popular light curve analysis. For saving storage space, we design an encoder to compress data. In addition, the usage of memory buffer can also improve the insert performance.

3.4 Query Engine

This component needs to support both the real-time analysis on short-term data and complex analysis on long-term data. Short-term data is kept into AstroCache being a disturbed key-value in-memory store, such as Redis cluster [6] and long-term data is kept into AstroBase being a disturbed on-disk store, such as Hbase [2]. Real-time analysis focuses on the scientific phenomena recognized by detector and complex analysis focuses on large-scale star data. We design a query engine integration framework which can transparently execute queries from different data sources. The framework uses Web service (i.e., Job server module) as the query interface and supports both SQL-like and API to access AstroServ. By query parser, query engine can choose the correct sub-engine to run query. As shown in Fig. 6, our on-line query engine runs approximate query efficiently and can meet the minimum accuracy, but our off-line query engine is accurate. When the query engine is launched, we first load the metadata into memory to improve the query performance and every query job will be run distributedly.

3.5 Other Components

Metrics and Log System. AstroServ as a distributed database follows the "master-slave" mode, so that we monitor services on master and slave nodes, respectively. Different logs are benefit from profiling our system. The major metrics include the size of data in both AstroCache and AstroBase, the Eset's size, the catalog size, the ingest and query latency, etc.

Applications. They mainly obtain data from the core components and help users analyze data in an easier way. Currently, our applications include inter-active analysis interface and data mining tool. Interactive analysis interface is a web UI which is ability to graphically display alerting signals and analysis results on an earth model. Data mining tool is a machine learning algorithm integration framework which can recognize long-timescale scientific phenomena from long-term historical data.

4 Experiments

4.1 Experimental Setup

We evaluate how well AstroServ works by using four metrics: insertion latency, query latency, detector recognition precision and storage usage. We use GWAC [19] as an example to test our system, in which each observation unit can collect 175,600 objects per 15 s and an observation experiment lasts 1,920 times (about 8 h).

We simulate GWAC by using the distributed data generator. It follows the "master-slave" mode, where each sub-generator simulates an observation unit. A sub-generator produces a catalog file which is a relational table with 175,600 lines and 25 columns per cycle. In our experiment, we assign every sub-generator to run on one machine. In addition, we simulate scientific event signals by setting the Eset size to subject to the geometric distribution and the locations of scientific events to subject to the uniform distribution. The duration of each scientific event is also random. Finally, we simulate 19 observation units on 20 machines (one master), each of which has 12 CPU cores (CPU frequency is low, only 1.6 GHz per core) and 96 GB RAM.

We also build AstroServ's cluster on the same 20 machines where each slave processes data produced by a sub-generator on the local machine. We use Go and C++ to implement detector service and real-time service, respectively. Redis cluster 3.2.11 as the key-value in-memory store where we launch 120 storage nodes (i.e., master and slave nodes are half of each) and each slave node backs up data of a master node. Hbase 1.3.1 is as the on-disk store. Spark 1.6.3 [4] is used for query processing. We will divide the experiment into three parts as follows.

4.2 Results

Detect Service on Streaming. For each slave node, the average detector latency on 17,5600 rows of data is 3.97 s and the variance is 1.6 s. Noting that the Go's garbage collection may cause the detector latency to be longer. Although it has no much impact on the overall performance, but it needs to be further optimized. We also test the detector's accuracy. We use GWAC's specific parameters to generate 3,240 time-series as test data and add the scientific phenomena for each time-series. Our's detector accuracy can achieve 73.5%. In other words, we have ability to detect 2,381 scientific phenomena.

Real-Time Data Service. We ingest 19 catalog files per 15 s by real-time data service lasting 1,920 times (8 h). The average insert latency is 2.35 s which is less than our requirement 15 s. In addition, our ingest latency is stable and the variance is 0.12 s. The size of 8 h data is 1.15TB, and AstroCache only uses 460GB main memory. It reduces the main memory overhead of our system. As a comparison, Redis cluster without our optimization will consume 2.34TB main memory. The main reason is that (1) we only manage the hot data and scientific phenomenon data instead of the entire data set and (2) we use partition data instead of star data as the storage unit which can cut unnecessary data structures to maintain more keys.

Long-Term Data Service. We prepare one-night of data in advance to test the continuous insertion performance. This component consumes 3.5 h to ingest these catalogs into AstroBase and about 6 s for each catalog. The latency is less than our requirement 8 h. As a comparison, Hbase without our optimization needs 13 h to ingest one-night of data and about 24.4 s for each catalog. Our optimization can improve 3.7× insertion performance.

Query Engine. For on-line query engine testing, we refer to the LSST telescope discovery ability [9] and decide to simulate the generation of 200,000 scientific phenomena one night. This number is more than it at the real case to test the worst performance of on-line query engine. The worst query latency on 6.7 billion rows of data is 2.72 s being less than 15 s. For off-line query engine testing, we preliminarily test it on 15.71TB (120 billion rows of data). We assume that the size of result set is less than 1 million rows of data. When the result set is 1.2 million rows of data, the average query latency is 26 s.

5   Related Work

We survey the related work involving (1) astronomical data management system and (2) distributed data store.

Astronomical Data Management System. In time-domain astronomy, catalogs collected by telescopes had to be stored into a long-term database for complex analysis. SkyServer [18] for SDSS was built on Microsoft SQL Server. It was primarily responsible for long-term storage (since 2000), complex query and the primary public interface catalogue data from SDSS. SciServer [7] was a major upgrade of SkyServer. SciServer was a collaborative research environment for large-scale data-driven science, but it still worked on long-term data. Qserv [20] was a distributed shared-nothing database to manage the LSST catalogs over 10 years. It was designed into a MySQL cluster by a proxy server and a distribution file system xrooted. In addition, PostgrelSQL was used for storing the original data for Skymapper. These databases did not consider the efficient real-time time-series data management, due to the low time resolution of HSR sky survey.

Distributed Data Store. The in-memory stores had high throughput and low latency, and several solutions were considered before we embarked on Astro-Cache development. In-memory distribution file systems, such as Alluxio [15], was similar to HDFS [5] on disk. It followed the write-once, read-many approach for its files and applications, but sometimes we had to update data. In-memory distribution messaging systems, such as Apache Kafka [3], did not support random read. In-memory relational databases, such as MonetDB [14], have been tested for the insert performance through GWAC's catalogs [19]. The experiment showed the insert latency was not stable, and sometimes an insert operation could not be finished within 15 s, because of periodically pushing data onto disk. In addition, it did not support query for scientific phenomenon data. In-memory distributed key-value stores, such as Redis cluster [6], have been tested by us through GWAC's catalogs without our optimization. However, the network latency was the main bottleneck because of too many keys. The total amount of memory, which was consumed to keep one-night GWAC's catalogs into Redis cluster, was also unacceptable (more than data size), even if data was compressed. The on-disk stores are suitable for long-term historical data storage. However, some popular solutions are not applicable to our scenario. Apache

Cassandra [1] is a distributed key-value database, but it is suitable for large-scale data storage. HDFS as a distributed file system, can support large-scale data storage but it does not support the necessary index and data organization schema. Hbase without our optimization is also tested by us through GWAC's catalogs. The insert latency is unacceptable in our cluster. In addition, Hbase does not support the scientific analysis methods.

6 Summary

AstroServ is a distributed database to analyze and manage large-scale full life-cycle astronomical data for short-timescale and large field of view sky survey. Detect service on streaming can finish the detect of 3.5 million stars within 3.97 s. Real-time data service can finish the ingest of 3.5 million rows of data within 2.35 s. Long-term data service can finish 6.7 billion rows of data within 3.5 h. The query latency on AstroCache is within 2.72 s and it on AstroBase is within 26 s. The overall performance is improved under our optimization. In future, our query will use polystore framework to automatically adjust data distribution for better performance.

Acknowledgement. This research was partially supported by the grants from the National Key Research and Development Program of China (No. 2016YFB1000602, 2016YFB1000603); the Natural Science Foundation of China (No. 91646203, 61532016, 61532010, 61379050, 61762082); the Fundamental Research Funds for the Central Universities, the Research Funds of Renmin University (No. 11XNL010); and the Science and Technology Opening up Cooperation project of Henan Province (172106000077).

References

1. Apache cassandra. https://cassandra.apache.org/
2. Apache hbase. http://hbase.apache.org/
3. Apache kafka. http://kafka.apache.org/
4. Apache spark. http://spark.apache.org/
5. Hdfs. http://hadoop.apache.org/
6. Redis. https://redis.io/
7. Sciserver. http://www.sciserver.org/
8. The skymapper transient survey. https://arxiv.org/abs/1702.05585
9. Becla, J., Lim, K.-T., Monkewitz, S., Nieto-Santisteban, M., Thakar, A.: Organizing the LSST database for real-time astronomical processing. In: Astronomical Data Analysis Software and Systems, pp. 114–117 (2008)
10. Feng, T., Du, Z., Sun, Y., Wei, J., Bi, J., Liu, J.: Real-time anomaly detection of short-time-scale GWAC survey light curves. In: Proceedings of IEEE International Congress on Big Data, pp. 224–231 (2017)
11. Frieman, J.A., et al.: The sloan digital sky survey - ii: supernova survey: technical summary. Astron. J. **135**(1), 338–347 (2008)
12. Gorski, K.M., Wandelt, B.D., Hansen, F.K., Hivon, E., Banday, A.J.: The healpix primer. Physics (1999)

13. Huber, M., Chambers, K.C., Flewelling, H., Smartt, S.J., Smith, K., Wright, D.: The pan-starrs survey for transients (PSST). In: IAU General Assembly, vol. 22 (2015)
14. Idreos, S., Groffen, F., Nes, N., Manegold, S., Mullender, S., Kersten, M., et al.: MonetDB: Two decades of research in column-oriented database architectures. Q. Bull. IEEE Comput. Soc. Techn. Comm. Database Eng. **35**(1), 40–45 (2012)
15. Li, H., Ghodsi, A., Zaharia, M., Shenker, S., Stoica, I.: Tachyon: reliable, memory speed storage for cluster computing frameworks. In: Proceedings of the ACM Symposium on Cloud Computing, pp. 1–15 (2014)
16. Nieto-Santisteban, M.A., Thakar, A.R., Szalay, A.S.: Cross-matching very large datasets. In: National Science and Technology Council (NSTC) NASA Conference (2007)
17. Nugent, P., Cao, Y., Kasliwal, M.: The palomar transient factory. In: Proceedings of SPIE, pp. 939702–939702 (2015)
18. Szalay, A.S., et al.: The SDSS skyserver: public access to the sloan digital sky server data. In: Proceedings of ACM SIGMOD International Conference on Management of Data, pp. 570–581 (2002)
19. Wan, M., et al.: Column store for GWAC: a high-cadence, high-density, large-scale astronomical light curve pipeline and distributed shared-nothing database. Publ. Astron. Soc. Pac. **128**(969), 114501–114532 (2016)
20. Wang, D.L., Monkewitz, S.M., Lim, K.T., Becla, J.: Qserv: a distributed shared-nothing database for the LSST catalog. In: High Performance Computing, Networking, Storage and Analysis, pp. 1–11 (2011)
21. Yang, X., et al.: A fast cross-identification algorithm for searching optical transient sources. Astron. Res. Technol. **10**(3), 273–282 (2013)

AstroBase: Distributed Long-Term Astronomical Database

Kenan Liang[1], Wei Guo[1], Lizhen Cui[1(✉)], Meng Xu[2], and Qingzhong Li[1]

[1] ShanDong University, JiNan, China
clz@sdu.edu.cn
[2] Shandong Technology and Business University, Yantai, China

Abstract. China's self-developed GWAC is different from previous astronomical projects. GWAC consists of 40 wide-angle telescopes, collecting image data of the entire sky every 15 s, and requires data to be processed and alerted in real time within 15 s. These requirements are due to GWAC. Committed to discovering and timely capturing the development of short-time astronomical phenomena, such as supernova explosions, gamma blasts [1], and microgravity lenses, hoping that GWAC will be able to make timely warnings and early warnings in the early stages of these astronomical phenomena. The astronomical researchers are provided with detailed information on the event by scheduling a deep telescope to record the entire process of astronomical time development. Second, the GWAC project requires observations for up to 10 years of storage. These long-term stored data are provided to astronomers to help the astronomers get new discoveries after the technology and means are updated, as well as astronomy for astronomers. Big data mining provides support.

Keywords: GWAC · Astronomical · Storage

1 Introduction

With the development of computer technology, modern astronomical observations use computers to analyze and process data on a large scale, and at the same time promote the richness of astronomical observation technology. These astronomical observation techniques combined with the development of big data technology, the real-time analysis and management of massive astronomical big data has become possible. The astronomical big data of these explosions also poses extremely high challenges and requirements for the analysis and management of astronomical data. In 2000, the Sloan Digital Sky Survey (SDSS) telescope in New Mexico collected more data in just a few weeks than in the history of astronomy. By 2010, information The file has been as high as 1.4 242B. However, the Large Synopticc Survey Telescope (LSST), which is expected to be used in Chile in 2019 [2], will receive the same amount of data in five days. In China, the Large Sky Area Multi-Object Fiber Spectroscopic Telescope (LAMOST) and the Ground-Based Wide-Angle Camera Array (GWAC) surveys generate massive amounts of data every day.

J. Li et al. (Eds.): BigSDM 2018, LNCS 11473, pp. 56–66, 2019.
https://doi.org/10.1007/978-3-030-28061-1_7

The early astronomical data management system used file data management. Early data management relied on observation methods and observation equipment. The amount of data to be managed was not large, and only data analysis and archiving were required. However, with the observance of observation methods and observation equipment, the amount of observed data is often difficult to cope with using traditional methods. Some projects use cloud computing to manage and analyze massive astronomical data. For example, the pulsar detection project in Swinburn, Australia [1] uses the cloud computing platform to analyze and process the data collected by the Parks radio telescope at a rate of 1 GB per second, which involves a large number of calculations and finally processes the results. And some intermediate results are archived and saved. Nowadays, astronomical researchers are no longer satisfied with the needs of the previous analysis and processing, but need to be able to query the data they need according to the needs of astronomers. Currently, some astronomical projects use the management of relational databases, such as the SDSS Sky Survey. The data is processed periodically and stored in a relational database with a sample period of 71.7 s as a sampling period for use by astronomical researchers and astronomers.

However, the GWAC designed in China is different from the previous astronomical project. The GWAC 40 wide-angle telescope collects the image data of the entire sky every 15 s, and requires the data to be processed and alerted in real time within 15 s. These requirements are due to GWAC's [3] efforts to discover and Timely capture the development of short-term astronomical phenomena, such as supernova explosions, gamma blasts, and microgravity lenses. It is hoped that GWAC will be able to make timely warnings and early warnings in the early stages of these astronomical phenomena. A deep telescope to record the entire process of astronomical time development [7] to help astronomical researchers obtain detailed information on the event. Second, the GWAC project requires observations for up to 10 years of storage. These long-term stored data are provided [5] to astronomers to help the astronomers get new discoveries after the technology and means are updated, as well as astronomy for astronomers. Big data mining provides support.

Massive astronomical data presents a series of challenges. The traditional astronomical data management system can not meet the needs of astronomers for real-time processing, storage, query and astronomical phenomena discovery of massive astronomical data. For the real-time processing of GWAC massive data processing and long-term management requirements, the traditional relational database is limited by the single-node limitation and cannot meet the processing requirements of a large amount of streaming data. Subsequent distributed systems such as Hadoop/HDFS cannot meet real-time queries and analysis. NoSQL databases [6] such as MongoDB have poor support for indexes and cannot meet the demand for efficient connection of large data volumes.

At present, researchers have designed a real-time data management system for GWAC, which initially realizes the real-time data processing and astronomical phenomenon discovery. However, there are still some problems in the real-time system. The data processed in real time is not only related to the astronomical phenomenon. None of them were saved, and astronomers could not view the historical data observed by the telescope, nor could they query the massive data. Not only that, the astronomical phenomenon discovery module of the real-time system uses an algorithm with low accuracy and low recall rate in order to ensure the performance of the system as much as possible, which leads to some astronomical phenomena that cannot be discovered, and can not guarantee the accuracy of the discovered astronomical phenomena. Sex.

Therefore, with the existing problems of the needle, we combined with the application background of the GWAC telescope to design an off-line analysis and management system for astronomical data to solve the above series of problems [4]. The system is mainly divided into three parts: storage manager, query engine and astronomical phenomenon mining [8]. The storage manager mainly implements the rapid storage of massive astronomical data, and organizes the data structure in the database according to the cluster state and the data interface. The query engine mainly implements a professional query interface for astronomy. The astronomical phenomenon mining part mainly realizes mining and analyzing astronomical phenomena from massive data.

2   Related Work

2.1   GWAC Telescope Array

The ground-based wide-angle camera array GWAC under construction in China consists of 40 wide-angle telescopes with a diameter of 18 cm. Each telescope is equipped with a 4k * 4k charge coupled device (CCD) detector. The entire camera array covers a range of 5,000 square degrees and a time sampling period of 15 s. The observed observations of the fixed sky area targets for each observation night were as long as 10 h. From the observation of the size of the field of view and the sampling frequency of the observation time, the ground wide-angle camera array has a special advantage in time domain astronomical observation. The huge amount of data and high-speed sampling rate pose great challenges to the management and processing of data. The star table data indicators of the ground wide-angle camera array are: (1) The star table data has about $1.7 * 10^5$ records per image, and the entire camera array produces $6.8 * 10^6$ (data volume is about 1.3 GB) records in 15 s, each There are about 2400 * 40 = 96000 pictures in the evening, which requires about 2 TB of storage overhead. (2) With a 10-year design cycle, GWAC will produce a super-large-scale catalogue of the order of 3–6PB.

2.2   Astronomical Data Management Related Work

At present, the main functions of astronomical databases at home and abroad are still concentrated on electronic archiving, search and download, and mainly through three stages.

(1) In the rise phase, the astronomical database at this time is mainly based on the data storage of the file system. The most famous one is the set of identifications, measurements, and bibliography for astronomical data (SIMBAD) of the Centre Data Center CDS (centre de Donnes stellarires). Manage astronomical data, archive, sort, and organize data, and provide cross-recognition and file catalog retrieval for global catalogs.

(2) The relational database implements the astronomical data management stage, represented by VizieR and SDSS, which provide star catalog services. By the end of the 1990 s, SIMBAD services could not meet more complex query requirements,

and CDS developed a more powerful ViziR. System, ViziR underlying dependency data model, supports ID and location based search, and does not have the maximum search radius requirement, has faster response speed [9], but the search is less customized, in addition, another professional astronomical data management service for Sloan Digital Sky Survey SDSS self-developed database. SDSS's astronomical database Skyserver is based on Microsoft's SQL Server custom development, with features such as fast query, batch download, SQL retrieval and visual graphical interface. This stage of astronomical data management began to customize the scientific application of various astronomical data on the basis of the database to meet the special retrieval needs of astronomical data.

(3) The upcoming large-scale astronomical database stage, represented by the US large-diameter panoramic LSST and SKA (square kilometer array) []. Some emerging astronomical fields such as gamma-ray bursts and supernova explosions have more urgent requirements for time-domain astronomical observations, which directly lead to the explosive growth of astronomical data. The US LSST design records 3 images of 1 billion pixels per 15 s. The amount of data collected every night is about 15–30 TB, and it can be visited once every 3 days. It is expected to receive observation missions [10] in 2022. The Australian SKA program generates more than 12 terabytes of data per second. The original image produced in a single day is 1 EB, and the first phase of construction is expected to begin in 2020. The above-mentioned large-scale astronomical observation project has created a huge challenge to the current data management framework. High-throughput, large-scale storage and fast search have become the main problems. It is worth mentioning that Meng et al. [7] have carried out some research work on the current GWAC data management scenario, and proposed a management scheme based on MonetDB database. The GWAC [3] data generator gwac_dbgen has been developed to simulate the real data format and magnitude of a CCD continuously generated. In addition, based on the simulation data of the generator, the cross-authentication algorithm in the MonetDB database is implemented using SQL to avoid data movement. However, when the accumulated data size is large, the scalability of MonetDB is poor and the storage time is not stable enough.

3 Design Concepts

3.1 Scalability

Over time, the system will accumulate a large amount of astronomical data, and the observatory will increase the number of observing equipment. We have designed an easily scalable model. This model can add data storage devices when the amount of data is large, or add data processing devices when adding astronomical observing devices.

3.2 Low Latency

GWAC is a short-cycle astronomical observing device, which requires data processing to be completed in a short time or it will cause data congestion [12]. We designed a high-performance astronomical data processing model and astronomical phenomena detection model.

3.3   Parallelism

The observation data acquired by each telescope in the GWAC telescope array is unshared, and the data processing flow of each telescope can be completely separated, so we use a parallel architecture for astronomical data processing.

4   System Architecture

4.1   Overall Architecture

The huge amount of observed data gives the system two challenges: (1) the system needs to have extremely high throughput. (2) The system needs high-performance real-time processing and analysis functions. In response to the above challenges, we propose a real-time astronomical data processing framework based on the characteristics of GWAC data. GWAC consists of forty telescopes, and the observation data of each telescope is not related to each other [14]. Therefore, we use forty servers for each telescope. Parallel real-time processing and analysis of observation data, one server performs overall control and status monitoring on the real-time layer system. The system architecture of the real-time layer is as follows (Fig. 1).

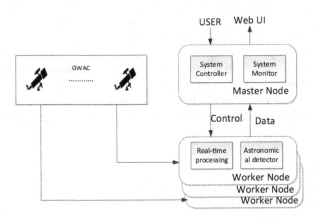

Fig. 1. Ovreall architecture.

In the system, the user performs overall control and monitoring through the master node. Each worker node processes the observation data acquired by one telescope. After real-time processing of the observation data and astronomical phenomenon discovery, the worker node returns the hot data such as astronomical phenomena. To the master node, the master node then tells these hotspot data to be displayed to the user through the front end.

4.2 Distributed Data Storage Scheme with Mixed Memory and Hard Disk

There are three parts of the data that astronomers pay more attention to, template star catalog data, observation star catalog data, and astronomical event data. The template star table data mainly records the celestial bodies in the sky, where each line represents the basic information of a celestial body. Observational star catalog data primarily records specific data for each observation of a celestial body in the sky, with each row representing information observed once by a celestial body. The astronomical event data mainly records every astronomical phenomenon in the sky, where each row represents the specific information of an astronomical event of a celestial body. The data descriptions of the specific three kinds of star tables are as follows.

There are three parts of the data that astronomers pay more attention to, template star catalog data, observation star catalog data, and astronomical event data. The template star table data mainly records the celestial bodies in the sky, where each line represents the basic information of a celestial body. Observational star catalog data primarily records specific data for each observation of a celestial body in the sky, with each row representing information observed once by a celestial body. The astronomical event data mainly records every astronomical phenomenon in the sky, where each row represents the specific information of an astronomical event of a celestial body. The data descriptions of the specific three kinds of star tables are as follows.

In order to satisfy the persistent storage of massive data and a series of query operations including join query, we designed a distributed storage scheme of mixed storage of memory and hard disk based on GWAC data background. For hot data such as template star table and astronomy for query access. The event star table, we store it in the Redis cluster, we store it in the HBase cluster for the observation star catalog with less access to the query. This storage method guarantees the reliability of the persistent star table data, and also makes the template star catalog in Redis and the join operation of the observation star table in HBase have better performance (Fig. 2).

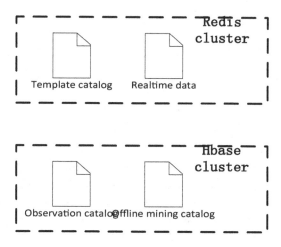

Fig. 2. Data storage schema.

4.3   Astronomical Data Offline Management and Mining Framework

The offline layer system mainly faces three challenges: (1) the system needs to be scalable, and the massive star table data makes the system have to periodically increase the storage node to provide space for subsequent data. (2) High-performance data storage and mining solutions. Since the real-time layer system occupies a large amount of system CPU and memory resources every night, the data storage and mining of the offline layer system can only be performed during the day. (3) The huge amount of data also brings a high delay to data query. How to combine storage and query to give a reasonable offline data storage solution. We designed astronomical data offline management and mining framework based on Hadoop\HBase\Spark cluster for the above problems. The architecture diagram of the offline layer system is as follows. The following mainly from offline data (Fig. 3).

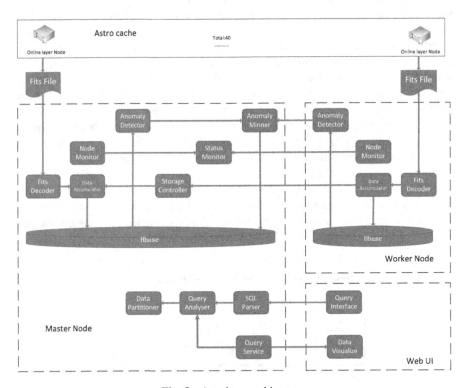

Fig. 3. Astrobase architecture.

For the data processed by the real-time layer, the offline layer also uses a parallel way to store data into the database. At the same time, combined with the characteristics of astronomical query, we pre-calculate and store the intermediate data required by some queries during the storage. Query performance. For the massive star table data stored in HBase, we use the time series data analysis algorithm to mine it based on Spark.

4.3.1 Parallel Data Entry Mechanism

In order to improve the warehousing performance as much as possible, for the 40 nodes processed in the real-time layer, we will perform the warehousing operation in parallel. HBase can effectively support the simultaneous storage of multiple nodes. The experiment shows that the GWAC nightly observation data is In the daytime, it can be completed in about 200 min.

4.3.2 Calculating the Integration Data

In order to provide high-performance query services for massive astronomical data, we propose the idea of calculating the integration of data, that is, in the daily storage process, the intermediate data (such as average value, maximum value, minimum value, etc.) required for some queries are stored in the process. Calculate in order to reduce the performance overhead of the query.

4.3.3 Data Blocking

In massive astronomical data, astronomers are always more interested in some of the data. From the perspective of data columns, queries such as star brightness values and star space coordinates are more frequent. In view of such data characteristics, we propose the idea of data partitioning, which divides the data column into hotspot columns and non-hotspot columns, and stores the hotspot columns and non-hotspot columns separately, thereby improving query performance.

5 Experiments

We have done a series of tests on the storage and query performance of the system. The test results show that our system has good storage speed and query performance.

We performed a series of tests on the performance of astrobase on a 20-node cluster, including data warehousing performance and query performance. The servers used in the experiment are all ubuntu16.4 system environments, and the deployed software environment is hadoop2.8.1, hbase1.3.1. The storage space of each server is about 1 TB, and the memory space of each server is 32 GB.

5.1 Astrobase Data Storage Performance Experiments

We simulated the data generated by 20 telescopes and tested the gwac data warehousing performance on 20 nodes. The experimental results show that the nightly observation data can complete the data warehousing operation in about 3.5 h, which can meet the basic needs of gwac.

5.2 Astrobase Query Performance Experiments

In order to improve the query performance of GWAC astronomical data, we adopt the idea of computing and integrating data. In the data storage process, some hot query data (such as average value, maximum and minimum value) are pre-calculated, and daily operation results are recorded. When the user queries, these calculation data can be directly called to improve the overall query performance.

We designed an intermediate calculation data table, saved the intermediate results in a table, and provided a query for the intermediate calculation data. We currently record the average of the daily light curve and store it in astroBase.

Intermediate Data Calculation Delay

The calculation of intermediate data is mainly carried out in the data warehousing process, and the calculation process will have a slight impact on the warehousing performance. The following figure is a graph of the storage performance when the calculated data is not used and the calculated data is used. We tested the performance impact of intermediate data calculations when storing 3000 star catalog files (Fig. 4).

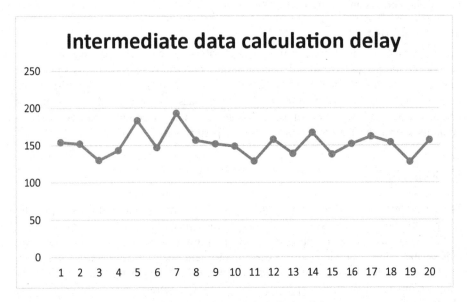

Fig. 4. Intermediate data calculation delay

As can be seen from the above figure, the calculation of the intermediate data will have a performance impact of about 150 s on the daily data storage. Compared with the overall storage time of about 4 h, the impact is not very large.

Intermediate Data Query Performance

The use of computational integration data can significantly improve query performance. We tested the query latency with and without this scheme. The chart is as follows (Fig. 5).

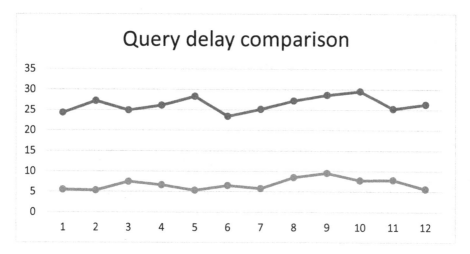

Fig. 5. Query delay comparison

It can be seen that before the calculation is integrated into the data scheme, the query delay is about 27 s. After the calculation is integrated into the data scheme, the query delay is reduced to about 6.4 s, which significantly improves the query performance.

6 Conclusion

Experiments show that astrobase can meet the long-term management needs of astronomers for astronomical data. Astrobase can realize long-term storage of billions of rows of data and provide high-performance index for data query.

Acknowledgement. This work is partially supported by National Key R&D Program No. 2016YFB1000602.

References

1. Yuan, D., Yang, Y., Liu, X., et al.: On-demand minimum cost benchmarking for intermediate dataset storage in scientific cloud workflow systems. J. Parallel Distrib. Comput. **71**(2), 316–332 (2011)
2. Boncz, P., Grust, T., Keulen, M.V., et al.: MonetDB/XQuery:a fast XQuery processor powered by a relational engine. In: ACM SIGMOD International Conference on Management of Data, pp. 479–490. ACM (2006)
3. Wan, M., Wu, C., Wang, J., et al.: Column store for GWAC: a high-cadence, high-density, large-scale astronomical light curve pipeline and distributed shared-nothing database. Publ. Astron. Soc. Pac. **128**, 114501 (2016)
4. Apache kafka. http://kafka.apache.org
5. Apache hbase. http://hbase.apache.org

6. Redislab redis. http://redis.io
7. Meng, W.: GWACdbgen. http://github.com/wanmeng/gwac_dbgen
8. Jian, L.I., Cui, C.Z., Bo-Liang, H.E., et al.: Review and prospect of the astronomical database. Prog. Astron. **31**(1), 1–16 (2013)
9. SDSS Skyserver. http://skyserver.org/
10. Zaharia, M., Chowdhury, M., Franklin, M.J., et al.: Spark: cluster computing with working sets. In: Usenix Conference on Hot Topics in Cloud Computing, p. 10. USENIX Association (2010)
11. Apache hbase. http://hadoop.apache.org
12. Meng, X., Meng, X., Meng, X., et al.: Spark SQL: relational data processing in spark. In: ACM SIGMOD International Conference on Management of Data, pp. 1383–1394. ACM (2015)
13. Chandola, V., Banerjee, A., Kumar, V.: Anomaly detection: a survey. ACM Comput. Surv. (CSUR) **41**(3), 1–58 (2009)
14. Ahmad, S., Purdy, S.: Real-Time Anomaly Detection for Streaming Analytics (2016)

An Efficient Parallel Framework
to Analyze Astronomical Sky Survey Data

Xiaobin Ma[1], Zhihui Du[1(✉)], Yankui Sun[1], Andrei Tchernykh[2], Chao Wu[3],
and Jianyan Wei[3]

[1] Tsinghua National Laboratory for Information Science and Technology,
Department of Computer Science and Technology, Tsinghua University,
Beijing, China
duzh@tsinghua.edu.cn
[2] CICESE Research Center, Ensenada, Mexico
[3] National Astronomical Observatories, Chinese Academy of Sciences, Beijing, China

Abstract. Big data has been an important analysis method anywhere
we turn today. We hold broad recognition of the value of data, and products obtained through analyzing it. There are multiple steps to the data
analysis pipeline, which can be abstracted as a framework provides uni-
versal parallel high-performance data analysis. Based on ray, this paper
proposed a parallel framework written in Python with an interface to
aggregate and analyze homogeneous astronomical sky survey time series
data. As such, we can achieve parallel training and analysis only by defining the customized analyze functions, decision module and I/O interfaces,
while the framework is able to manage the pipeline such as data fetching,
saving, parallel job scheduling and load balancing. Meanwhile, the data
scientists can focus on the analysis procedure and save the time speeding
this program up. We tested out the framework on synthetic data with
raw files and HBase entries as data sources and result formats, reduced
the analyze cost for scientists not familiar with parallel programming
while needs to handle a mass of data. We integrate time series anomaly
detection algorithms with our parallel dispatching module to achieve
high-performance data processing frameworks. Experimental results on
synthetic astronomical sky survey time series data show that our model
achieves good speed up ratio in executing analysis programs.

Keywords: Big data analysis · Deep learning algorithm ·
Performance optimization

1 Introduction

In recent years human society generates data increasingly, the large proportion
of which are time series data, such as ECG [1], data center system metrics, traffic flow data, financial data and other industrial monitor data. For example, a
search engine can monitor its search response time (SRT) by detecting anomalies in SRT time series in an online fashion [7]. These time-resolved values can

© Springer Nature Switzerland AG 2019
J. Li et al. (Eds.): BigSDM 2018, LNCS 11473, pp. 67–77, 2019.
https://doi.org/10.1007/978-3-030-28061-1_8

reflect the characteristics of a metric through time axis. We are able to sketch the outlines of a dataset, but handling outpouring data requires lots of computing resources, paralleled high-performance computing emerged as a promising approach for data processing. But most data scientists are not familiar with database programming or parallel programming, or they are not proficient in implementing high-performance programs which can meet the need for high-performance computing and digesting data in quantities in modern society. What is inspiring is that time series data are formatted data that they are homogeneous, so processing procedures of these data have many steps in common that could be summarized as a general parallel strategic module to speed up the programming and synthesize the result from different sources. The I/O of the data and the analysis operation is the only variable between two different analysis tasks. Other components like the scheduling policy, data cache pool, and auxiliary utilities can be reused.

The major contributions of our paper are as follows: Firstly, we present a parallel data analysis framework based on parallel programming modules and data analysis modules. We provide simple interface allows programmers not familiar with parallel programming knowledge to implement high-performance data analysis programs dealing with big data [10] simply.

Secondly, we implemented a data cache pool based on producer-consumer module for distributed shared memory task scheduling. Based on these ideas, we proposed a modularized data computing pipeline, a database and disk I/O optimized parallel data processing framework composed of data acquisition module, data persistence module, training module and data processing module. By organizing them with a parallel programming framework, we build a high-performance data processing framework with strong scalability. In this paper, we introduce the architecture of the parallel data processing pipeline framework.

In this paper, we introduced the background of astronomical survey data analysis and discussed the related works in Sect. 2. Section 3 presents the overall design and implementation of our framework dealing with the need for analysis task. Section 4 introduced the application of our framework in production. Finally, Sect. 5 concludes this paper.

2   Background and Related Work

Ground-Wide Angle Cameras (GWAC) [9] in China is a telescope build for observing the sky continuously, which can generate 1-Dimensional star lightness time series data every dozen seconds. Exploiting anomaly incidents in these accumulated curves are critical for observing astronomical exploration like exoplanet or supernova explosion. These astronomical phenomena would cause anomalies on light curves with certain patterns, which could be a spike or frequency, phase shift on the data values. Detecting anomalies means discovering transient astronomical objects or capture historical astronomical incidents, is of great significance in the field of astronomy. Data analysis plays an important role in the latest researches and industrial fields. But online real-time anomaly detection attaches more importance to time-efficiency, it requires low alert delay after the anomaly appeared. On the online system, the input data from real-time cameras to the module is not complete that it only consists of the data before the current time. Offline data analysis allow a more complex module to process these data with higher complexity and resource consumption. Moreover, the view of the module is more complete, the input to the module can be the data of the day, containing the whole data points of the anomaly, such that the method we use and the feature we focus on would be much different from the real-time data analysis module.

Existing big data processing framework like MapReduce [8] is prepared for computer engineers manipulating a cluster with hundreds or thousands of nodes. Hadoop is a distributed data management architecture allows users to manage big data with high throughput by a bunch of parallel computing instructions, which aims at large scale computing tasks and high reliability. Data scientist without much understanding of parallel programming cannot make the best of such heavyweight programming model, and the abstraction of this programming module is not easy for data scientists to implement scientific computation algorithms. Analogously, Spark [17] is a memory-based big data computing architecture, which mainly aims at the programming module and needs a distributed or high-performance file system to manage I/O and the data flow. And its elusive architecture is, similarly, hard for analyst without parallel programming basics to write a data analysis programming fast and efficiently. In the meantime, data analysis software like R [14] or SPSS [5] are not scalable enough to connect various data sources or implement parallel acceleration modules, to provide the user a full and complete programming package to implement an analysis system. While MATLAB and other data analysis software are not fully paralleled for big data and large training tasks, although they have integrated with many statistic algorithms, efficiency is the other important point dealing with such massive datasets. So it is urgent to develop a parallel programming framework with a simplified interface for high-performance big data analysis.

Ray [12] is a high-performance distributed execution framework targeted at large-scale machine learning and reinforcement learning applications. It provides us with a unified interface that can express both task-parallel and actor-based computations, supported by a single dynamic execution engine. To meet the performance requirements, Ray employs a distributed scheduler and a distributed and fault-tolerant store to manage the system's control state. It is a high performance framework. Python [4] is a scripting language absorbed the advantages of both mathematical computation and system level programming, which has better potentials to implement parallel programs while with great richness of scientific computing libraries like machine learning libraries as *Keras* [2], *Scikit-learn* [13] and math libraries such as *Numpy* [3]. Secondly, Python is with better portability and is cross-platform while scientists can execute this program among all supported systems. Thirdly, Python is compatible with different databases and network libraries, we can easily connect our framework with various databases for file systems to a collection. By contrast, C++ and other languages which are more close to the system level is not friendly to data analysis for data scientists, although they can achieve great parallel performance. On the contrary, the R language is not feasible for achieving parallel model though it has powerful scientific computing libraries. So we pick Python to achieve this data analysis framework which can exploit best parallel performance and provide strong script libraries to scientific analysis tasks.

3   Architecture

The whole pipeline consists of four components, a data cache pool, an analysis module, a result write-back interface and a dispatcher. We implemented the dispatcher which determined the interfaces among these components and the pipeline logic. The data cache pool reads from the data source constantly to feed the data to the dispatcher for analysis, the framework is implemented by a controller and a Python lib *Queue*, a producer-consumer model, which ensured the thread safety among the threads dispatched by the dispatcher component. The dispatcher controls all the slave threads who carry out the data fetching, analysis and write back procedures, while it just invokes the corresponding functions and has no idea about how these components work, the dispatcher connects the data flow and dispatches these threads.

3.1   Task Scheduling

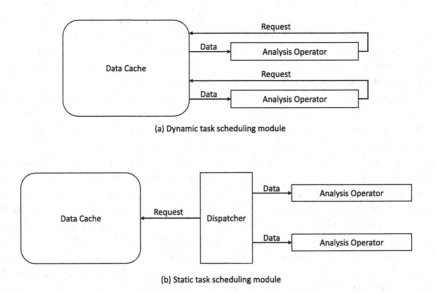

(a) Dynamic task scheduling module

(b) Static task scheduling module

Fig. 1. Task scheduling strategies

The framework supports two strategies for task scheduling: static scheduling and dynamic scheduling. Static scheduling generates certain jobs at the same time, and distributes each task to one thread respectively, the slaves will process and analyze the data and wait for others to finish, they will be assigned with new task together once all the threads finished the analysis operation. On the contrary, the request for data is done by each thread under the scenario of dynamic scheduling, and once a slave finished its analysis operation, it will request for a new one immediately (Fig. 1).

By default, the framework works in static scheduling mode, because there won't be much difference applying the same analysis model to different data with similar format but different content and static scheduling could aggregate the I/O and decrease the time spend on communication. We implemented the dispatch logic that manages the data flow. While the user needs to define the interface between the raw data source and data cache pool and get known of the format we used to store the data.

3.2 Analysis Pipeline and Parallel Neural Network Training

Figure 2 illustrates the architecture of our framework. From the programmer's view, they need to define the function of each link in the pipeline like how the model gets data, how to analyze them to obtain the result and how to save the results.

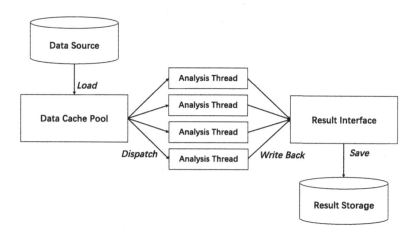

Fig. 2. Analyzer architecture

The cache pool controller will fetch data from the predefined data source once the amount of data in the pool is less than a threshold. We keep a cache pool to feed all the threads the time series data, and the write back is done respectively by these slaves. Both training and processing jobs can be scheduled by this framework.

For deep neural network training tasks, the ray is talented with a framework can be configured to use both GPUs and CPUs to proceed the training jobs, named *Ray Agent Policy Optimizer* as Fig. 3 shows. At a high level, Ray provides an Agent class which holds a policy for environment interaction. Through the agent interface, the policy can be trained, checkpointed, or an action computed. Slaves are managed by Ray to update the parameters and configure automatically and thus to speed up the training procedure. As for other training jobs, the parameters might not able to be updated by more than one task or accelerated by parallel computing, to ensure the correctness of the training algorithm and the capability of our framework to all algorithms the module would use, the framework can only digest the data serially for typical analysis algorithms. Meanwhile, RLlib [11] is integrated with Ray for reinforcement learning that offers both a collection of reference algorithms and scalable primitives for composing new ones.

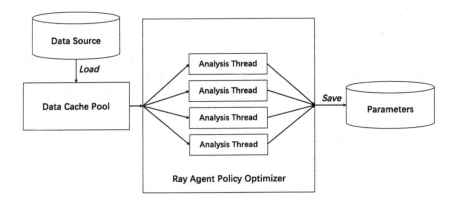

Fig. 3. Distributed deep network training architecture

As Fig. 2 shows, the parallel data processing module is designed to achieve high-performance analysis, the programmer is expected to implement following functions to complete the whole data analysis flow: (1) Function defining the procedure we collect the data from the database for raw CSV file, as the connection between the manager of the data cache pool. (2) The function implements the entire analysis procedure, with interface *getData* to get single data from data pool as an input indexed by the name of single data series, while with interface *writeResult* to save corresponding analysis result to the storage system. (3) The function *writeResult* for saving the result, each of the slaves will call this function to write back the analysis result, *.csv* file and database are supported in our framework.

3.3   System Implementation

The framework consists of four components, *DataFetcher*, *Dispatcher*, *DataWriter* and *Utils*, which each of them is a Python class. As Fig. 4 shows, *Dispatcher* is the motherboard of the framework and organize all other classes.

DataFetcher defines the procedure the data cache pool read from the data source, and the logic we maintain the cache pool. The programmer needs to implement the function *_fetchData* to read from the raw data source, and we implemented the function *init* to initialize the cache and the function *getData* to read from the cache. *Dispatcher* is the class that pulls all these basic components together into a framework. The parallel architecture of ray defines an *Actor* which is the minimum unit executing the analysis logic.

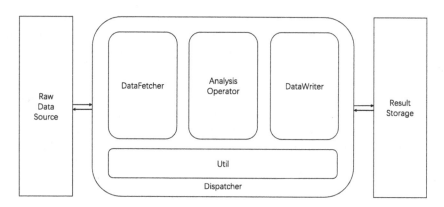

Fig. 4. Class hierarchy and analyzer implementation

For the data cache pool, a daemon thread is created to manage an HBase [16] and keep it filled, after we invoke the *getData* function, which can make full use of the I/O of the disk or network, overlap the time computing and I/O. The data read request from slaves can be returned directly by memory access instead of through network communication or disk access which could costs tens or hundreds of time more than memory operations. Currently, the data requests are made in a distributed way that managed by a synchronized queue that could feed these data to the slaves. As Eq. (1), the size of cache pool should be big enough for all the threads to get a job data and finish corresponding processing for one time. Thus once any one of these slaves found the data remaining in the cache pool is not adequate, and started the read request from the raw data resource, fittingly, all other slaves finished their job and request data from cache, at this extreme circumstance, all these slaves are able to get one job and start their task without any block. And the data could be ready for the threads finished processing the data from the last batch, ensuring the pipeline filled.

$$cache\_pool\_size > \frac{read\_latency \times number\_of\_threads}{analysis\_module\_processing\_time} \tag{1}$$

Secondly, the analysis framework, theoretically, can avoid all communications between two analysis thread slaves by fetching enough data items from the cache pool, gathering all the prerequisite data for one result in a single thread. The data flow is denoted by *job* which include two fields named *name* and *data*. Function *getData* will return a list of a dictionary which each element of the array is a job. The *name* of a job is a unique identifier used for job scheduling and result write back. And the analysis operator will focus on the content of the dictionary. What's inside of the *data* need to be determined by the data scientist. In our implementation, it is an array which each of the elements is the magnitude of the brightness of a star, and the array is a time series data. If the algorithm needs a certain amount of time series data as a batch, we can pass a parameter to the function *getData* to obtain multiple data at the same time.

Finally, the writeback cannot be avoided, it is done by the slaves respectively, parallel writeback could save more time than creating an individual thread to handle this job or aggregate all the writeback request to a single agency thread in real time, cause it will turn the parallel module into a serial one and the write-back of results could be the bottleneck and slows the whole module down, not as the cache pool does. In most cases, the analysis result is always more simple than the input, the analysis is a procedure summarizing and obtaining the feature of a data set. As an example, the size of the average value of an array with millions of elements is one, so each analysis output of a data is much smaller than the input, and the output could be very small compared to the input. Therefore, we can exploit the best performance of disk or network by distributed result write back, no matter saving the result to a remote database server or local disk. If the result data is not too big, a more flexible approach is to store all the result in memory managed by the class *DataWriter*, asynchronous write back to a database server or disk is another solution. In a word, our framework is able to implement various write back policy meeting different task and storage conditions.

4   Offline Time Series Data Mining Application

4.1   Anomaly Detection of Astronomical Sky Survey Data

Astronomical sky survey data generated by GWAC is a group of time series data which each of a data item is a one-dimensional key-value set. The magnitude of light of the star is indexed with a time stamp. To analyze these formatted data, we make the best of the pipeline and the framework is able to process them efficiently.

Based on the previous framework, we can implement almost any offline big data mining algorithms including deep neural networks with our framework to achieve high-performance analysis procedure, as long as the data are homogeneous and can be processed with the same method. Take Seasonal Hybrid Extreme Studentized Deviate (S-H-ESD) [15] as an instance, it is a time series anomaly detection algorithm based on a statistical hypothesis test. The algorithm decomposes time series data into seasonal, trend and residual components.

Periodic variation of the time series is extracted by the seasonal component; Non-periodic variation of the time series is extracted by the trend component, and the residual is the time series remaining after the previous two components are extracted. The S-H-EAS algorithm estimates the long-term trend of a time series and detects anomalies using median absolute deviation. This approach doesn't require enormous parameters to be trained or warming up and is scalable for being parallelized.

We integrate S-H-ESD anomaly detection module with our framework as an analysis component, which digests the time series data preprocessed from raw data source, when it found anomaly in a time series, we record the anomalous values together with its anomaly score, here we use the value of *Median Absolute Deviation* (MAD) to denote the intensity of the anomaly, this score is saved to the anomaly result database directly. The *Anomaly* object in this application is a key-value pair whose key is a string by concatenating *region_id* of a star, start time, end time and the star id in the corresponding region.

Because this algorithm detects anomaly by testing the distribution of time series data, so the anomaly is discovered by a single data point, thus the anomaly is reported point by point. We added a function in *DataWriter* to aggregate some continuous anomalies into a larger one, in order to lighten the load of data I/O and organize these anomalies better simultaneously.

As an example, the following code demonstrate the main function of a program applying the S-H-ESD anomaly detection algorithm to a dataset, loading data from csv files and write back to HBase database:

```
from Dispatchers import Dispatcher
from Fetchers import CsvFetcher
from DataWriters import HBaseWriter
from Detectors import ShesdD

dt = ShesdD()
df = CsvFetcher("./data/all")
dw = HBaseWriter(host = "127.0.0.1", port = 50000)
dp = Dispatcher(dt,df,dw,4)
dp.init()
dp.run()
```

4.2   Train a Deep Network in the Distributed Setting with Astronomical Sky Survey Data

The deep network can be trained in a distributed way, we might need to ship the model parameter between process or machines, compute a gradient on slaves and update the parameter on the master node. In a traditional way, we define the architecture of the deep neural network on each node or process memory respectively, and feed them the data to update the gradients. And then gather them to a single node to calculate the average value of these parameters, thus to achieve distributed training.

Ray works with TensorFlow [6] by extracting gradients from the module and shipping the parameters to achieve distributed training. Benefit from the computing framework we applied, it is very convenient to implement parallel training with Python, and we are able to train deep network in parallel using Ray by passing parameters between processes. Integrated with our high performance I/O cache pool, the user can define the deep network while train them with few codes to write.

5 Conclusion

We find that (1) This framework can accelerate analysis algorithms, achieve almost linear acceleration ratio and simplify the analysis programming, helping data scientist without parallel programming basics to implement high performance, high throughput data analysis module. (2) Queue managed by Python is strong and scalable enough that can be used to scheduling these tasks, and implement high performance distributed data cache pool. Computing and I/O overlapping is an efficient approach to decrease the cost caused by insufficient disk or network bandwidth. (3) This framework is modularized which is applicable for different data sources and storage format. This programming architecture is of high significance in abstracting the big data analysis pipeline, optimization for this module is always ongoing depending on various application scenarios.

Acknowledgements. This research is supported in part by Key Research and Development Program of China (No. 2016YFB1000602), "the Key Laboratory of Space Astronomy and Technology, National Astronomical Observatories, Chinese Academy of Sciences, Beijing, 100012, China", National Natural Science Foundation of China (Nos. 61440057, 61272087, 61363019 and 61073008, 11690023), MOE research center for online education foundation (No. 2016ZD302).

References

1. ECG. https://en.wikipedia.org/wiki/Electrocardiography
2. Keras. https://en.wikipedia.org/wiki/Keras
3. Numpy. https://en.wikipedia.org/wiki/NumPy
4. Python. https://en.wikipedia.org/wiki/Python_(programming_language)
5. SPSS. https://en.wikipedia.org/wiki/SPSS
6. Abadi, M., et al.: Tensorflow: a system for large-scale machine learning. In: OSDI 2016, pp. 265–283 (2016)
7. Chen, Y., Mahajan, R., Sridharan, B., Zhang, Z.-L.: A provider-side view of web search response time. In: ACM SIGCOMM Computer Communication Review, vol. 43, pp. 243–254. ACM (2013)
8. Dean, J., Ghemawat, S.: MapReduce: simplified data processing on large clusters. Commun. ACM **51**(1), 107–113 (2008)
9. Godet, O., et al.: The Chinese-French SVOM mission: studying the brightest astronomical explosions. In: Space Telescopes and Instrumentation 2012: Ultraviolet to Gamma Ray, vol. 8443, p. 844310. International Society for Optics and Photonics (2012)

10. Walker, S.J.: Big Data: A Revolution That Will Transform How We Live, Work, and Think (2014)
11. Liang, E., et al.: Ray RLLib: a composable and scalable reinforcement learning library. CoRR, abs/1712.09381 (2017)
12. Moritz, P., et al.: Ray: a distributed framework for emerging AI applications (2017)
13. Pedregosa, F., et al.: Scikit-learn: machine learning in Python. J. Mach. Learn. Res. **12**(Oct), 2825–2830 (2011)
14. R Core Team: R: a language and environment for statistical computing (2013)
15. Vallis, O., Hochenbaum, J., Kejariwal, A.: A novel technique for long-term anomaly detection in the cloud. In: HotCloud (2014)
16. Vora, M.N.: Hadoop-HBase for large-scale data. In: 2011 International Conference on Computer Science and Network Technology (ICCSNT), vol. 1, pp. 601–605. IEEE (2011)
17. Zaharia, M., Chowdhury, M., Franklin, M.J., Shenker, S., Stoica, I.: Spark: cluster computing with working sets. In: Usenix Conference on Hot Topics in Cloud Computing, p. 10 (2010)

Real-Time Query Enabled by Variable Precision in Astronomy

Yongjie Du, Xiaofeng Meng$^{(\boxtimes)}$, Chen Yang, and Zhiqiang Duan

School of Information, Renmin University, Beijing, China
xfmeng@ruc.edu.cn

Abstract. As sky survey projects coming out, petabytes and exabytes of astronomical data are continuously collected from highly productive space missions. Especially, in time-domain astronomy, Short-Timescale and Large Field-of-view (STLF) sky survey not only requires real-time analysis on short-time data, but also need precise astronomical data for special phenomena. Additionally, it is important to find a partition method and build an index based on that for effective storage and query. However, the existing methods cannot simultaneously support real-time and variable-precision query in astronomy. In this paper, we propose a novel astronomical real-time and variable precision query method based on data partitioning with Hierarchical Equal Area isoLatitude Pixelation (HEALPix for short). Our method calculates the time through model and predict precision by machine learning, which can accurately predict the partition level number of HEALPix which can effectively reduce the cost of time for query by layer and layer. The method can meet the user's requirements of real-time and variable-precision query. The experimental results show that our method can optimize previous query strategies and reach a better performance.

Keywords: Real-time · Variable precision · HEALPix · Astronomical data

1 Introduction

In the 21st century, with the emergence of various latest observation techniques, the astronomical field has ushered in the era of information explosion. And the main challenge of the era is the management of astronomical big data. For example, GAIA [2] and LSST [3] are expected to produce PB-level data. At the same time, the investigation and analysis of astronomical phenomena is the basis of astronomical discovery. By analyzing a large amount of astronomical data, astronomers can more effectively discover astronomical phenomena, such as gamma-ray bursts, supernovae and microgravity lenses. However, we still face many challenges: as the scale of astronomical data increases dramatically, (1) how astronomers find the data they need in the shortest time, it is crucial to find astronomical phenomena in time. (2) many astronomical operations are computationally intensive, complex and costly. Thus, it is important to maintain balance between real-time and variable precision that is a method finding an approximate solution in computer. For example, when an astronomical phenomenon appeared, the astronomer may query database to find the star. However,

© Springer Nature Switzerland AG 2019
J. Li et al. (Eds.): BigSDM 2018, LNCS 11473, pp. 78–85, 2019.
https://doi.org/10.1007/978-3-030-28061-1_9

different astronomers have different needs. An astronomers may be focused on real-time. At the same time, another astronomer can pay close attention to precision. Therefore,

In this respect, with the growing of astronomical surveys, we need a scalable solution to study complete spatial data exploration. In order to solve the above challenge, a good solution is to propose a variable precision query optimization method based on HEALPix [1], which can not only meet the short time scale requirements of astronomical data query, but also satisfy the variable precision queries. However, due to the complexity of astronomical data, there are also many problems to be addressed: (1) How do we provide astronomers with expressive queries? (2) How do we storage and query effectively? (3) How do we reduce the complexity of astronomical operations?

So far, there are some complex astronomical query systems and spatial indexes. However, these systems have some limitations.

(1) Lack of query patterns that can satisfy both time and precision.
(2) Limited performance for partition mode queries.
(3) The query engine based on partitioning algorithm does not have good optimization measures or methods.

Therefore, we redesign query optimization strategy based on HEALPix partition algorithm so that it can meet the user's time query and the variable precision query requirements. These optimizations can provide an effective, fast and accurate large-scale astronomical data query solution.

The rest of the paper is organized as follows. Section 2 is background. Section 3 is related work. Section 4 introduces our work. Section 5 describes the experimental results. Section 6 summarizes our work and presents directions of future work.

2 Background

2.1 GWAC

The Ground Wide Angle Camera (GWAC) that built in China, is consisted of 20 wide angle telescopes with 18 cm aperture. Each telescope is equipped with 4k × 4k charge coupled device (CCD) detector. The entire camera array covers 5,000 degrees^2 and the sampling period is 15 s. The observed time of the fixed sky area is as long as 8–14 h at each observation night. GWAC can generate 3.5 million rows of data pre 15 s, 6700 million rows of data at one night (8 h) and 2.24PB in 10 years. The GWAC has special advantages in time-domain astronomical observation that reflect from the field of observation view and the sampling frequency of the observation time. Therefore, it is crucial to ensure real-time of store and query for such a large amount of data.

2.2 HEALPix

HEALPix is one of the most popular astronomical data indexing methods for astronomical data. In this paper, we use the partition characteristics of HEALPix to satisfy our needs and optimize the partition coverage method to achieve variable precision query requirements.

HEALPix divides the sky space into 12 basic pixels (cells). The nesting numbering scheme of HEALPix has a layered design method which the basic pixels are recursively subdivided into four equally sized pixels on a spherical coordinate system. These subspaces are organized into trees and the amount of subdivision (the height of the tree) is given by the Nside parameter.

We adopt partition method of HEALPix for the following reasons:

(1) HEALPix divides the sphere into equal areas, which is good for calculating precision.
(2) HEALPix uses the mapping technique of the celestial space to divide the sky. Each small unit has a unique ID. We can effectively index and retrieve data through the ID.
(3) Data linearization using HEALPix preserves the locality of the data, which can help us organize similar points in the same partition and continuous partition. Thus, optimized query is accessing similar object groupings and avoiding to search unrelated partitions.

Previous work has proposed methods for sky spherical division, such as simple mesh sky division and HTM algorithm [4]. The work of comparing the two algorithms is carried out in [5]. Both HTM and HEALPix use a fixed number of cells to hierarchically divide the sky area. The difference between them is that area divided by HEALPIX is equal. In our article, we use HEALPIX instead of HTM for the reason.

3 Related Work

Previous work has been used in various fields, Hao [6] uses variable precision to accelerate 3D rendering. User input minimum precision, computer output the result to meet user accuracy requirement through arithmetic calculations. Schulte [7] introduces a variable precision interval arithmetic processor, which can detect and correct implicit errors in finite precision numerical calculations by variable precision method. It also provides the method to solve problems that cannot be effectively solved using traditional floating point calculation. Li [8] uses variable precision for computational estimation. In the SKA-SDP workflow, FFT calculations take up a lot of overhead. It can improve power efficiency through variable precision calculation. However, no one query uses variable precision query in the astronomy.

4 Implementation

In our method, we have two query process strategies. First, we regard the time as the first consideration. Then we consider the precision more important. When user inputs right ascension, declination, radius, time, precision, flag (Ra, Dec, R, T, P, F for short) for query, we firstly decide what the Flag is. When flag is 0, we will do the first strategy. When flag is 1, we will do the second strategy.

There are two kind of strategies that process the query. The first strategy determines that the level can meet the given time requirement, the second step is to determine whether the accuracy satisfies the user query. As long as the current accuracy is greater than user input accuracy, the model outputs the partition number of coverage area. When the query accuracy does not meet the requirements, we will turn to the next level to judge. If the condition is not corresponded, there will be an error message for user.

We split the factors that affect time into layers and radius. As showing in the left figure of Fig. 1, the time taken for query is firstly increased and then decreased as the radius of the query increasing, that is due to the query mechanism of HEALPix. So we can formalize the relationship between time and radius as

$$t_1 = a \times r^2 + b \times r + c \qquad (1)$$

Where t_1 is the time of query, r is the radius of query and a, b, c is the parameters. The left figure of Fig. 1 shows the relationship between the time and Nside that is the layer of HEALPix. It can be seen from the figure that the time of query increases exponentially with Nside. So we can formalize the relationship between time and Nside as

$$t_2 = p \times 2^{q \times l} \qquad (2)$$

Where t_2 is the time of query, l is the number of layers and p, q is the parameters.

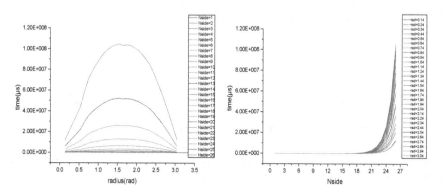

Fig. 1. The left is relationship between radius and time. The right is relationship between radius and number of layers.

In summary, the time of query may be affected by the radius and the number of layers. So, we combine the above two formulas into one formula as follows:

$$t = q_0 + q_1 (q_2 \times r^2 + q_3 \times r + q_4) \times 2^{q_5 \times l} \tag{3}$$

Where t is the time of query, r is the radius of query, l is the number of layers and $q_0, q_1, q_2, q_3, q_4, q_5$ is the parameters.

The second strategy determines that the level can meet the given precision requirement. Then we judge whether the query time of the layer satisfies the user's needs. As long as the time is less than the user's requirement, the model outputs the partition number of coverage area. When the time does not meet the requirements, we will judge next level again. If the condition is not corresponded, there will be an error message for user.

We calculate the area of crown. Then we can get all the number of cell which is the intersection of crown and quadrilateral of HEALPix. Since each area of quadrilateral is equal, the total square of the spherical surface is 41252.96 and the number of quads is fixed per layer. So we can formulate the precision as follows:

$$P = \frac{A_1}{A} = \frac{A_1}{N * S_i} \tag{4}$$

P is the precision of query; A is the crown area; N is the number of cell that is intersection of crown and quadrilateral of HEALPix; S_i is per area of ceil in the ith layer.

We project the spherical surface onto a plane. Size of it is 4 and the corresponding spherical surface is 41252.96 degrees^2. The spherical crown is an approximately square which projected onto the plane. We can calculate the area as follows:

When radius of the circle is less than $\pi/2$, the length of side can be expressed as:

$$y = |\cos(\pi \div 2 - rad)| \tag{5}$$

When radius of the circle is more than $\pi/2$, the length of side can be expressed as:

$$y = 1 + |\cos(rad)| \tag{6}$$

The crown area that is projected to the plane is formulated as $A_1 = y^2$. The area of each small grid can be formulated as $S_i = 4/np_i$ that is projected onto the plane for per layer. The number of blocks in the ith layer is recorded as np_i.

The processing flow of variable precision query is showed as the algorithm 1, that is composited by the model of time and precision. The Ra and Dec is used to determine center point of the query. The R is the radius of query. The Pre and T is the precision and time. The F is a flag bit that is used to determine time priority or precision priority.

The HT and HP is the prediction time and precision.

Algorithm 1. Variable precision query the tables.

Input : Ra , Dec , R , Pre , T, F
Result : cell number
1: If F=0
2:    If T>HT
3:       If P<HP
4:          output cell number
5:       else check the next level
6:    else output error
7: else If F=1
8:    If P<HP
9:       IF T>HT
10:          check the next level
11:       else output cell number
13: else output error
14: return cell number

---

5 Experiment

In this section, we evaluate our two strategies firstly. Then we provide an overall evaluation of the method. We obtain data about time, radius and number of layers through experiments. We model the first method by several sets of data and fit the second method by machine learning. Metrics that we evaluate our methods are query time and accuracy. Table 1 shows that the prediction of time surpasses the precision of prediction. There is lower cost of time and loss of precision by fitting polynomial. For predicting precision, the random forest is better than other methods. The main reason is too few features.

We obtained the correspondence among the number of layers, the radius and the precision through experiments. Then we fit our experimental results through SVM [9], KNN [10] and random forest [11]. The experiment shows as the Table 1, that the random forest predict effectively and can accurately locate the number of layers.

Table 1. The evaluation of our methods.

Strategy	Method	Time (seconds)	Precision
Time prediction	Polynomial fitting	0.00089	99.3%
Precision prediction	SVM	0.00125	20.4%
	KNN	0.00461	43.1%
	Random forest	0.00172	99.1%

The baseline is the original method that queries layer and layer. Comparing with the method, we found that our algorithm can locate the number of layers required and reduce the unnecessary consumption of time. As showing in the Fig. 2, when the level is less than 7, our method is as effective as the traditional method. As the number of layers increasing, our method is significantly better than the traditional method.

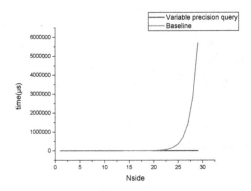

Fig. 2. The cost of time between our method and original method.

6 Summary

In this article, we present a variable-precision real-time astronomical query method. This method can apply HEALPix that is one of the astronomical partition algorithms to astronomical queries through an efficient query mechanism. The result outperforms the previous method. This method can simultaneously satisfy the user's real-time and variable precision query. Not only accuracy but also performance is better in the method. This approach provides a way to speed up query for many astronomical queries. In the future, we will organize the partitioning algorithm of the astronomy to construct a variable precision query framework for astronomers.

Acknowledgement. This research was partially supported by the grants from the National Key Research and Development Program of China (No. 2016YFB1000602, 2016YFB1000603); the Natural Science Foundation of China (No. 91646203, 61532016, 61532010, 61379050, 61762082); the Fundamental Research Funds for the Central Universities, the Research Funds of Renmin University (No. 11XNL010); and the Science and Technology Opening up Cooperation project of Henan Province (172106000077).

References

1. HEALPix. http://HEALPix.sourceforge.net/index.php
2. GAIA. http://sci.esa.int/gaia/
3. LSST. https://www.lsst.org/
4. Szalay, A.S., Gray, J., Fekete, G., et al.: Indexing the sphere with the hierarchical triangular mesh. arXiv preprint arXiv:cs/0701164 (2007)

5. O'Mullane, W., Banday, A.J., Górski, K.M., Kunszt, P., Szalay, A.S.: Splitting the Sky-HTM and HEALPix. In: Banday, A.J., Zaroubi, S., Bartelmann, M. (eds.) Mining the Sky, pp. 638–648. Springer, Heidelberg (2000). https://doi.org/10.1007/10849171_84

6. Hao, X., Varshney, A.: Variable-precision rendering. In: ACM Symposium on Interactive 3D Graphics, 19–21 March 2001, Research Triangle Park, NC (2001)

7. Schulte, M.J., Swartzlander, E.E.: A family of variable-precision interval arithmetic processors. IEEE Trans. Comput. **49**(5), 387–397 (2000)

8. Li, M., et al.: Evaluation of variable precision computing with variable precision FFT implementation on FPGA. In: IEEE 2016 International Conference on Field-Programmable Technology (FPT), Xi'an, China, 18 May 2017

9. Burges, C.J.C.: Geometry and invariance in kernel based methods. In: Scholkopf, B., Burges, C.J.C., Smola, A.J. (eds.) Advances in Kernel Methods—Support Vector Learning, pp. 89–116. MIT Press, Cambridge (1999)

10. Bromley, J., Sackinger, E.: Neural-network and k-nearest-neighbor classifiers. Technical report 11359- 910819-16TM, AT&T (1991)

11. Breiman, L.: Random forests. Mach. Learn. **45**, 5–32 (2001)

Continuous Cross Identification in Large-Scale Dynamic Astronomical Data Flow

Zhiqiang Duan, Chen Yang, Xiaofeng Meng[(✉)], and Yongjie Du

School of Information, Renmin University, Beijing, China
xfmeng@ruc.edu.cn

Abstract. In modern astronomy, Short-Timescale and Large Field-of-view (STLF) sky survey produce large volume data and face a great challenge in cross identification. Furthermore, transient survey projects are required to select the candidates fast from large volume data. However, traditional cross identification methods didn't satisfy the observation of transient survey. We present a fast and efficient cross identification system for large-scale astronomical data streams. By receiving a high-frequency star catalog and maintaining a local star catalog, the system partitions the star catalog and cross identification with the object catalog. A coding strategy is used to manage the unique ID of the all-sky star. After processing data, all the results are stored in Redis and generate the light curve. Our experiment shows that the method could meet the strict performance requirements and good recognition accuracy on fast real-time sky survey project. Additionally, our system shows good performance in low latency large volume astronomical data processing and our system has been successfully applied in the Ground-based Wide Angle Camera (GWAC) online data processing pipeline.

Keywords: Cross identification · Pipeline · Data flow · Astronomical data processing

1 Introduction

With the development of astronomical telescopes, the volume of astronomical data has been continuously expanded. At the same time higher requirements and challenges have been put forward for the processing and storage of a large amount of astronomical data. The research on optical transient sources has become a major topic in the field of astronomy. The transient source is a sudden short-term aperiodic astronomical phenomenon. And the time span consists of seconds to year, including supernovae, gamma bursts, and microgravity lenses [3]. The study of transient sources is of great importance to astronomy and physical phenomena in the universe, so the observation and research of transient sources has become the focus of the astronomical community, such as the US-supported science project Large Weather Survey Telescope (LSST), Catalina Real-Time Transient Survey (CRTS) [8] and so on. China's current construction of the transient source search equipment Ground-based Wide Angle Camera (GWAC) is composed of 36 wide-angle cameras with a diameter of 18 cm. Each wide angle camera is equipped with a 4k × 4k CCD detector and the field reached

© Springer Nature Switzerland AG 2019
J. Li et al. (Eds.): BigSDM 2018, LNCS 11473, pp. 86–91, 2019.
https://doi.org/10.1007/978-3-030-28061-1_10

5,000 square degrees. An observation image is generated per 15 s and each observation image contains approximately 175,600 objects. The device is of great significance for the search of unknown short-time transient sources.

Due to the characteristics of the transient source, the observation transient source astronomical device has the characteristics of large field of view and high sampling rate, which poses a great challenge to the processing and management of the star catalog. Taking GWAC as an example, GWAC produces an observation image about every 15 s, which has at least 3.5 million rows of data, about 6.7 billion rows of data per night and about 2.24 PB of data for the next decade.

Our contribution is the following three points:

1. Continuous processing of astronomical data. Our algorithm receives the astronomical star catalog continuously and ensures cross identification within a short time. In addition, maintaining the template star catalog and saving the results include intermediate data into Redis [1].
2. Manage the unique ID of the all-sky star. Efficient and accurate cross identification, maintaining a unique template star catalog and giving the star a unique ID.
3. High throughput and low latency pipeline. Designed an efficient pipeline that combines cross identification, traffic normalization and maintenance of the star catalog into one component to find scientific phenomena as quickly as possible.

The rest of the paper is organized as follows. Section 2 is related work. Section 3 introduces our work. Section 4 describes the experimental results. Section 5 summarizes our work and presents directions of future work.

2 Related Work

Our related work focuses on the astronomical data management system. In astronomy, the data collected by the telescope must be stored in a long-term database for complex analysis. SkyServer [2, 5] is based on the Microsoft SQL Server mainly used to be responsible for Sloan Digital Sky Survey (SDSS). And SkyServer is used to long-term storage and management of data. Qserv [7] is a distributed shared-nothing database for the LSST catalog and a key feature of Catalina Real-time Transient Survey (CRTS) is that it is the first fully open synoptic sky survey: all detected transients are published immediately, with No proprietary period at all, using several Internet-based mechanisms, this open-data approach benefits the entire astronomical community.

These studies play an important role in multi-band astronomy. They are mainly aimed at the cross identification of massive data. And the real-time requirements for the intersection of the stars are not demanding [4]. But the cross match of transient sources requires high real-time requirements and the order of magnitude is at least 10^5 [6]. Therefore, for the transient source survey project such as GWAC, real-time fast cross match and data processing algorithms are needed. Our algorithm is mainly for the data processing of the GWAC optical camera array in the Sino-French astronomical satellite SVOM project. In the large field view of GWAC, five million stars can be observed simultaneously. Considering the time series observation mode, it will get about 5 Millions of light curves with a 15-second sampling interval.

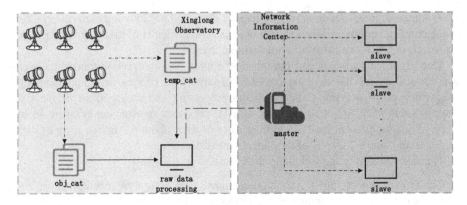

Fig. 1. Scientific data process pipeline

3 Architecture

As shown in Fig. 1, our data source is from the Xinglong Observatory and the GWAC camera array collects data at regular intervals. The collected raw data is subjected to three steps. The first step uses the original data to construct and form a template star catalog of the current sky area. The second step performs a point source extraction operation on the original data to generate an observation star catalog for the observation. The third step adjusts the observation star catalog according to the template star catalog and process the data that cannot be used. Then GWAC sends data to the primary node of the cluster. And the primary node selects a suitable child node through the scheduling algorithm to transfer the data to the child node. The child node deals with the data and finally stored into the database.

3.1 Scientific Data Process Pipeline

Our algorithm mainly receives and processes three types of catalogs and generates data products. The temp catalog is an irregularly generated star catalog (excluding peripheral catalog) in the template image. It contains 14 basic astronomical tables shown in Table 1 (exclude No. 15 16). And the object catalog is an observation catalog generated every 15 s. It contains the time of the current observation, the camera number, the sky area number and the basic data of the 16 astronomical tables in Table 1. The object catalog needs to be aligned with the temp catalog and X, Y need to be converted to the coordinates of the template catalog. The Temp catalog is a master template table that is maintained on the local machine and used for fast cross identification. And it is used to quickly determine the star ID and maintain consistent updates. The core part of our algorithm is consisted a cross identification of star catalog and a unique way of identifying the star ID.

And we use the HEALPix [11] to partition the catalog. HEALPix is one of the most popular astronomical data indexing methods for astronomical data. And uniquely determine a star ID using the camera number, the sky zone number, the HEALPix partition number and its number in the current catalog. By using this encoding method, it can be well adapted to the mode of our current multi-camera with multi-day zone. At the same time, it is simpler and convenient to query and inset data.

Table 1. Basic data set

No.	Col name	Font size and style
1	NUMBER	Star number
2	ALPHA_J2000	Right ascension
3	DELTA_J2000	Declination
4	X_IMAGE	X coordinate
5	Y_IMAGE	Y coordinate
6	MAG_APER	Magnitude
7	MAGERR_APER	Magnitude error
8	THETA_IMAGE	Image parameter
9	FLAGS	Identification
10	ELLIPTICITY	Ellipticity
11	CLASS_STAR	Classification identifier
12	BACKGROUND	Background light
13	IMAGE	Image contour parameter
14	VIGNET	Image quality parameter
15	PIXXO	X original coordinates
16	PIXYO	Y original coordinates

3.2   Catalog Cross Identification

In the science of astronomy, it is common to record the physical quantities in astronomical catalogs. The same star has different designations in different catalogs. For example, we have α and β two catalogs and α_id, β_id are the designations of the same star in α and β respectively. It is often necessary to know β_id given α_id.

The biggest challenge of the cross match for the large field of view massive data astronomical survey project is to find optical transient sources and how to process these data in real time. According to its data characteristics, matching method of the star template is more suitable for data processing of optical transient sources. The main speed bottleneck of the star template matching method lies in the cross identification of the star catalog because a large number of cross match processes are needed.

In the astronomical catalog data, the same stars have different names in different star tables. We need to match these stars in data processing and management. The offline cross match algorithm is often a template for maintaining an all-sky area. For example, USNO-B [9], USNO-B is an all-sky catalog that presents positions, proper motions, magnitudes in various optical passbands, the catalog's size is about 80 GBytes. We cannot use such a template catalog of all-day zones for cross match because this catalog cannot meet our real-time requirements for finding transient sources.

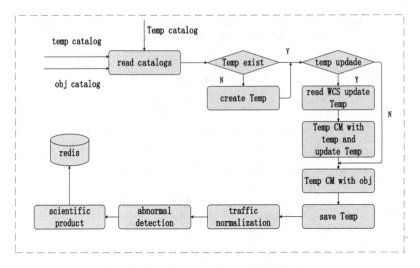

Fig. 2. Data process and cross identification

As shown in Fig. 2, we propose a fast and continuous cross match algorithm which does not need to maintain a large all-sky area star table and cross match by maintaining a local template star table of a partial sky area. For the cross identification of the catalog, our algorithm is mainly divided into two parts. Firstly, we can quickly filter out the required reference stars by partitioning the template catalogue and quickly filtering out the reference stars of the target celestial bodies from the template catalogue. Then there is a fast position matching operation between the target object and a small number of selected reference stars. After finding the target star, update the data of the target star and the template catalog of the current sky zone. As the data continues to be processed, the cross identification of the target star will be more accurate and faster.

4   Experiments

We use GWAC's real data as a test set. Each catalog contains approximately 29,700 stars of data. In our experiment, we placed the subroutine in the Docker [10] for testing. By using Docker, we can better schedule our programs and resources. We simulated the data transmission mechanism of GWAC and simulated 16 telescopes each telescope sent an observation star catalog in 15 s. And we ingest catalog files per 15 s lasting 960 times (4 h).

We have 20 computers (including one master node and 19 child nodes) and each computer including 12 CPU with 1.6 GHz per core and 96 GB RAM. Our data processing is about 0.72 s per cross match and the entire time (include insert the database) is about 2.82 s. In addition, our algorithm only uses 4.8 GB main memory of per child node and the latency is stable

5   Summary

Our system is a distributed big data processing framework for analyzing and managing astronomical data about optical transient sources. The entire real-time processing flow can satisfy the short-time and large-field characteristics of the astronomical transient source project. A complete processing flow is about 2.82 s. Our system maintains the template star catalog corresponding to each sky zone without maintaining a star catalog of the all-sky zone and the processing speed is able to be accepted. With the processing data increases, the processing efficiency and accuracy will continue to increase. In the future, our child node scheduling will be more efficient and get better performance.

Acknowledgement. This research was partially supported by the grants from the National Key Research and Development Program of China (No. 2016YFB1000602, 2016YFB1000603); the Natural Science Foundation of China (No. 91646203, 61532016, 61532010, 61379050, 61762082); the Fundamental Research Funds for the Central Universities, the Research Funds of Renmin University (No. 11XNL010); and the Science and Technology Opening up Cooperation project of Henan Province (172106000077).

References

1. Redis. https://redis.io/
2. Sciserver. http://www.sciserver.org/
3. Wu, C., Ma, D., Tian, H.-J., Li, X.-R., Wei, J.-Y.: Study and development of a fast and automatic astronomical-transient-identification system. Acta Automatica Sinica **43**(12), 2170–2177 (2017)
4. Nieto-Santisteban, M.A., Thakar, A.R., Szalay, A.S.: Crossmatching very large datasets. In: National Science and Technology Council (NSTC) NASA Conference (2007)
5. Szalay, A.S., et al.: The SDSS skyserver: public access to the sloan digital sky server data. In: Proceedings of ACM SIGMOD International Conference on Management of Data, pp. 570–581 (2002)
6. Wan, M., et al.: Column store for GWAC: a high-cadence, high-density, large-scale astronomical light curve pipeline and distributed shared-nothing database. Publ. Astron. Soc. Pac. **128**(969), 114501–114532 (2016)
7. Wang, D.L., Monkewitz, S.M., Lim, K.T., Becla, J.: Qserv: a distributed shared-nothing database for the LSST catalog. In: High Performance Computing, Networking, Storage and Analysis, pp. 1–11 (2011)
8. Djorgovski, S.G., et al.: The Catalina Real-Time Transient Survey (CRTS). arXiv preprint arXiv:1102.5004 (2011)
9. http://tdc-www.harvard.edu/catalogs/ub1.html
10. https://www.docker.com/
11. https://healpix.sourceforge.io/

Scientific Data Management and Application in High Energy Physics

Gang Chen and Yaodong Cheng[⊠]

Institute of High Energy Physics (IHEP), Chinese Academy of Sciences,
Beijing, China
chyd@ihep.ac.cn

Abstract. High energy physics experiments have been producing a large amount of data at PB or EB level, and there will be ambitious experimental programs in the coming decades. The efficiency of data-intensive researches is closely related to how fast data can be accessed and how many computational resources can be used. Changes in computing technology and large increases in data volume require new computing models. This paper will give an overall introduction to scientific data management technologies and applications in high energy physics. The current data management framework and workflow will be investigated at first. These include data acquisition, data transfer, data storage, data processing, data sharing and data preservation. Then some ongoing research and development on data organization, management and access will be introduced. Finally the EventDB, an event-based big scientific data management system will be introduced. The test on more than ten billion physics events shows the query speed is greatly improved than traditional file-base data management system.

Keywords: Scientific data management · EventDB · EOS ·
High energy physics

1 Introduction

Nowadays, high energy physics (HEP) experiments have been producing a large amount of data at PB or EB level, and distributed computing technologies has been widely used in high energy physics field. To provide computing support to LHC experiments, the Worldwide LHC Computing Grid (WLCG) [1], the one of the largest distributed computing systems in the world, is built and operated. To process the BES-III experiment's data the distributed computing infrastructure [2] based on DIRAC middleware [3] has been setup and became operational since 2013. Other large experiments such as Belle II [4], Icecube [5] and so on also build their computing platforms to process the large amount of data.

Moreover, there will be ambitious experimental programs in the coming decades. The High-Luminosity Large Hadron Collider (HL-LHC) [6] will be a major upgrade of the current LHC. HL-LHC will investigate the properties of the Higgs boson and other

© Springer Nature Switzerland AG 2019
J. Li et al. (Eds.): BigSDM 2018, LNCS 11473, pp. 92–104, 2019.
https://doi.org/10.1007/978-3-030-28061-1_11

related particles. It is scheduled to begin taking data in 2026 and to run into the 2030s, some 30 times more data than the one LHC has currently produced will be collected by ATLAS and CMS. China is upgrading current neutrino experiment (dayabay) to the new generation experiment JUNO [7], and the cosmic ray observatory (ARGO-YBJ) to LHAASO [8]. They will produce 10 times more data than the old ones.

High energy physics is typical data-intensive application. The scientific research is closely related to how fast data can be accessed and how many computational resources can be used. Changes in computing technology and large increases in data volume require new computing models [9], compatible with budget constraints. The integration of newly emerging data analysis paradigms into our computing model has the potential to enable new analysis methods and increase scientific output. In this paper we will focus on data management technologies and application in high energy physics.

This paper is structured as followings. Section 2 summarizes the HEP data processing workflow. Section 3 introduces current data management technologies and ongoing activities in these fields including data acquisition, data storage, data transfer, data processing and data preservation. In Sect. 4 an event-based big scientific data management system - EventDB is introduced. Finally Sect. 5 discusses the conclusions and future work.

2 HEP Data Processing Workflow

An HEP data processing workflow can generally be divided into data acquisition (DAQ), event reconstruction, and data analysis stages, and the results of each stage will be saved to storage system separately. Detector systems are generally responsible for the collection of data, filtering and building of events. The data saved to the storage system at this stage are called "raw data". These raw data are processed during the event reconstruction stage to produce physics-related information including particle identification and energy measurements etc. This step will take much time and may actually produce a larger data set than the original raw data. Data analysis is usually performed for an individual physical interaction process. The process includes data selection, statistical summarization, creation of plots, and production of the final result. Data analysis workflows are generally more demanding of I/O than compute power.

Moreover, Monte Carlo simulations and event generations are very important for an HEP experiment [10]. Simulation is the process to mock up the physics theory and response of detectors. The structure of a simulation package can be split into two main components including the physics modeling and the detector simulation. The reconstruction package will be the same to both the simulation and the real data flows.

A typical HEP data processing workflow is depicted in Fig. 1.

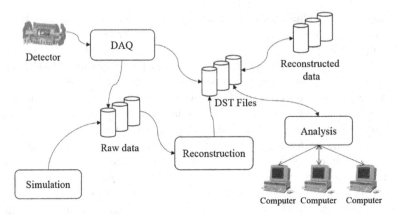

Fig. 1. Data processing workflow in high energy physics.

3   Data Management Framework

As mentioned in Sect. 2, each step of data processing may face big challenges with the rapid development of HEP experiments. These challenges cover many topics such as data acquisition, data transfer, data storage, data processing, data sharing and preservation. We summarize the basic technologies and main specific challenges each topic will face, describe current practices and some activities.

3.1   Data Acquisition

Scientific data are generally produced by detector or physics simulation. Data acquisition systems [11] are critical components in high energy physics experiments. They are embedded in an environment of custom electronics. With high performance computing platforms and affordable high-speed networking equipment, the volume of acquired data can be significantly reduced. For example, ATLAS detector on LHC produce tens of terabytes per second and the trigger system consists of a hardware Level-1 (L1) and a single software-based high-level trigger (HLT) [12]. This two-stage system reduces the event rate from 40 MHz to 100 kHz at L1 and to an average recording rate of 1 kHz at the HLT. The data rate is reduced to about 1.5 GB/s by the trigger system, which means that about 47 PB of raw data is sent and kept in the offline storage system.

Monte-Carlo event generators are also critical to modern particle physics, providing a key component of the understanding and interpretation of experiment data. The event simulation also produce "raw data" similar to the detector output. Collider experiments also have a need for theoretical QCD [13] predictions at very high precision.

Both simulation and detector will produce a large amount of data which are sent to offline storage system. These data will then be transferred, processed or analyzed by physicists around the world.

3.2   Data Storage and Access

The LHC experiments currently provide and manage about an exabyte of storage, approximately half of which is archival, and half is traditional disk storage. Other experiments that will soon start data taking have similar needs, e.g., Belle II has the same data volumes as ATLAS about tens of hundreds of PB per year. The HL-LHC storage requirements per year are expected to increase by a factor close to 10, which is a growth rate faster than can be accommodated by projected technology gains.

Currently there are three types of storage systems widely used in HEP community, including clustered file system, application-layer storage system and hierarchical storage system. These three kinds of systems don't have strict differences, and they are all adopting distributed storage technologies. A clustered file system is a file system which is shared by being simultaneously mounted on multiple servers. Clustered file systems usually implement client kernel modules which are POSIX compatible. Lustre file system [14] is one of typical clustered file system, which supports many requirements of leadership class HPC environments. IHEP has deployed Lustre since 2008, and currently provides more than 10 PB disk capacity.

An application-layer storage system is similar to a clustered file system, but lack of some file system syntax features such as client kernel module, complete POSIX compatibility and so on. EOS [15] and dCache [16] are typical ones. EOS has been developed by CERN since 2010, which is a software solution for central data recording, user analysis and data processing. EOS supports thousands of clients with random remote I/O patterns with multiprotocol support including HTTP, WebDAV, CIFS, FUSE, xrootd, gsiFTP, etc. EOS provides over 250 PB of raw disk space on more than 1.2k nodes and 50k disks at CERN. The architecture of EOS is depicted in Fig. 2.

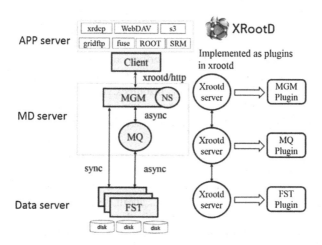

Fig. 2. EOS architecture.

Hierarchical storage management (HSM) is a data storage technique that automatically moves data between high-cost and low-cost storage media, for example between hard disks and tapes. CASTOR [17], Enstor [18] and HPSS [19] are widely used in HEP community. Currently about 350 PB LHC data is stored on tapes managed by CASTOR at CERN.

Changes in computing technology and large increases in data volume bring huge challenges for data storage system. There are some ongoing activities on Data Organization, Management and Access (DOMA) to solve these challenges [20], including:

Sub-file granularity, for example event-based, data management systems are implemented for all applications to select and analyze events aiming at offering an advantage over current file-based granularity.

Organize data using column-wise, versus row-wise to improve performance of each kind of access, or design efficient data storage and access solutions that support the use of map-reduce or Spark-like analysis services.

Place data in different storage layers intelligently, or cache date remotely to use computing resources effectively.

Study how to minimize HEP infrastructure costs by exploiting varied quality of service from different storage technologies.

Optimize data access latency globally with respect to the efficiency of using CPU, at a sustainable cost.

3.3   Data Processing and Distributed Computing

Real raw data and simulated data will be reconstructed in real time after they are sent to offline storage system. It is important that event reconstruction algorithms continue to evolve so that they are able to efficiently exploit future computing architectures, and deal with the increase in data rates without loss of physics. Scientific questions are then answered by analyzing the data obtained from suitably designed experiments. The physics analysis also need to compare measurements with predictions from models and theories. Such comparisons are typically performed long after data taking, but can sometimes also be executed in quasi real-time on selected samples of reduced size. The data processing requires huge computing resources. In past decades, HEP has successfully setup global distributed computing system such as WLCG (Worldwide LHC Computing Grid) [21]. The WLCG infrastructure has been divided into "Tiers" as shown in Fig. 3.

WLCG is made up of four layers, or "tiers"; 0, 1, 2 and 3. Each tier provides a specific set of services. Tier 0 is the CERN Data Centre, responsible for the safekeeping of the raw data (first copy), first pass reconstruction, distribution of raw data and reconstruction output to the Tier 1 s. Tier1 are thirteen large computer centers with sufficient storage capacity. Tier 2 s are typically universities and other scientific institutes, which can store sufficient data and provide adequate computing power for specific analysis tasks. Individual scientists will access these facilities through local (also sometimes referred to as Tier 3) computing resources. In total, WLCG currently provides experiments with resources distributed at about 170 sites, in 42 countries, which pledge every year the amount of CPU and disk resources they are committed to delivering. These sites are connected by 10–100 Gb links, and deliver approximately 500k CPU cores and 1 EB of storage, of which 400 PB is disk. More than 200M jobs are executed each day [22].

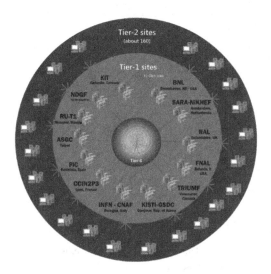

Fig. 3. WLCG Tiered computing model

There are a number of changes that can be expected in the next decade that must be taken into account. There is an increasing need to use highly heterogeneous resources, including the use of HPC infrastructures which can often have very particular setups and policies; volunteer computing which is restricted in scope and unreliable, but can be a significant resource; and cloud computing, both commercial and research. All of these resources can be more dynamic than directly funded HEP computing sites. In addition, diversity of computing architectures is expected to become the normal, with different CPU architectures and more specialized GPUs and FPGAs.

3.4 Distributed Data Management and Transfer

Distributed computing model allows jobs to run at different sites, so there is continual challenge of integrating heterogeneous systems and sites into data management infrastructure. Distributed data management and transfer play an important role in HEP computing environment. Here we will introduce it using an example of WLCG/EGEE data management [23] as shown in Fig. 4.

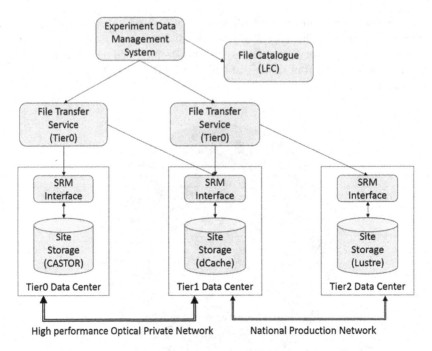

Fig. 4. Overview of data management components in WLCG/EGEE

The WLCG infrastructure provides a set of data management services and interfaces for the LHC experiments and their data models. The File Transfer Service (FTS) [24] allows for the scheduling of the transfer of data files between sites. It allows applications to access files using abstractions such as the "Logical File Name" (LFN), a human-readable identifier of a file in the Grid. The Storage Resource Manager (SRM) provides a web service interface to storage, offering the basic functionality necessary to add files, extract them, or delete them. Storage systems are deployed in different sites to provide storage capacity and SRM interface. These storage systems has been mentioned in Sect. 3.2, for example EOS, CASTOR, dCache, luster, etc. The WLCG experiments also have developed their own data management frameworks. For example, a transfer service such as PhEDEx [25] developed by the CMS collaboration, could have evolved to fulfill the role of the FTS. The ATLAS collaboration developed Rucio [26] which is the Distributed Data Management (DDM) system in charge of managing all ATLAS data on grid.

Some HEP experiment detectors are located far away from data center, for example IceCube South Pole neutrino observatory. The data produced by detectors has to be transferred to data center reliably and efficiency in real time. Spade [27] is a decentralized mechanism for data management that has been used in High Energy and Light Source physics experiments.

High performance optical private network is very important to transfer massive data between different sites. For example, LHCOPN connects CERN with the Tier-1 centers and a mixture of LHCONE [28] and generic academic networks connect other sites. In the network domain, there are new technology developments, such as Software Defined Networks (SDNs), which enable user-defined high capacity network paths to be controlled via experiment software, and which could help manage these data flows. These new technologies require considerable R&D to prove their utility and practicality. In addition, the networks used by HEP are likely to see large increases in traffic from other science domains.

3.5   Data Preservation and Sharing Data Preservation and Sharing

Data from high-energy physics (HEP) experiments are collected with significant financial and human effort and are mostly unique. An inter-experimental study group on HEP data preservation and long-term analysis (DPHEP) was convened as a panel of the International Committee for Future Accelerators (ICFA). The group was formed by large collider-based experiments and investigated the technical and organizational aspects of HEP data preservation. A report [29] was released in 2012 addressing the general issues of data preservation in HEP. According to the report, data produced by the HEP experiments are usually categorized in four different levels: Level 1 data comprises data that is directly related to publications which provide documentation for the published results; Level 2 data includes simplified data formats for analysis in outreach and training exercises; Level 3 data comprises reconstructed data and simulations as well as the analysis level software to allow a full scientific analysis; Level 4 covers basic raw level data (if not yet covered as level 3 data) and their associated software and allows access to the full potential of the experimental data.

Data in different levels will be opened and shared to different use cases, and these cases are summarized in Table 1 [29].

Table 1. Various preservation models, listed in order of increasing complexity.

Preservation model	Use case
1. Provide additional documentation	Publication-related information search
2. Preserve the data in simplified format	Outreach, simple training analyses
3. Preserve the analysis level software and data format	Full scientific analysis based on existing reconstruction
4. Preserve the reconstruction and simulation software and basic level data	Full potential of the experimental data

Currently each of the LHC experiments has adopted a data access and/or data preservation policy, all of which can be found on the CERN Open Data Portal [30]. All of the LHC experiments support public access to some subset of the data in a highly reduced data format for the purposes of outreach and education. The Durham High Energy Physics Database (HEPData) [31] has been built up over the past four decades as a unique open-access repository for scattering data from experimental particle physics papers. It comprises data points underlying several thousand publications. Over the last two years, the HEPData software has been completely rewritten using modern computing technologies. In the future, HEP community will continue to work together to extend and standardize the final data and analysis preservation scheme via HEPData, Rivet [32] and/or other reinterpretation tools.

4   Event-Based Management System

The traditional high energy physics data processing technology uses file as a basic data management unit, and each file contains thousands of events. The benefit of file-based method is to simplify the complexity of data management system. However, one physical analysis task is only interested in very few events, which leads to some problems including transferring too much redundant data, I/O bottleneck and low efficiency of data processing. To solve these problems, some experiments begin to develop and use event-based management system.

The EventIndex [33] is the complete catalogue of all ATLAS events, keeping the references to all files that contain a given event in any processing stage. It replaces the TAG database, which had been in use during LHC Run 1. For each event it contains its identifiers, the trigger pattern and the GUIDs of the files containing it. Major use cases are event picking, feeding the Event Service used on some production sites, and technical checks of the completion and consistency of processing campaigns.

EventDB is a system supported and developed by 'Big Scientific Data Management Systems' project in China. In this system, event data is still stored in ROOT file while a large amount of events are indexed by some specified properties and stored in in NoSQL database. Moreover, IndexDB also provides event-oriented cache and transfer services to improve the physics analysis efficiency. The architecture of EventDB is depicted in Fig. 5.

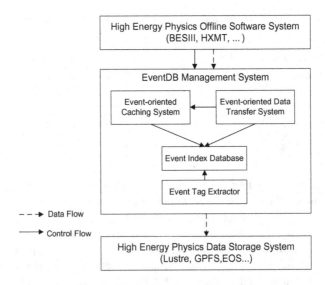

Fig. 5. The architecture of EventDB

There are three components including EventIndexer, EventAccess and EventExtr-actor. EventIndexer is an event index database. EventAccess includes two components: Event-oriented Data Transfer System and Event-oriented Caching System. EventExtractor is an event tag extractor, it will scan all of DST files and extract event attributes into DB.

We have set up a test bed including two sites between Beijing and Chengdu to evaluate the performance of the system. The distance of the two sites is about 2000 km. The network latency is about 35 ms, and the bandwidth is 1 Gpbs. Currently EventDB already support BESIII and HXMT experiments, and more than hundreds of billions of events are indexed. We chose four use cases to perform the performance test as shown in Table 2. In the test, 11237 BESIII events was selected and analysed from 99130 events in total.

Table 2. Test cases of EventDB.

Case	Site	Event indexer	Event cache
1: Original analysis	Local	No	No
2: Original analysis with EventIndexer	Local	Yes	No
3: Original analysis with EventIndexer and Event Cache	Local	Yes	Yes
4: Original analysis with EventIndexer on remote site	Remote	Yes	Yes

The test results showed that the analysis time was 9.3 s with event indexer, about 30% faster than file-based model. When using event cache, the analysis time was reduced to 2.16 s greatly. For remote sites, the analysis time was 11.68 s if using event indexer and event cache, which was even better than original analysis on local site (Fig. 6).

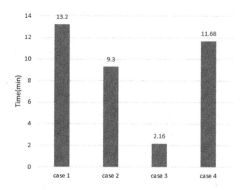

Fig. 6. The performance of EventDB vs original file based storage system.

5   Conclusions

Future challenges for High Energy Physics in the domain of data management are not simply an increase of data volume. The significantly increased computational requirements will also place new requirements on data access. Specially, the use of new types of computing resources (cloud, HPC, etc.) that have different dynamic availability and characteristics will require more dynamic data management systems. Applications employing new techniques, such as training for machine learning, will likely be employed to meet the computational constraints and to extend physics reach. The data management systems will need to meet these changes.

One of data and storage management solutions is trending towards heterogeneity. This means that storage becomes more cost-effective as it becomes available, for example from a cloud provider. With the increase of wide area network bandwidth and the employment of dynamic cloud resources, the data placement optimizations can play important role to better manage data between different sites, whilst maintaining the overall CPU efficiency. It will be different from traditional distributed data access. Others solutions, such as event-based data management, will improve the efficiency of physics analysis in a scalable, cost-effective manner. And the "data-lake" approach concentrates data in fewer, larger locations and making increased use of opportunistic compute resources located further from the data.

Acknowledgements. This work was supported by the National key Research Program of China "Scientific Big Data Management System" (No.2016YFB1000605) and National Natural Science Foundation of China (No. 11675201 and 11575223).

References

1. WLCG Homepage. http://wlcg.web.cern.ch/. Accessed 27 Oct 2018
2. Belov, S., Suo, B., Deng, Z.Y., et al.: Design and operation of the BES-III distributed computing system. Procedia Comput. Sci. **66**, 619–624 (2015)
3. Ayllon, A.A., Salichos, M., Simon, M.K., et al.: FTS3: new data movement service for WLCG. J. Phys. Conf. Ser. **513**(3), 032081 (2014)
4. Takanori, H., Belle, I.I.: Computing at the Belle II experiment. J. Phys: Conf. Ser. **664**(1), 012002 (2015)
5. Karle, A., Ahrens, J., Bahcall, J.N., et al.: IceCube—the next generation neutrino telescope at the South Pole. Nucl. Phys. B-Proc. Suppl. **118**, 388–395 (2003)
6. Apollinari, G., et al.: High-Luminosity Large Hadron Collider (HL-LHC): Technical Design Report V. 0.1. CERN Yellow Reports: Monographs. CERN, Geneva (2017). https://cds.cern.ch/record/2284929
7. Djurcic, Z., Li, X., Hu, W., et al.: JUNO conceptual design report (2015). https://arxiv.org/abs/1508.07166
8. He, H.H., LHAASO Collaboration: Design highlights and status of the LHAASO project. In: Proceedings of the 34rd ICRC (2015)

9. Butler, M., Mount, R., Hildreth, M.: Snowmass 2013 Computing Frontier Storage and Data Management. arXiv preprint arXiv:1311.4580 (2013)
10. Perret-Gallix, D.: Simulation and event generation in high-energy physics. Comput. Phys. Commun. **147**(1), 488–493 (2002)
11. Gutleber, J., Murray, S., Orsini, L.: Towards a homogeneous architecture for high-energy physics data acquisition systems. Comput. Phys. Commun. **153**(2), 155–163 (2003)
12. Nakahama, Y.: The atlas trigger system: Ready for run-2. J. Phys: Conf. Ser. **664**(8), 082037 (2015)
13. Ratti, C., Thaler, M.A., Weise, W.: Phases of QCD: lattice thermodynamics and a field theoretical model. Phys. Rev. D **73**(1), 014019 (2006)
14. Schwan, P.: Lustre: building a file system for 1000-node clusters. In: Proceedings of the 2003 Linux Symposium, pp. 380–386 (2003)
15. Peters, A.J., Janyst, L.: Exabyte scale storage at CERN. J. Phys: Conf. Ser. **331**(5), 052015 (2011)
16. Fuhrmann, P., Gülzow, V.: dCache, storage system for the future. In: Nagel, W.E., Walter, W.V., Lehner, W. (eds.) Euro-Par 2006. LNCS, vol. 4128, pp. 1106–1113. Springer, Heidelberg (2006). https://doi.org/10.1007/11823285_116
17. Presti, G.L., Barring, O., Earl, A., et al.: CASTOR: a distributed storage resource facility for high performance data processing at CERN. In: MSST, vol. 7, pp. 275–280 (2007)
18. Devision, C.: Fermi National Accelerator Laboratory, "Enstore mass storage system". http://www-ccf.fnal.gov/enstore/design.html
19. Watson, R.W., Coyne, R.A.: The parallel I/O architecture of the high-performance storage system (HPSS). In: MSS, p. 27. IEEE (1995)
20. Alves Jr., A.A., Amadio, G., Anh-Ky, N., et al.: A Roadmap for HEP Software and Computing R&D for the 2020 s. arXiv preprint arXiv:1712.06982 (2017)
21. Bonacorsi, D., Ferrari, T.: WLCG service challenges and tiered architecture in the LHC era. In: IFAE 2006, pp. 365–368. Springer, Milano (2007). https://doi.org/10.1007/978-88-470-0530-3_68
22. I Bird. The Challenges of Big (Science) Data. https://indico.cern.ch/event/466934/contributions/2524828/attachments/1490181/2315978/BigDataChallenges-EPS-Venice-080717.pdf
23. Stewart, G.A., Cameron, D., Cowan, G.A., et al.: Storage and data management in EGEE. In: Proceedings of the Fifth Australasian Symposium on ACSW Frontiers, vol. 68, pp. 69–77. Australian Computer Society, Inc. (2007)
24. Baud, J.-P., Casey, J.: Evolution of LCG-2 Data Management CHEP, La Jolla, California, March 2004
25. Barrass, T., Newbold, D., Tuura, L.: The CMS PhEDEx system: a novel approach to robust grid data distribution. In: AHM 2005, 19–22nd September 2005, Nottingham (UK) (2005)
26. Garonne, V., et al.: Rucio - the next generation of large scale distributed system for ATLAS Data Management. J. Phys.: Conf. Ser. **513**, 042021 (2014)
27. Patton, S., Samak, T., Tull, C.E., et al.: Spade: decentralized orchestration of data movement and warehousing for physics experiments. In: 2015 IFIP/IEEE International Symposium on Integrated Network Management (IM), pp. 1014–1019. IEEE (2015)
28. Martelli, E., Stancu, S.: Lhcopn and lhcone: status and future evolution. J. Phys: Conf. Ser. **664**(5), 052025 (2015)
29. Akopov, Z., Amerio, S., Asner, D., et al.: Status report of the DPHEP Study Group: Towards a global effort for sustainable data preservation in high energy physics. arXiv preprint arXiv:1205.4667 (2012)

30. CERN Open Data Portal. http://opendata.cern.ch/. Accessed 27 Oct 2018
31. Maguire, E., Heinrich, L., Watt, G.: HEPData: a repository for high energy physics data. J. Phys: Conf. Ser. **898**(10), 102006 (2017)
32. Buckley, A., Butterworth, J., Grellscheid, D., et al.: Rivet user manual. Comput. Phys. Commun. **184**(12), 2803–2819 (2013)
33. Barberis, D., Zárate, S.E.C., Cranshaw, J., et al.: The ATLAS EventIndex: architecture, design choices, deployment and first operation experience. J. Phys: Conf. Ser. **664**(4), 042003 (2015)

EventDB: A Large-Scale Semi-structured Scientific Data Management System

Wenjia Zhao, Yong Qi$^{(\boxtimes)}$, Di Hou, Peijian Wang, Xin Gao, Zirong Du,
Yudong Zhang, and Yongfang Zong

Department of Computer Science and Technology,
Xi'an Jiaotong University, Xi'an, China
qiy@xjtu.edu.cn

Abstract. During the process of scientific research, the amount of data collected from scientific experimental devices has reached hundreds of PB per year. So how to use these data efficiently to produce some scientific findings is a hot problem. There are many challenges in the use of these scientific big data, such as the storage, processing and sharing of the data. In this paper, we propose a data management system, EventDB, for scientific big data. EventDB provides data management function for massive semi-structured scientific data; In EventDB, we propose IndexDB to provide a faster data retrieval, cross-domain access to provide a better data sharing and operator libraries to provide higher performance data analysis. Our preliminary experiments show that our system has improved performance by more than 6 times in data retrieval.

Keywords: Scientific big data · Data storage · Data retrieval · HBase

1 Introduction

With the development of technology, more and more large scientific devices have been applied to scientific experiments, such as the Europe Large Hadron Collider (LHC) [12] and the Beijing Electron-Positron Collider (BEPC) [13] in high-energy physics; Sloan Digital Sky Survey (SDSS) [14] and Five-hundred-meter Aperture Spherical radio Telescope (FAST) [15] in astronomy and so on. The use of these large scientific devices has made the data generated during scientific experiments larger and larger. Particles collide in the LHC detectors approximately 1 billion times per second, generating collision data about 1PB per second and the data permanently archived in its tape libraries has over 200PB [2]. As time goes on, the amount of data will grow larger and larger.

The data generated by many large scientific experimental devices has semi-structured data feature. For example, high-energy physics and astronomical observations. The data format commonly used in high-energy physics data analysis is root [3], which is made up by many Events. The experiment data exists as

This research is supported by the National Key R&D Program of China under grant No. 2016YFB1000604.

J. Li et al. (Eds.): BigSDM 2018, LNCS 11473, pp. 105–115, 2019.
https://doi.org/10.1007/978-3-030-28061-1_12

attribute value of Event. The Event attributes are different, either the number of attribute or the value format of attribute may be different. The data format of attributes are not structured. The format commonly used in astronomical observation data analysis is fits [4], which also has the same characteristics. The semi-structured data characteristics make these data unfavorable for managing with RDBMS. Currently, the data is usually stored on disk or tape as a file.

Although the amount of data generated in scientific experiments is large, scientists are only interested in a small part of the data. For example, in high-energy physics, the Events used by physicists in an analysis are usually less than 1% of the original data, or even one parts per million [5]. Because of the large amount of the data for analysis, the scientists pick the data based on certain criteria and then analyze the picked data. It takes a lot of time to screen out a small part of the original data. At present, the file-based storage way needs the file to be opened and read in the screening process, resulting in a large amount of disk I/O. However, only a part of the read data will be used in the analysis, and many others are actually useless.

At the same time, the experiment data may be used by scientists located in different organizations or regions. The different organizations or regions have different storage size and processing capacity. In high-energy physics, the World-wide LHC Computing Grid (WLCG) is a global computing infrastructure whose mission is to provide resources to store, distribute and analyze the data generated by the LHC [6]. Due to the concentration of initial data and the huge differences environment in network between organizations we need an efficient, transparent system to support physical analysis issued in different organizations. However, WLCG doesn't support event-level data transfer and is not satisfied with our requirements.

As we described before, we think that the traditional scientific big data processing system has some problems, it is not limited to the following aspects. Firstly, the problem of data storage scientists often care about a small part of the overall data. This small amount of data is screened from the original data with multiple times. Therefore, the way of data management directly affects the performance of data screening. Secondly, the problem of data processing. In the analysis and processing of existing scientific big data, it is done on the retrieved small data set. In fact, the data analysis can be run on a distributed platform along with retrieving data; it can improve the speed of data processing. At last, the problem of data sharing. Scientific experimental data is open to scientists all over the world under certain conditions, and data analysis should be available to execute at any site. To deal with these challenges existing in scientific big data, we design a system, EventDB, which mainly contains three components, IndexDB, Operator Library, Cross-domain access. To improve the speed of screening useful data, we propose IndexDB. In order to speed up the data analysis, we propose Operator Library. At the same time, we propose Cross-domain access to provide better data sharing.

The rest of paper is structured as follows. Section 2 introduces some background knowledge. Section 3 describes the related works. Section 4 describes our system design. Section 5 shows the evaluation to our system. At last, Sect. 6 concludes our work.

2   Background

2.1   Data Processing

Figure 1 depicts the process of data analysis in a common scientific experiments. Firstly, we record the raw data came from detectors, sensors, etc. Then, we use some tools to reconstruct the raw data to build useful data that can be analyzed. Because of the large size of the data, scientists usually retrieve the data according to certain conditions. Further, they will do simulations or further analysis on the screened data to obtain scientific findings. Taking high-energy physics as an example, the processing of high-energy physics data mainly includes data reconstruction, analysis, and simulation. The information generated by the high-speed collision of the particles is filtered by the online capture system of the sensor and transmitted to the offline system for storage. These data are called raw data and stored in RAW file format. The raw data consists of individual events, which are saved as files to the disk storage system and stored in the tape library for a long time; these raw data need to be reconstructed before being analyzed. The reconstructed data is saved as a DST (Data Summary Tapes) file. Physicists analyze the events in these DST files through an analysis framework to produce ROOT files that describe histograms or curves. In addition, Monte Carlo simulation software is often used to verify the analysis results. A data analysis usually needs to screen out interesting events from billions of events and then performs statistical analysis. These events are further analyzed to generate statistical charts or to perform fitting analysis.

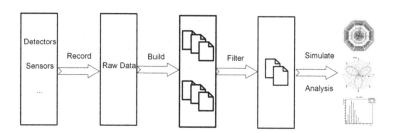

Fig. 1. The data processing flow of scientific big data

2.2   HBase and Fuse

HBase. HBase is an open-source, non-relational, distributed scalable, big data store [18]. It is developed as part of Apache Software Foundation's Apache

Hadoop project and runs on top of HDFS (Hadoop Distributed File System). HBase has very effective MapReduce integration for distributed computation over data stored within its tables and provides a fault-tolerant way of storing large quantities of sparse data. HBase supports coprocessors, which was inspired by Google's BigTable coprocessors. With coprocessor, HBase can push the computation to the server in where it can operate on the data directly without communication overheads. This method can give a dramatic performance improvement over HBase's already good scanning performance.

Filesystem in Userspace. Filesystem in Userspace (FUSE) is a software interface for Unix-like computer operating systems that lets non-privileged users create their own file systems without editing kernel code [19]. FUSE consists of a bridge-like module in kernel and a libfuse library in userspace. Many existing file systems are based on fuse development, such as Lustre, GlusterFS, WebDrive and so on.

3   Related Work

With the amount of scientific experiments data increasing, many scientific fields have proposed countermeasures. In high-energy physics, Becla *et al.* [10] and Düllmann *et al.* [11] propose Objectivity/DB database system to store the all data. Goosens in CERN, Cranshaw in ANL and *et al.* [7] use the Oracle database to store the index information of the ATLAS experiment. With database optimization techniques such as horizontal and vertical partition, they build TAGDB, which can support one billion-level Event index. Sánchez *et al.* [8] use HBase to build an Event index database, EventIndex, for ATLAS. Lei *et al.* [9] propose a high-energy physics data storage and process system, which uses HBase to store the Event data and uses MapReduce to implementing parallel process. In astronomy, Wiley *et al.* [17] use MapReduce framework to implement a scalable image-processing pipeline for the SDSS imaging database. Brahem *et al.* [16] present AstroSpark, a distributed data server for astronomical data. AstroSpark extends Spark, a distributed in-memory computing framework, to analyze and query huge volume of astronomical data. These works only focus on only one part of the processing analysis. Our work points out several problems in the analysis of scientific big data, and provides a holistic solution.

4   The Architecture

In order to efficiently use the large amount of data generated by scientific experiments, we design a system, EventDB, which improves the processing performance of scientific big data. Figure 2 shows the three main components of EventDB. EventDB consists of three main components, namely IndexDB, Operator library and cross-domain access. Next we will introduce these three components according to previous three problems existing in scientific big data analysis.

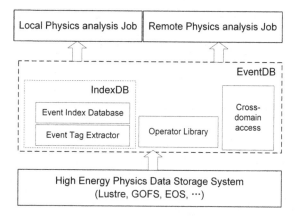

Fig. 2. The three main components of EventDB

4.1   Data Storage

At present, the experiment data still stores in the form of files. The data screening is very inefficient based on this form. In order to get an eligible data set quickly we build IndexDB, which provides an inverted index table of the data. Inverted index is an index data structure storing a mapping from content to its locations. Objects in scientific experiment data usually have many data attributes. We use the inverted index table to map the attribute value to the location including the filename, the offset in the file. Although the basic idea sounds simple, but it is not simple at all. We met many difficult problems to solve. Such as How to import data quickly facing so many events data? How to place the index table for faster access? And so on. To import data quickly, we write the record to HFile concurrently without using HBase APIs. To reduce the size of the inverted index table, we reduce the dimensions of the data attributes of the object; we build a multi-level index and do some data compression using bitmap. To speed up the query, we adjust the data layout using our algorithm to improve the data locality.

Taking the high-energy physics as an example, in order to quickly obtain the data set required by physicists, Liu [1] has extracted a series of feature quantities (called TAGs) and stored them in separate ROOT files and the TAGs as an index has been applied to BESIII. However, the tag information stored in the file is still slow when searching for the tag itself. We extracted TAGs from the data and use the TAGs information to create an inverted index for the events. We store the inverted index table into the HBase table as the key-value form. Figure 3 shows an example of a table in HBase.

The Rowkey contains the runID, each data attribute and the value. Each line represents a list of files containing the attribute and the offset in the file. Because the construction of the primary key in HBase uses a lexicographically ordered index structure and is usually cached in memory, it has good query efficiency. Through the inverted index of key-value table, we can quickly locate

Rowkey	EventIndex
~8026#Neutral#003	pips1.dst-0
~8026#Neutral#004	pips1.dst-0, pips1.dst-5, pips2.dst-8
~8026#Neutral#005	pips1.dst-0, pips1.dst-5, pips2.dst-9
~8026#Neutral#006	pips1.dst-0, pips1.dst-5, pips2.dst-10, ⋯
~8026#Neutral#007	pips1.dst-0, pips1.dst-5
~8026#Neutral#008	pips1.dst-0, pips1.dst-5, pips2.dst-12
~8026#Neutral#009	pips1.dst-0, pips1.dst-5, pips2.dst-13, ⋯

Fig. 3. The inverted index of Events stored in HBase

the file and the offset of the instance, which greatly improves the efficiency of Event screening.

IndexDB not only provides indexing function, but also supports to store original data directly. For astronomy, the size of each object is small, so it can be stored in IndexDB, I call this way ADSD (All raw Data Stored in the Database). ADSD can further improve the performance of data access, but it requires a very large storage overhead when the amount of raw data is too large.

4.2   Data Processing

In the current scientific big data analysis, the analysis usually starts after the screening ends. In order to accelerate the process of data analysis, we propose an Operator Library, which is a package for the common operations of the scientific analysis. With the coprocessor provided by HBase, the analysis can be started directly at each node running the screening, without waiting to complete all the screening results.

Taking the high-energy physics as example. The data processing of existing high-energy physics is usually submitted to the offline system for data analysis after completing the screening. With the help of the IndexDB, our system has the ability to quickly scan. Due to the distributed nature of HBase, after each node completes the filtering of a given condition, we can analyze the data in advance at the node, and finally combine the analysis results at the master node. This does not require the screening to be completed before data analysis. The most common data analysis tool is ROOT in high-energy physics. Our Operator Library contains the calculation operations commonly used in root, such as *Mag(), Dot(), Angle()* and so on. These operators are implemented on HBase through coprocessor. So the scientists just need to write their analysis program with the API provided by us, the analysis can be done at each node running the filtering. With Operator Library, we can get the analysis results faster.

In addition to providing some basic operators, Operator Library also supports users to define operators themselves. The scientists only need to write code following our specification, we can load the new operator dynamically and support the new operator to use in the analysis.

4.3   Data Sharing

In general, the experiment data will be used by many scientists, who come from different scientific research institutions all over the world, so a cross-organizational and cross-regional mechanism is necessary to ensure efficient access to experimental data.

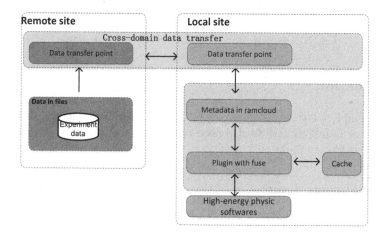

Fig. 4. The architecture of Cross-domain access

To provide object-level data sharing, we provide a cross-domain access system. The cross-domain access system stores the metadata information of the experiment data file at the local site using Ramcloud, which syncs the metadata of data files located in remote site. RAMCloud is a storage system that provides low-latency access to large-scale datasets. To achieve low latency, RAMCloud stores all data in DRAM at all times [24]. When the file or the object is accessed on the local site, the Plugin firstly queries the related metadata from the Ramcloud, then the remote site's data will be fetched by the FUSE module in the form of a file block or an object and cached on the local site. Through this system, the data analysis program can run seamlessly anywhere. The granularity of data transfer is file blocks level or object level, thus greatly reducing the amount of useless data transfer. Taking the high-energy physics as an example, cross-domain access seamlessly supports existing high-energy physics analysis and provides event-level access support. The architecture is shown in Fig. 4.

With EventDB, the pattern of scientific big data analysis has changed. Figure 5 shows the changes. In the traditional way, analysis job should commit to the data owner site and run as a batch job, which commits to a job scheduling system, such as condor [20], openpbs [21], etc. After introducing EventDB, each site can commit job on local, the cross-domain access can fetch the data from the data owner to insure the job running. At the same time, with the IndexDB and operator library, we reduce a large amount of I/O requests and improve the processing performance.

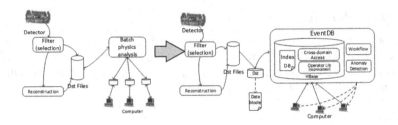

Fig. 5. The new processing flow of scientific big data

5   Evaluation

We just completed the preliminary evaluation, the following evaluations are mainly about IndexDB and we still need to do more test on cross-domain access and operator library parts.

5.1   Experiments Setup

We build HBase on 10 Dell servers, with Intel Xeon E5-2630v4 CPU, 64G RAM, 256 GB SSD and 20TB 10000 rpm disk, running Red Hat 4.4.7. All nodes are inter-connected by a 1000M Ethernet switch. The version of core softwares are: HBase v1.2.0, JDK v1.8.0. The data set in the experiment uses the running data of BESIII, which contains a total 2800 files and 500 billions events. The total size of the data is about 450 TB. We use 7 tags to construct the event index.

Importing Data. In EventDB, the first step is to build the inverted index table into HBase. Due to the large amount of the data, the construction speed also is important to the data users. New experimental data needs to add into HBase every once in a while. We do lots of optimization about importing data. Currently, the speed of importing event can be reached 200 millions per second. It takes only about three days to import a 500 billions Events. From the data disclosed by EventIndex [23], it requires 660 s to import 100 millions events with 128 parallel threads, and about 38 days to import 500 billions events. We are far faster than him. We think the reason is that EventIndex build record based on message and HBase API, our system write records to the HFile directly.

Events Query. BESIII Offline Software System(BOSS) software suite is used for BESIII simulation and analysis [22]. We simulate the scientists' query operation used in BOSS. We compare the time spent by using BOSS and using EventDB in executing same query operations. Figure 6 shows the comparison. In Fig. 6, the time consumption of the single user query using EventDB is just 1/10 of BOSS, some cases even is 1/100 of BOSS. Because of the query conditions are very complicated; it causes that the final results only contain little Events. The BOSS still need search from large number of files and EventDB can quickly get the results from the IndexDB. We also do some tests on the multiple users to EventDB. In general, the average query efficiency increases more than 6 times and our throughput can reach 2 millions Events per second.

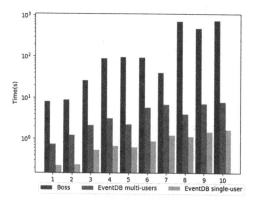

Fig. 6. The query performance comparison of Boss and EventDB

Functionality. We compare our system with the other scientific big data system in high-energy physics and astronomy. We choose EventIndex used in high-energy physics and the astronomy work which uses MapReduce for the SDSS imaging database. Table 1 shows the difference of these systems. The EventIndex only supports to store the data as files, EventDB can also store data in HBase. The EventIndex only supports data screening and do not speed the data processing, EventDB can speed up the data processing using the implemented API. EventDB also provides the Cross-domain access to support data sharing. Comparing with MapReduce for SDSS, it only tries to use mapreduce pattern to speed up the data analysis; it doesn't provide the ease to use APIs; it also doesn't provide the solution for data sharing across the organization or regions. Comparing with other two systems, our system provides better performance and more comprehensive functions.

Table 1. The solution to three challenges in different systems.

	EventIndex	MapReduce for SDSS	EventDB
Data storage	File	SQL	(file + Index) or (ADSD)
Data processing	Traditional way	Mapreduce jobs	Coprocessor
Data sharing	-	-	Cross-domain access

6   Conclusion

In our paper, we summarize several problems in the management of scientific big data with semi-structured features, and propose an EventDB system. EventDB overcomes difficulties in scientific big data through IndexDB, Operator Library, and Cross-domain access. Compared with the traditional scientific big data management system, it has the characteristics of more efficient data retrieving, faster data analysis, and better data sharing.

References

1. Liu, B.: High performance computing activities in hadron spectroscopy at BESIII. J. Phys.: Conf. Ser. **523**(1), 012008 (2014)
2. Gaillard, M.: CERN Data Centre passes the 200-petabyte milestone (2017)
3. Brun, R., Rademakers, F.: ROOT–an object oriented data analysis framework. Nucl. Instrum. Methods Phys. Res. Sect. A: Accelerators Spectrometers Detectors Assoc. Equipment **389**(1–2), 81–86 (1997)
4. Ponz, J.D., Thompson, R.W., Munoz, J.R.: The FITS image extension. Astron. Astrophys. Suppl. Ser. **105**, 53–55 (1994)
5. Cheng, Y., et al.: Data management challenges and event index technologies in high energy physics. J. Comput. Res. Dev. **54**(2), 258–266 (2017)
6. Girone, M., Shiers, J.: WLCG operations and the first prolonged LHC run. J. Phys.: Conf. Ser. **331**(7), 072014 (2011)
7. Cranshaw, J., Goosens, L., Malon, D., McGlone, H., Viegas, F.T.A.: Building a scalable event-level metadata service for ATLAS. J. Phys.: Conf. Ser. **119**(7), 072012 (2008)
8. Sánchez, J., Casaní, A.F., de la Hoz, S.G.: Distributed data collection for the ATLAS EventIndex. J. Phys: Conf. Ser. **664**(4), 042046 (2015)
9. Lei, X., Li, Q., Sun, G.: HBase-based storage and analysis platform for high energy physics data. Comput. Eng. **41**(6), 49–55 (2015)
10. Becla, J.: Improving performance of object oriented databases. BaBar case studies. In: CHEP Proceedings, Padova, Italy (2000)
11. Düllmann, D.: Petabyte databases. ACM SIGMOD Rec. **28**(2), 506 (1999)
12. Large Hadron Collider, European Organization for Nuclear Research. http://lhc. web.cern.ch/lhc/
13. Beijing Electron-Positron Collider, institute of High Energy Physics Chinese Academy of Sciences. http://bepclab.ihep.cas.cn/
14. Sloan Digital Sky Survey. http://www.sdss.org
15. Nan, R., et al.: The five-hundred-meter aperture spherical radio telescope (FAST) project. Int. J. Modern Phys. D **20**(06), 989–1024 (2011)
16. Brahem, M., Lopes, S., Yeh, L., Zeitouni, K.: AstroSpark: towards a distributed data server for big data in astronomy. In: Proceedings of the 3rd ACM SIGSPA-TIAL Ph.D. Symposium, p. 3. ACM, October 2016
17. Wiley, K., et al.: Astronomy in the cloud: using mapreduce for image co-addition. Publ. Astron. Soc. Pac. **123**(901), 366 (2011)
18. Apache HBase. https://hbase.apache.org/
19. Filesystem in Userspace. https://github.com/libfuse/libfuse

20. Tannenbaum, T., Wright, D., Miller, K., Livny, M.: Condor: a distributed job scheduler. In: Beowulf Cluster Computing with Linux, pp. 307–350. MIT Press, November 2001
21. OpenPBS. http://www.openpbs.org
22. Li, W.-D., Mao, Y.-J., Wang, Y.-F.: Chapter 2 the BES-III detector and offline software. Int. J. Modern Phys. A **24**(Supp01), 9–21 (2009)
23. The ATLAS EventIndex and its evolution based on Apache Kudu storage. https://indico.jinr.ru/getFile.py/access?contribId=199&sessionId=10&resId=0&materialId=slides&confId=447
24. Ousterhout, J., et al.: The RAMCloud storage system. ACM Trans. Comput. Syst. (TOCS) **33**(3), 7 (2015)

An Operator Library System for HEP Events Processing Based on Hbase Coprocessor

Zirong Du[1(✉)], Di Hou[2], and Yong Qi[2]

[1] School of Software Engineering, Xi'an Jiaotong University,
Xi'an, China
455755543@qq.com
[2] School of Electronic and Information Engineering,
Xi'an Jiaotong University, Xi'an, China

Abstract. At present, data generated by high-energy physical devices has reached the PB or EB level as it is constantly updated and running. Therefore, the technical requirements for physical analysis are constantly increasing with the mass of physical events generated by high-energy physical colliders. The physical analysis for high-energy physics events refers to the selection of thousands of meaningful events from massive physical events. The analysis process relies on different attribute parameters and different operators to obtain the final selection criteria. Because the traditional method of high-energy physics events analysis is less efficient and more complex, it is very important to design an efficient and convenient operator library for high-energy physics events processing. This paper proposes a new operator library system based on hbase coprocessor for high-energy physics events processing. It uses storage structure and endpoint coprocessor of the hbase as the framework of the physical analysis operator library, uses protocol buffer to customize the client and server RPC communication and encapsulates the operators in the object-oriented data analysis framework ROOT to hbase coprocessor. In this way, the complexity of the high-energy physics events analysis is reduced and the efficiency of the calculation is improved. In the experiment, we take the zc3900 analysis process as an example and apply it to the 10 nodes experimental cluster. The test results show that the proposed system is more efficient and avoids the complexity of the events processing.

Keywords: Scientific big data · High-energy physics · Event analysis · Operator library · Hbase coprocessor

1 Introduction

High-energy physics is a frontier branch of physics, also known as particle physics. Its scientific goal is to study the smallest unit and interaction law of constituent materials. It is a subject based on experiment, and it is also a subject based on the close combination of experiment and theory [1, 2]. With the development of science and technology, the scale and complexity of high-energy physics are constantly improving. The increase in the scale and complexity of the experiment means increases in the amount of data and the difficulty of data manipulation, but the calculation mode of high-energy

J. Li et al. (Eds.): BigSDM 2018, LNCS 11473, pp. 116–124, 2019.
https://doi.org/10.1007/978-3-030-28061-1_13

physics is basically unchanged. Its calculation mode is that the modern data acquisition system collects and rapidly filters the experimental data to form the original experimental data, and then the data storage system records the original data for subsequent data analysis and processing [3]. The accuracy of high-energy physics experiments depends on the statistics of the data. The construction and operation of a new generation of high-energy physics experiment equipment has produced data of PB and EB magnitude. Massive amounts of data require super-large scale storage, computing resources and network resources, which poses great challenges to data management technologies such as data collection, storage, transmission, sharing, analysis and processing [4].

Event screening and analysis is an important part of data analysis and processing in high-energy physics. We call the data calculation process in the data analysis process as an operator, which is the theoretical basis for the analysis and prediction of physicists [5, 6]. With the rapid development of high-energy physics, the complexity of event analysis for high-energy physics is getting higher and higher, and the scheduling, combination, communication, and mapping of algorithms applied to big data analytics platforms are becoming more and more complex. It is very complicated and redundant to analyze each type of operator and apply it to the big data management platform of high-energy physics. The operator library is a collection of high-energy physics basic operators. It can be combined into various required algorithms according to the needs, giving full play to the advantages of the big data platform. It can not only effectively solve the execution time of operators, but also give full play to the parallel characteristics of distributed systems to improve the calculation efficiency of operators and achieve the optimal operation cost. Operator library is a set of operators with certain functions and a standard library of the whole high-energy physics complex algorithm. A complete operator library can lay a good application of big data management platform in high-energy physics data. The operator library plays a very important role in the big data management platform for high-energy physics events. Decomposition and mapping based on high-energy physics operators have become one of the research hotspots of high-energy physics. It is urgently needed by physicists to assemble the basic operators of high-energy physics into a complete standard operator library with certain functions. Therefore, this paper proposed the design and implementation of events processing operator library based on Hbase coprocessor for high-energy physics.

The rest of the text is organized as follows. The second part introduces the related work of the system, including the ROOT framework and Hbase coprocessor. The third part describes in detail the design and implementation of events processing operator library based on Hbase coprocessor for high-energy physics, including operator library, event analysis process and system analysis. The fourth part describes the system performance evaluation. Finally, the fifth part is Conclusions.

2 Related Work

At present, the massive data produced by high-energy physics experiments are widely dependent on the storage and processing of ROOT framework [7]. The application of Hbase in high-energy physics experiments is relatively small, but the combination of

Hbase and high-energy physics experiments has become a research hotspot. Therefore, this section is based on the application of the ROOT framework and Hbase coprocessor in high-energy physics.

2.1 ROOT Framework

ROOT is an object-oriented framework system used primarily for high-energy physics data analysis. The ROOT framework provides the basic tools and services for analysis of high-energy physics, such as multi-dimensional histogram, curve fitting, data modeling, and simulation. It is programmed in C++ and contains 60 libraries and 19 modules. It uses an object-oriented structure and provides methods for serialization and deserialization. It can quickly convert between memory objects and disk files [7, 8].

High-energy physics is based on ROOT files when relying on the ROOT framework for data analysis. The ROOT file has a strict storage format. The file begins with 100 bytes to record the basic information of the file and the position of the pointer. There are two fields indicating the offset of the first data object and the offset of the file end for the file. There are consecutive data objects behind the file header. Each data object contains two parts, which are the size of the object and the offset of the object in the file. Therefore, each data object in the ROOT file can be accessed sequentially from the beginning of the file. This sequential access mode does not meet the requirements of large-scale data analysis. So ROOT provides a tree structure to support more flexible file access methods, which can selectively access certain branches of the event. A tree structure contains one or more branches, each branch contains a list of leaves, each leaf represents an attribute class of the branch and its variable is a simple type [9].

2.2 Hbase Coprocessor

Hbase is a column-oriented storage database based on Hadoop, and its design idea comes from BigTable of Google. The storage structure of Hbase is a multi-dimensional orderly mapping, commonly referred to as column-oriented or column-cluster storage. The storage structure does not care about the data type. This structure enables HBase to process structured, semi structured or even unstructured data. Its underlying layer uses Hadoop distributed file system HDFS as the storage system, which makes Hbase have good fault-tolerant performance and horizontal extension [10]. Compared with the traditional database, Hbase only provides row key indexing and does not provide flexible filtering query function. Therefore, in some query cases, full table scanning is required. And it does not have the computational power of some commonly used aggregate functions.

The Hbase coprocessor provides a mechanism for developers to run custom code directly on region server to manage data. Users can use the coprocessor to write code running on the Hbase Server side. HBase supports two types of coprocessors, which are Endpoint and Observer. The Endpoint coprocessor is similar to the stored procedure in the traditional database. The client can call these Endpoint coprocessors to execute a piece of Server-side code and return the results of the Server-side code to the client for further processing. The most common usage is to the aggregation operation. Another kind of coprocessor is called Observer, which is similar to the trigger in the traditional database. This coprocessor will be called by the server side when certain events occur [10, 11].

3 Design and Implementation of the System

The system in this paper is based on the basic operator in the ROOT framework. The protocol buffer is used to customize the RPC service. Based on the Hbase coprocessor programming framework, the base operator is encapsulated on the Hbase server and the aggregation operation is implemented on the client. In this way, the data analysis process with high efficiency and low bandwidth is realized.

3.1 Operator Library

The operator library is a basic operator commonly used in the process of high-energy physics data analysis. The complete operator library can reduce the redundant operation of high-energy physics in the event analysis process. In order to make the process of event analysis more convenient and efficient, we try our best to abstract the basic operators in the ROOT framework and combine them into a complete operator library. In the ROOT framework, we can divide operators into MathCore, MathMore, and Physics Vectors. MathCore contains some basic implementations of numerical algorithms, such as probability density functions, cumulative distribution functions, quantile functions, trigonometric functions, and statistical functions. MathMore is a supplement to MathCore, including polynomial calculation, elliptic integral, hypergeometric function and coupling coefficient. Physics Vectors is for describing vectors in 2, 3 and 4 dimensions and their rotation and transformation algorithms, including transposed matrices, symmetric matrices, vector products, cross vectors, and so on.

This paper takes the analysis process of zc3900 as an example. We split the basic operator in the zc3900 event analysis process and found that the analysis process uses the basic operator list in the ROOT framework as follows. They are TVector3.Mag2(), TVector3.Mag(), TVector3.Dot(), TVector3.Angle(), TLorentzVector.Mag2(), TLorentzVector.Mag() and TLorentzVector.M(). So we reproduce these basic operators in the experiment and add them to the operator library for use in the zc3900 analysis process.

We integrate the abstracted operators into the operator library, which provides a more convenient and effective combination calculation method for specific event analysis. The system encapsulates the operator library on the Hbase coprocessor after assembling into a complete operator library. Since the operator is the calculation of the event attribute value, the Endpoint coprocessor is used. That is, the operator library is published on the Hbase server side, and the aggregation of different Region calculation results is implemented on the client side. During the implementation of the custom Hbase endpoint coprocessor, protocol buffer is used to customize the RPC service to achieve communication between different Hbase nodes. The working principle of the operator library mounted on the Hbase endpoint coprocessor is shown in Fig. 1.

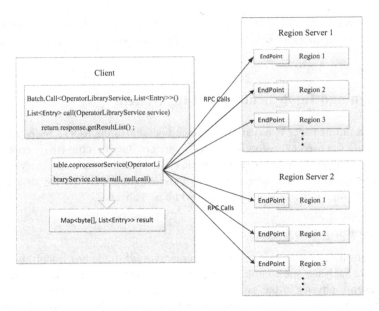

Fig. 1. Schematic diagram of the operator library mounted on the Hbase endpoint coprocessor.

In the process of implementing the operator library using the Hbase endpoint coprocessor, we prepared a proto file to define the RPC interface and then use the protocol buffer to generate the java file corresponding to the proto file. Then we implement the server class. It inherits an abstract class generated by the protocol buffer, which is the inner class of the proto corresponding java file. And it implements the two necessary interfaces CoprocessorService and Coprocessor for Hbase coprocessor. The concrete operator is then written into the class as a method to form the final server-side class. The generated interface file and implementation class file are packaged into a jar and placed on hdfs. Finally, the client class is implemented, and the operators in the operator library can be directly invoked to aggregate the calculation results in different regions to solve the final result.

3.2 Event Analysis Process

Event analysis refers to the process of event filter and reconstruction of original data generated by high-energy physics experiments. In the process of data analysis for high-energy physical, different event filter conditions are adopted for different event analysis. The event filter condition consists of different operators in the operator library. Each event is judged to satisfy the filtering condition according to the calculation results of the combined operator, and then the signal histogram is generated to show the event distribution. Taking the zc3900 analysis process as an example, the event filter flowchart of the operator library based on the Hbase coprocessor is shown in Fig. 2. It can be seen that we can directly combine the operators of the operator library to calculate whether each event meets the filtering conditions in a distributed and concurrent way. And then we can pick out the events that meet the conditions to generate the signal graph. Because it is based on Hbase to scan the data in full table, the IO overhead is much smaller than that of directly reading the ROOT file. In addition, the efficiency has been greatly improved due to the synchronized calculations on different Regions.

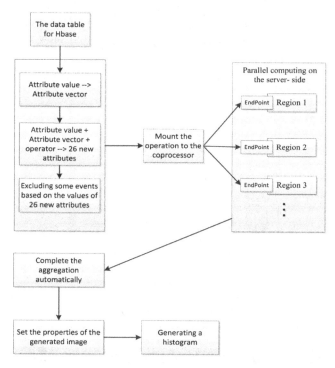

Fig. 2. Flow chart of zc3900 analysis process based on Hbase coprocessor operator library

3.3   System Analysis

The demand for data processing of high-energy physics can no longer be satisfied with traditional computing methods. The huge data requirements for computing environment of high-energy physics are gradually improved, including massive data storage capacity, superior computing power, high-speed data transmission, and large-scale distributed collaborative processing mechanism. Therefore, this paper designs a operator library system based on Hbase coprocessor for event processing of high-energy physical. And it uses the Hbase coprocessor for event filtering.

In this system, we designed the operator library and implemented the basic operator in the ROOT tool as much as possible, so the system can reduce the amount of code and provide convenience. In addition, we make full use of the high concurrency and high availability of Hbase to carry out the calculation process by multiple nodes simultaneously. Therefore, the efficiency of operator library is optimized.

In the process of implementing events processing operator library based on Hbase coprocessor for high-energy physics, we first analyze the high-energy physics analysis tool ROOT to abstract its base operator and encapsulate it into the operator library. Then the operator library is mounted on the Hbase coprocessor. The user can directly call the operator in the operator library to calculate the event data and directly return the calculation results to the client. Its internal calculation process is invisible to the user. Finally, we use the interface as a user. The framework design of the system is shown in Fig. 3. We take the zc3900 analysis process as an example to show the interface, as shown in Fig. 4.

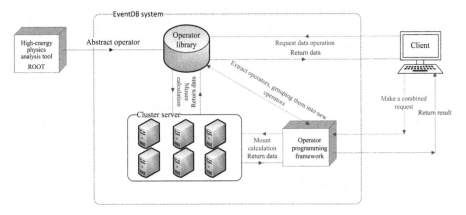

Fig. 3. Frame design diagram of operator library.

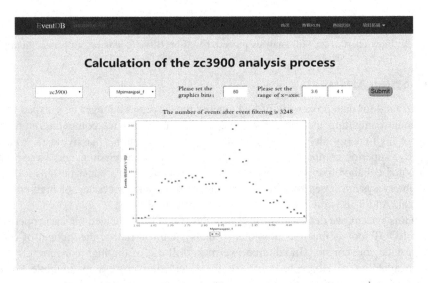

Fig. 4. User interface of zc3900 analysis process

4   System Performance Evaluations

The performance evaluation uses the data used in the zc3900 analysis process as the evaluation object, and the filter condition is the track set in advance. Figure 5 shows the comparison of the amount of code between the system we designed and the ROOT tool in the zc3900 analysis process. It can be seen that the system we designed has less code and simpler operation in the process of event filtering. Figure 6 shows the comparison of the calculation time between the system we designed and the ROOT tool in the zc3900 analysis process. It can be seen that the system we designed has improved the efficiency by 16% for the zc3900 analysis process. Therefore, the system we designed improves the computational efficiency of event filter in data analysis of high-energy physics and reduces the redundancy and complexity of user operations.

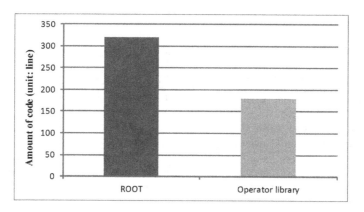

Fig. 5. The amount of code between the operator library and the ROOT tool was compared in the zc3900 analysis process.

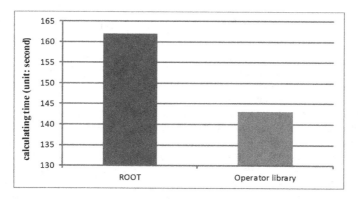

Fig. 6. The calculation time between the operator library and the ROOT tool was compared in the zc3900 analysis process.

5   Conclusions

Based on the high-energy physics and the zc3900 case analysis process and combined with the advantages of Hbase, this paper studies and designs a new system of events processing operator library based on Hbase coprocessor for high-energy physics. It makes the event filter process highly parallel. The test results show that the system has faster data processing speed, more convenient operation mode and more friendly interface than traditional systems. The Hbase endpoint coprocessor can place logical calculations on the server side, allowing users to perform their own operations on the server side. However, the system has not been verified in the large-scale production environment. There are still many areas to be studied and improved in the study of the operator library based on Hbase coprocessor.

References

1. Chen, Y., Shi, J., Chen, G.: A survey of high energy physics computing system. e-Sci. Technol. Appl. **5**, 3–10 (2014). https://doi.org/10.11871/j.issn.1674-9480.2014.03.001
2. Cheng, Y.D., Shi, J.Y., Chen, G.: Design and optimization of storage system in HEP computing environment. Comput. Sci. **42**, 54–58 (2015). https://doi.org/10.11896/j.issn.1002-137X.2015.01.012
3. Wang, L., Chen, Y.D., Chen, G.: Design and performance optimization of metadata server in mass storage system. Comput. Eng. **38**, 1–3 (2012). https://doi.org/10.3969/j.issn.1000-3428.2012.02.001
4. Huo, J., Lei, X., Li, Q., Sun, G.: High energy physics data processing system with parallel heterogeneous clusters. Comput. Eng. **41**, 1–5 (2015). https://doi.org/10.3969/j.issn.1000-3428.2015.01.001
5. Lei, X., Li, Q., Sun, G.: Hbase-based storage and analysis platform for high energy physics data. Comput. Eng. **41**, 49–55 (2015). https://doi.org/10.3969/j.issn.1000-3428.2015.06.010
6. Dong-Song, Z., Jing, H., Dong, L., Gong-Xing, S.: High energy physics data analysis system based on mapreduce. Comput. Eng. **40**, 1–5 (2014). https://doi.org/10.3969/j.issn.1000-3428.2014.02.001
7. Brun, R., Rademakers, F.: ROOT—an object oriented data analysis framework. Nuclear Instrum. Methods A **389**, 81–86 (1997). https://doi.org/10.1016/S0168-9002(97)00048-X
8. Antcheva, I., Ballintijn, M., Bellenot, B., Biskup, M., Brun, R., Buncic, N., et al.: ROOT—A C++ framework for petabyte data storage, statistical analysis and visualization. Comput. Phys. Commun. **180**, 2499–2512 (2009). https://doi.org/10.1016/j.cpc.2009.08.005
9. ROOT a Data analysis Framework. https://root.cern.ch/
10. George, L.: HBase: The Definitive Guide: Random Access to Your Planet-Size Data. O'Reilly Media Inc., Sebastopol (2011)
11. Vashishtha, H., Stroulia, E.: Enhancing query support in HBase via an extended coprocessors framework. In: Abramowicz, W., Llorente, I.M., Surridge, M., Zisman, A., Vayssière, J. (eds.) ServiceWave 2011. LNCS, vol. 6994, pp. 75–87. Springer, Heidelberg (2011). https://doi.org/10.1007/978-3-642-24755-2_7

Event-Oriented Caching System for ROOT Data Analysis in HEP

Baoan Liu[1,2], Can Ma[2(✉)], Bing Li[2], and Weiping Wang[2]

[1] University of Chinese Academy of Sciences, Beijing, China
[2] Institute of Information Engineering, Chinese Academy of Sciences, Beijing, China
macan@iie.ac.cn

Abstract. In the field of high-energy physics, Event is the basic data unit, referring to a particle collision or interaction among particles. At present, advanced physical experimental devices can produce a large amount of Event data up to PB level. While Compared to these massive data generation, data storage system based on files at the moment is out of date. Event data are mostly random accessed, but searching a few specific Event in large files is an inefficient job. Therefore this paper proposes an event-oriented data storage technology, caching frequently accessed data in HBase. This paper serializes the Event data in the ROOT files and dumps them to intermediate files, and then a large number of intermediate files are transferred into the HBase by the method of bulkload and cluster resources. Eventually, using the data of Beijing Spectrometer (BESIII) Experiment, we conduct a data access experiment. The result shows that the new storage schema can improve the performance of data random access by more than 4 times.

Keywords: Event-oriented · Caching system · Random access

1 Introduction

Event is the basic unit of high energy physics experimental data, referring to a particle collision or the interaction among particles. In a large physical experiment, the number of the produced Event data can be as large as trillion scale. At present, the Events are mainly stored and calculated in the form of ROOT [1,2] file. It is a standard format file in the field of high energy physics, which may contains millions of events, and each event may have hundreds of properties. These files are currently stored in some distributed file systems like Lustre [3]. Storage based on large files leads to huge I/O overhead and poor performance when retrieving events by some attributions' value. Meanwhile, the calculation nodes and storage nodes are separated and connected through high speed network. When it is required to access some events in a file, the whole file will be

Supported by the National Key Research and Development Program of China (2016YFB1000604).

loaded, however, most of which are useless. All these cause serious network and storage resources waste. Therefore event-oriented storage and calculation mode is a good solution to these problems.

On the other hand, according to the investigation and survey on high-energy physics data and the related data processing, most physicists merely focus on some certain events, which may be distributed in different files. Even in the same file, there is a skipping access between consecutive requests. Based on the two points above, the I/O performance is difficult to optimize while conducting data analysis in such storage mode.

In view of the problems faced by current high-energy physical data storage, many research institutes at home and abroad have studied the current mature big data tools, such as HDFS [4], MapReduce [5,6], HBase [7] etc, and applied them to high-energy physics [8]. Such as, the University of Iceland has conducted research and test on data parallel analysis of MapReduce replacing Parallel ROOT Facility [9]. ATLAS has studied the use of Hbase to store metadata. Cern studied the application of HDFS and MapReduce in the field of high-energy physics. Nebraska University tried to use HDFS as the storage system of the CMS experiment. Xiaofeng Lei's HBase platform [10] is instructive, but the mapreduce job to load data also has much I/O consume.

2   Related Work to the Event-Oriented Caching System

This paper proposes an event-oriented storage technology caching frequently accessed data into HBase. we use different data processing methods and improve data storage performance. While in specific implementation, through serializing the Event data ROOT files, generate the representation of intermediate file. Then by the spark cluster, bulkload these files into HFile, a standard storage format of HBase data [11]. This method is currently used as a caching tool for the original storage system. Physicists can import a batch of experimental data as a whole and analysis experiments for this batch of data will be accelerated. If some Events are unavailable in HBase, it will turn to the ROOT files automatically. Combined with the function of the database, the paper provides a function for prefetching continuous events. By setting the size of prefetching data, the performance of data sequential access can be improved as well. Experiment results show this new storage mode can help improve access efficiency of the physical offline analysis system.

3   The Traditional Data Storage Method

ROOT is a framework for data processing, born at CERN, at the heart of the research on high-energy physics. It processes PB-level experimental data generated by various high-energy physics experiments every year and stored these data in the form of ROOT files. Figure 1 is the structure of TFile [12], an abstraction of ROOT files. The header portion holds the format information of the file, ensuring that the file conforms to the ROOT specification without the suffix.

The header also stores some serialization characteristics of objects stored in the file so that the original ROOT object can deserialized when loading. Each object is assigned with an offset for indexing. Eventually, the actual data is stored in the data section and is compressed when saved, which reduces space overhead [13].

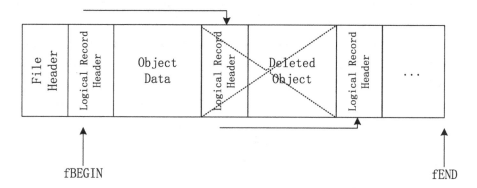

Fig. 1. Structure of TFile.

For high-energy physical data, the storage adopts a more complex data structure - TTree. TFile writes the TTree object directly to the file. TTree is a data structure implemented in the ROOT framework for random reading, as shown in Fig. 2 [14]. Event data is stored in the Branch part of the TTree. Each branch contains a leaf list for storing data of smaller granularity. One branch has a basket structure, which is a buffer used to implement I/O operations between files and memory. TTree is something like column-stored mode, branches of one tree hold the same attribution of different events. When a particular event needed to be load to the memory, depending on the nature of the cache, Basket will load the same attribution of other events in the vicinity, and the load resources will be wasted if the caching data is not truly needed. That is also a significant reason for proposing event-oriented cache schema.

High-energy data is stored in seven top-level objects, and the seven objects are TevtRecObject, TdstEvent, TdigiEvent, TevtHeader, ThltEvent, TMcEvent and TtrigEvent. These seven objects are stored in the Branch structure in Fig. 2. There is a lot of leaf data under the Branch of these event objects. Branches and leaf nodes are organized into one TTree. Data is organized hierarchically like a tree. Furthermore, an event data is not always containing all the seven Branches mentioned above.

For a generic ROOT object, the framework provides a complete mechanism for accessing. The seven event classes above, however, are actually defined by the physicists according to the requirements of high-energy data and extended based on rules of ROOT framework. The definition of the seven classes are in BOSS software. So it is necessary to integrate the resources of the two frameworks when using Event data. ROOT class for storing event objects [6].

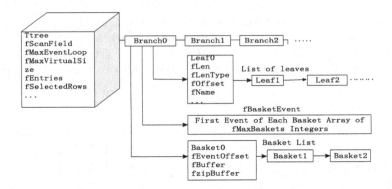

Fig. 2. Data structure of TTree.

This chapter mainly describes the storage format, data structure, related framework and technology of the original data storage mode.

4   Design and Implementation of the Event-Oriented Caching System

The granularity of traditional storage mode is too big for offline analysis software with event as a basic unit of processing. Although TTree improves the efficiency of random reading, considering the skip reading access mode of physicists, there are still massive useless event data being loaded into memory, resulting in serious I/O resource waste. The method proposed in this paper will extract the events frequently accessed from ROOT files, and cache them into the HBase. When the physical analysis software conducts data access, it first accesses the K-V database and then the original file system if not hit. The new storage mode with event granularity is more suitable for the accessing mode.

In order to dump event data into the database, a series of transformation is necessary. Meanwhile, the database needs to connect with the existing offline analysis software, so a new set of data access interface should be designed. The overall process is shown in Fig. 3. The data abstraction part extracts events from the ROOT files, and the data serialization part transforms the complex structure of event into a serialized string. The event import part compared two storage mode and use the method of bulkload and spark cluster to load the events into HBase. The data interface part provides a C++ interface for the traditional high-energy physical software to access the caching system. Depending on the custom of the physical experiments, this paper adopts the batch update strategy to refresh the entire caching system by the new data produced in the latest experiment. Besides, this paper implements a prefetch function by adding a cache strategy at the client side. Each section will be covered in detail next.

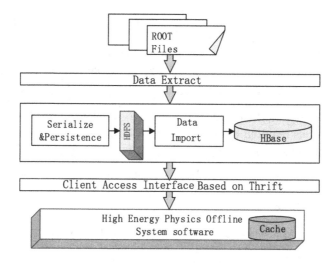

Fig. 3. Processing flow chart of the event-oriented caching system.

4.1 Event Abstraction

This section includes loading event data into memory and data Persistence. These Event classes are extended based on the rules of ROOT framework and defined in BOSS software-an offline analysis software based on Gaudi [15] framework. So it is necessary to integrate the two resources. The specific process of extracting event data into memory is shown in Fig. 4.

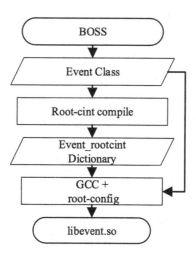

Fig. 4. Event data abstraction.

The event data classes belong to user-defined ROOT classes, and the definition file is in the physical analysis software known as BOSS. For the use of the custom ROOT classes, there is a set of specification in the ROOT framework. Operate on the source file using ROOTs root-cint tool to generate an interface dictionary. ROOT uses a compiler called Cling to compile C++. This compiler has been embedded inside the ROOT framework and it is not convenient for external development. Following the complicated compiling rules of ROOT framework, this paper grasp the way to compile using gcc. Besides the interface dictionary, another tool is also needed, root-config, to get varies compilation conditions required by ROOT [13]. To facilitate modular programming, this paper compiles the code and generate a dynamic library as a tool to extract event data.

4.2 Event Serialization

The design of event data class is complex, which can hardly be disassembled into basic data type to storage. On the other side, physical analysis is generally based on event, so if an event is disassembled into basic data types, extra work will be done to restore it. The main character of the event-oriented storage system is to store the event itself into the K-V system directly. The event class are previously persisted to ROOT files, involving complicated data structure, different from the K-V storage mode, instructive to our work indeed. In this paper, a new event serialization method is proposed referring to the storage schema of ROOT framework. Serialization refers to the conversion of objects to binary bytes. This concept often occurs, for instance, when data is transmitted over the network it needs to be serialized by sender and deserialized by the receiver. Also in persistence, objects stored in disk needs to be serialized first, and when objects are read, it needs to be deserialized back. Serialization solutions must be mature enough to prove their availability in productive practice. Meanwhile, it is very convenient to debug if the correctness of the serialization and deserialization data results is recognizable. This paper proposes a serialization method based on persistency and network transport of ROOT framework.

The serialization tool implemented in this paper is SerTMessage class, providing serialization and deserialization interfaces. Serialization interface takes TObject class as input parameter and outputs a binary byte stream. TObject is a base class of the entire framework, that is, the interface supports all the TObject descendants serialization in ROOT. The deserialized one is to get the byte stream output the original object back. This tool class inherits from TMessage class. TMessage is actually a tool for transferring objects in the network. It involves the specific process of serialization, but provides no related public interface.

As the superclass of TMessage, TBufferFile implements some read/write interface provided in TBuffer and also implements some serializable method of basic data types. However, it has limitations. With serialization, persistency and network transport bundled together, the underlying implementation is complex and cannot accept TObject type and get serialized byte stream, which is not a tool that can provide good interface.

TBuffer inherits from TObject, which is used for serialization and preserving the serialized structure in a character point variable. TBuffer provides read/write interface for serialization and deserialization, but has not been implemented [9].

SerTMessage is a combination of the above tools and provide available serialization function directly. It is noted that the serialized data is 1.5 times larger compared to the original file. The reason is that ROOT files is compressed during persistency. Considering this, compression algorithm should be cooperatively considered when we store these data.

4.3   Event Import

This section covers how to persist event data to K-V database from memory, including the design of the storage schema, selection of data import method and implementation of access interface.

Storage Schema. The serialized events can be stored into HBase directly. For hbase, if the muti-column cluster is used for storage, some more mapping data of the key and column cluster will generate. When the amount of data becomes large, it will take up a lot of storage capacity. Therefore, a single cluster storage mode is selected. Another issue to be considered is to separate or merge branches when storing event, which actually depend on the access way of the data. The original TTree structure provides a schema similar to column storage and it gives us direction to store the different branches separately. These schema is suitable if Branches are accessed mostly separately. If the whole Event is accessed more often, the merge schema is better otherwise. While both situation exist when physicists conduct analysis, and there is no clear proof that which one is more frequently. One the other side, it was found by statistics that if no compression algorithm is used when the serialized events are preserved into database directly, no matter what schema, the data will expand compared to the original file. When storing data with the schema of merging Branches together, the expansion ratio is slightly lower, but space will also increase by about 60%. Considering the large volume of high energy physics experimental data, a compression algorithm is necessary. Compared on various performance of data compression algorithm supported by HBase, which is shown in Table 1 [16], snappy compression algorithm is prefered. The effect and performance comparison of the two storage modes will be detailed in the experimental section.

Table 1. Efficiency comparison of three compression algorithms.

Compression algorithm	Compression ratio	Compression speed	Decompression speed
GZIP	13.4%	21 MB/s	118 MB/s
LZO	20.5%	135 MB/s	410 MB/s
Zippy/snappy	22.2%	172 MB/s	409 MB/s

Improvement of the Importing Method. There are many methods for data importing, and the following three are the commonly used at present: (1) The native Client API; (2) mapreduce job; (3) Generating HFile directly by using BulkLoad. The first two methods will communicate with RegionServers frequently, when massive data is stored at one time, a lot of RegionServer resources may be occupied, which impairs queries on other tables.

Data in HBase are stored in HDFS in form of HFile. So a more efficient and convenient method is to take the method of BulkLoad by Spark [20]. This method is suitable for batch writing. BulkLoad generates StoreFiles through a cluster job and loads them into the HBase directly, which will cause less I/O, CPU as well as and resources overhead.

With the method of BulkLoading and spark cluster to complete data migration, the efficiency and anti-pressure capacity are improved. Because the serialized event is an unrecognized binary string, it is impossible to distinguish the boundaries of two events in a file when importing data. Therefore, an extra operation of encoding is added after the serialization, so that the serialized events can be stored as a well-formed intermediate text file in hdfs.

4.4   Data Access Interface

HBase supports java language, but does not provide interface for C++. In the field of high-energy physics analysis, BOSS is a commonly used analysis software, which will obtain events from storage system and conduct analysis by interface provided in ROOT framework. Both BOSS and ROOT are developed in the C++ language, so a set of C++ interfaces to access the new storage system is required for platform consistency.

Thrift [17] is a software framework, which is used for the development of extensible and cross-language service that allows users to pass messages between systems implemented in one or more languages [18]. In this paper, the C++ interfaces are accomplished by the services provided by thrift. The Thrift framework [19] actually implements a kind of C/S mode. Through the Customized files of data structure and service interface, we compile and generate server-side and client-side codes to support cross-programming languages communication. In this paper, the client code is written by c++, and the server side is java.

Throughout the development environment, physical analysis software BOSS is strict with the versions of operating system and compiler. The version of operating system must be Redhat linux 6.5 and the compiler must be gcc-4.4.6. The compiler of thrift2 which is developed by C++ and C language, will encounter version-incompatibility problems when install the thrift environment by source code. This causes the available thrift versions are restricted in a scope. This paper uses thrift-0.9.2.

4.5   Cache Replacement Strategy

The main function of a caching system is to keep hot data in an efficient storage region and reduce the time delay for frequent data accessing. Thus, how to update the stored data is of great importance. That is, how to seek the hot data in and old data out. At the same time, due to the change of storage mode, the data access method of traditional physical analysis software should be also changed. In order to further improve the access efficiency, a function of prefetching has also been added in the client.

During a high energy physics analysis work, different physicists only care about certain part of the whole data set according to their research content. There might be no any overlap among the data sets. However, what all physicists have in common is that they are interested in the latest data, which is generated from recent large experiment. Therefore, the event-oriented caching system adopts the cache replacement strategy of batch update, that is, the latest experimental data are imported into caching system.

For the client cache, a bidirectional linked list is used to cache recently accessed event data. The pointer of each event is located by its key in HBase, and the query time is O(1). In general, the classical replacement strategy LRU is used. The new access data will be updated to the head of the linked list. When the cache overflows, event at the end of the linked list will be remove firstly. The bidirectional linked list and scan function of HBase can be used together. We use a configurable pre-fetch function along with scan function to obtain continuous data. Improving the pre-fetch number of events will improve the efficiency of data sequential access.

5   Evaluation and Analysis of Event-Oriented Caching System

5.1   Storage Mode Comparison

There are two storage modes proposed in 3.3.1. This paper tried the two storage schemas respectively. For the merge storage schema, it creates the class called TBossFullEvent, which inherits from the TObject class. This class includes the seven objects fields and a flag used to mark the branch of each object. The class graph is shown in Fig. 5. This schema stores the TBossFullEvent class and necessary extra properties.

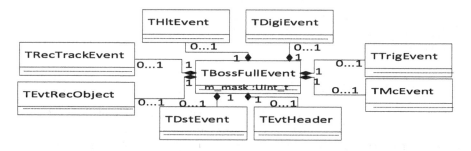

Fig. 5. The class structure of TBossFullEvent.

The table structure designed in HBase for this schema is simpler. The schema of the caching system is shown as Table 2. The key of each event is spliced by the filename and entry id. The filename is from ROOT file in which the event originally exists and the entryid is one property of the event. There is only one cluster named Event, which includes 3 fields: eventInfo, length, and d_length. EventInfo represents the final processed result of the entire event, length is the size of the serialized result, and d_length is the length of the encoding length.

Table 2. Event table merged in caching system.

Key	Event:eventInfo		
Filename:entryid	Event	Length	D_length
	SeriString1	size of SeriString1	size of coded SeriString1

Different from the merged one, the separated storage schema stores the seven Top ROOT objects independently. The key of each event is also the concatenation of the filename and entry id. There is only one cluster named Event, which includes at most 9 fields. The field of Slen_array and Dlen_array are required. And the other seven fields depend on whether each Branch exists. Slen_array is spliced by each length of the serialized string of each branch. And Dlen_array is spliced by each length of the coded string of each branch. The schema is shown as Table 3. Ls1 represents the length of serialized String1 and Lc1 represents the length of the decoding String1 and so on.

Table 3. Event table separated in caching system.

Key	Event:eventInfo				
Filename:entryid	Slen_array	Dlen_array	Branch1	Branch2	...
	ls1-ls2-...	lc1-lc2-...	SeriString1	SeriString2	...

The performance of the two in storage space and read-write time is compared by serializing and loading 15000 events into the caching system. The results are shown in Table 4, which conveys that the merged schema has advantage over the separated one.

5.2 Data Access Comparison

A series comparative experiment for random access to real Event data are conducted to verify the performance of the new storage schema. When conducting the random-access experiment, it preserved 190,000 events and randomly generated 10,000 event-ids. Contrast experiments are using the original file. In each

Table 4. Performance comparison of writing by different storage mode.

Storage mode	Size of all events	Average size of event	Time of loading	Occupied storage space
ROOT file	511.5 MB	34.91 K		313.65 MB
Merged storage	493.2 MB	33.67 KB	19.8 s	339.45 MB
Separated storage	494.5 MB	33.72 K	21.26 s	340.31 MB

Table 5. Performance testing of random reading.

Num of events (piece)	500	1000	2000	3000	5000	10000
Time of ROOT file	67.44 s	137.16 s	271.01 s	404.28 s	678.65 s	1363.83 s
Time of caching system	11.5 s	30.36 s	62.76 s	98.29 s	165.63 s	336.55 s

case, three experiments were conducted to get the average result. The performance of the event-oriented storage system and the traditional storage system for random access are shown in the Table 5.

Compared with the traditional storage system, the speed of the random data reading is increased by more than 4 times. Considering the cost of opening multiple files, the efficiency improvement should be more obvious if more ROOT files are loaded into the caching system.

6   Conclusion

In this paper, through a lot of analysis on characteristics of high-energy physics Event data, combining with the existing storage mode, we proposed a Event-based storage schema.

The main work of the paper is the processing of data transformation. Through in-depth study of ROOT, we realized Event extraction from ROOT files. The serialization and deserialization of the Event data is done based on ROOT framework. The communication between Database and analysis software is solved by using thrift. In the end, the performance has been improved by more than 4 times compared with the traditional data access method. At present, the new storage system has been applied to event data management and transfer system in high energy physics.

References

1. CERN ROOT [EB/OL]. https://root.cern.ch/. Accessed 29 Oct 2018
2. Rademakers, F., Brun, R., et al.: ROOT: an object oriented data analysis framework. Nucl. Instrum. Methods Phys. Res. **389**(1–2), 81–86 (2000)
3. Braam, P.J., Schwan, P.: Lustre: the intergalactic file system. In: Ottawa Linux Symposium, vol. 8, no. 11, pp. 3429–3441 (2002)

4. White, T.: Hadoop-The Definitive Guide 4e, 4th edn, pp. 1–4. O'Reilly Media, Newton (2015)
5. Zang, D.S., Huo, J., Liang, D., et al.: High energy physics data analysis system based on mapreduce. Comput. Eng. **40**(2), 1–5 (2014)
6. Glaser, F., Neukirchen, H., Rings, T., et al.: Using mapreduce for high energy physics data analysis. In: International Conference on Computational Science and Engineering, pp. 1271–1278. IEEE (2013)
7. Vora, M.N.: Hadoop-HBase for large-scale data. In: International Conference on Computer Science and Network Technology, pp. 601–605. IEEE (2012)
8. Pivarski, J., Lange, D., Jatuphattharachat, T.: Toward real-time data query systems in HEP (2017)
9. Glaser, F., Neukirchen, H., Rings, T., et al.: Using mapreduce for high energy physics data analysis. In: International Conference on Computational Science and Engineering, pp. 1271–1278. IEEE (2014)
10. Lei, X., Qiang, L.I., Sun, G.: HBase-based Storage and Analysis Platform for High Energy Physics Data (2015)
11. Apache HBase Reference Guide. http://hbase.apache.org/book.html. Accessed 29 Oct 2018
12. TFile Class Reference. https://root.cern.ch/doc/master/classTFile.html. Accessed 29 Oct 2018
13. ROOT user Guide. https://root.cern.ch/root/htmldoc/guides/users-guide/ROOTUsersGuide.html. Accessed 29 Oct 2018
14. TTree Class Reference. https://root.cern.ch/doc/master/classTTree.html. Accessed 29 Oct 2018
15. The Gaudi Project. http://proj-gaudi.web.cern.ch/proj-gaudi/. Accessed 27 Oct 2018
16. George, L.: HBase: The Definitive Guide. O'Reilly Media, Newton (2011)
17. Apache Thrift Homepage. http://thrift.apache.org/. Accessed 29 Oct 2018
18. Rakowski, K.: Learning Apache Thrift. Packt Publishing, Birmingham (2015)
19. Abernethy, R.: The Programmer's Guide to Apache Thrift. Manning Publications, Shelter Island (2015)
20. Apache Spark Homepage. http://spark.apache.org/. Accessed 29 Oct 2018

Automated and Intelligent Data Migration Strategy in High Energy Physical Storage Systems

Zhenjing Cheng[1,2](✉), Yaodong Cheng[2], Lu Wang[2], Qingbao Hu[2], Haibo Li[2], Qi Xu[1,2], and Gang Chen[2]

[1] University of Chinese Academy of Sciences, Beijing 100049, China
chengzj@ihep.ac.cn
[2] IHEP Computing Center, 19B Yuquan Road, Beijing 100049, China

Abstract. As a data-intensive computing application, high-energy physics requires to process and store massive data at the PB or EB level. It requires high performance data access and large volume of data storage as well. Some enterprises and research organizations are beginning to use tiered storage architectures, using tapes, disks or solid drives at the same time to reduce hardware purchase costs and power consumption. Tiered storage requires data management software to migrate less active data to lower cost storage devices. Thus an automated data migration strategy is very necessary. Data access requests are driven by the behavior of users or programs. There must be associations between different files that are accessed consecutively. This paper proposes a method to predict the heat of data access and use data heat trend as the basis criteria for data migration. This paper proposes a deep learning algorithm model to predict the evolution trend of data access heat. This paper discussed the implementation of some initial parts of the system. Then some preliminary experiments are conducted with these parts.

Keywords: High-energy physics storage · Tiered storage · Data migration · Data access heat · Deep learning

1 Introduction

The scale of High Energy Physics computing has been expanding as human exploring origins of the universe and basic material composition [1]. High-energy physics experiments storage produce and process PB level data. On one hand, it is difficult for traditional storage systems based on disks to support higher IOPS services. On the other hand, only a very small number of files would keep active for a fixed period of time. So IHEP computing center are considering to use tiered storage architectures [2], which include tape, normal disk or solid state drives to reduce hardware purchase costs and power consumption.

Limited by cost factors, an automated data migration strategy [3] is needed to migrate less active data to lower cost storage devices. At present, automatic data migration strategies, such as LRU, CLOCK, 2Q, GDSF, LFUDA, FIFO, etc., are all

© Springer Nature Switzerland AG 2019
J. Li et al. (Eds.): BigSDM 2018, LNCS 11473, pp. 137–145, 2019.
https://doi.org/10.1007/978-3-030-28061-1_15

strategies for the exchange of in-memory caches. They need to run in the kernel of operating system, so their exchange rules are relatively simple [4].

Data access requests are not completely random. They are driven by the behavior of users or programs so they would show access locality in file system [5]. This motivates file access prediction as the basis criteria for data migration.

1.1 Prediction Model

LS model [6] uses a file's most recent successor as the next one to be accessed. Essentially the basis of LS and the extended version, DLS is when a certain user reads some files sequentially, generally the past access sequence would be repeated more or less. On this basis, Aigui Liu et al. proposed a ULNS model which adds user information to produce more accurate subsequent file prediction [7]. The prediction accuracy is better than LS model, but it also relies heavily on historical file access order. It is not very efficient when there are massive files, or the access order laws are not so obvious. Palmer et al. used associative memory to identify patterns in the context of file access order. They used cache which called Fido to learn this pattern [8]. Griffioen et al. proposed a scheme for predicting file access using some probability maps. But those probability maps only record files whose access frequency is within a certain window size [9]. Tait et al. studied techniques for detecting laws of file access order in client-side cache management and used them to prefetch files from the servers. They assumed that most users' behavioral patterns would produce special file access patterns for data sets of specific applications [10].

The Stable Successor (or Noah) and Recent Popularity [11, 12] methods extended the Last Successor algorithm by attempting to filter out noise of the observed file access order. Stable Successor records subsequent accessed files for each file. Recent Popularity looks in the access order for successors of the file, then uses the Best-k-of-n algorithm to pick the next file most likely to be accessed.

1.2 File Access Heat

Since mass storage systems for high-energy physics experiments tend to have billions of files, it is difficult to achieve good results regardless of which file access prediction algorithm is used. Therefore this paper predicts file access heat, i.e. file access frequency. In general, performance metrics or load predictions for continuous time can use regression analysis [13]. To predict data access, regression analysis can also be used to predict and help generate migration decisions. Since migration decisions and the I/O performance of tiered storage are directly affected by predictions' accuracy, so we need to train a regression analysis model for every file in storage system. That is an impossible task to accomplish at this stage for the reality of high-energy physics storage systems with billions of massive files. Generally speaking, the best file migration strategy does not change within a certain interval of access frequency. This paper divides file access frequency into multiple bins. The model only needs to predict which bin a file's access frequency would fall in. It is worth the cost of not being able to predict precisely if more accurate prediction could be made on which bins a file belongs to. Therefore the prediction problem is re-presented as a classification problem. Only one classification model is needed for all files, as shown in Fig. 1.

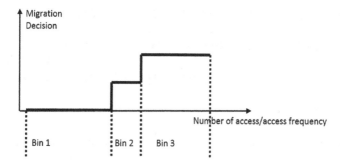

Fig. 1. Given a file, there are n possible migration decisions for the tiered storage where n is the number of tiers. This paper introduces three bins for we have three tires at present: tapes, disks and solid drives.

After some research, we found that more than 95% of the files in high-energy physics storage would not be accessed over three times in the next 7 days after being accessed once. Those files were defined as cold file/cold data. Nearly 3% of the files would be accessed between one to three times. Those files were defined as warm file/warm data. Nearly 1% of the files would be accessed more than three times. Those files were defined as hot file/hot data.

2 File Access Feature Construction

The file heat classifier prediction accuracy is very dependent on inputs, that is, training set. Generally speaking, file heat changes are affected by multiple factors, such as different data access methods used by different users/applications, and different high-energy physics data analysis software. Therefore, data heat prediction not only needs to consider data access frequency in the past, but also needs to consider data access mode of different users/applications. In the underlying file system it means different data access features. In this paper. The premise of our project is try to predict file access heat from the perspective of underlying file system. That is to say, for different files file system can perceive the difference of file metadata and different file access trace. From file system traces we can get file reading and writing, file pointer seeking operations and so on.

Lustre [14] and EOS [15] are the most widely used mass storage systems in high-energy field. Taking CERN as an example, EOS manages 15 instances, 2.6 Billion files, and nearly 250 PB of data. The amount of data read and written in EOS in the past month is nearly 100 PB. Institute of High Energy of the Chinese Academy of Sciences is also using Lustre and EOS to manage massive amount of data over 10 PB. To not affect file system performance, Linux's trace functionality is disabled. It forces us to obtain file access features in Lustre and EOS from other sources. The log system of EOS FST servers keep data access records in file units, including file opening, closing, reading, writing e.g. As follows,

```
log=a048f57a-6034-11e8-8f98-288023415e08&path=/#curl#/eos
/user/b/biby/yinlq/rootdata/QGSJET-FLUKA/Helium/1.e14_1.e
15/wcda003363.root&ruid=10408&rgid=1000&td=*CioA-gA.16391
02:551@vm088029&host=eos07.ihep.ac.cn&lid=1048578&fid=123
971808&fsid=25&ots=1527263977&otms=887&cts=1527263998&ctm
s=734&rb=0&rb_min=0&rb_max=0&rb_sigma=0.00&wb=8830528&wb_
min=63&wb_max=32768&wb_sigma=2225.83&sfwdb=8814629&sbwdb=
8814592&sxlfwdb=8781824&sxlbwdb=8814592&nrc=0&nwc=271&nfw
ds=3&nbwds=1&nxlfwds=1&nxlbwds=1&rt=0.00&wt=24.91&osize=0
&csize=8830565&sec.prot=unix&sec.name=root&sec.host=vm088
029.ihep.ac.cn&sec.vorg=&sec.grps=root&sec.role=&sec.info
=&sec.app=fuse
```

Since our project is still in the early stages of research, all file access records were stored in the column-oriented NoSQL database – Hbase [16]. Using Hbase makes it easier to select different features to train the prediction model.

After pre-processing, we get such file access vectors such as <access timestamp, file name, file size, file read/write times, sequential and random access ratio, read and write ratio, read and write bytes>. This paper compacted multiple vectors into a sequence of time series by hour, as shown in Fig. 2.

T: timestamp F_1:filename F_2: file size R_1:file read/write ratio S: sequential/random ratio R_2: file read/write bytes

Fig. 2. File access eigenvector organized by hour.

When building training samples, a fixed training time window should be set, such as 30 days. Model inputs are file access features in the past 30 days, as shown in Fig. 3.

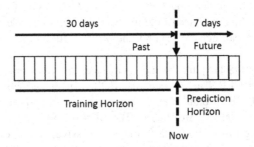

Fig. 3. If the training horizon is 30 days and the granularity is 1 h, the training horizon is represented by 720 records, which means 720 access eigenvectors.

Usually the width of each time record is 1 h, which means every file records one access feature vector every hour. Each training sample would contain 720 records and 720 access feature vectors. In the training process, the predicting model has equal ability to learn the access characteristics for different time since the width of the time record is fixed. In some cases, we want to use as much historical information as possible. So expanding the training horizon could help learn more file access information. More access features need to be saved in each training sample. This could lead to an increase in the complexity of the prediction model, and the difficulty of training can even lead to over-fitting problems. In this paper, the method of dynamical training time window is used, as shown in Fig. 4, and the width of each time record would become larger as time goes forward. It not only achieved the goal of using as much as historical information as possible, but also avoided the problem that the model was too complicated to train.

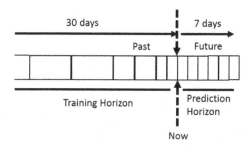

Fig. 4. The longer the time, the wider of a time record. Old data access information get consolidated as time moving forward.

3 Overall System Design

As shown in the foregoing, the obtained file access feature vector is a fixed-length time series which is very suitable for solving with Long-Short Term Memory (LSTM). LSTM [17] is an improved recurrent neural network which is capable of learning the long-term hidden dependencies. Each LSTM cell has three gates: input gate, output gate and forget gate. Those gates are used to control whether the information can pass through the cell or not. Therefore the LSTM network is capable of memorizing contextual information when mapping between input and output sequences [18]. The LSTM network structure is shown in Fig. 5.

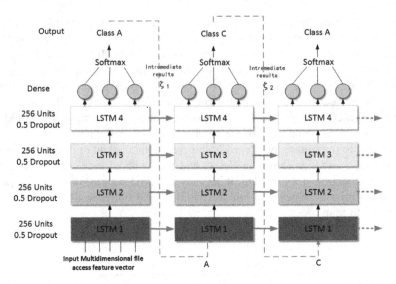

Fig. 5. LSTM network structure. Its composed of units of recurrent neural network (RNN). It's capable of learning long-term dependencies.

The tiered storage system designed in this paper has a fixed data migration cycle. The cycle begins with collecting file access records from the EOS log system, ends with migrating a set of files across different ties. The specific steps included in the data migration cycle are as follows.

Step1: collect file access records from the mass storage system of high energy physics. According to statistics, in the storage system used by the LHAASO collaboration, tens of thousands to hundreds of thousands of such file access records are generated every day. In this paper, HBase is used to store such unstructured access log data. File name and access time are used as the rowkey in HBase.

Step2: generate features from the access records. Useful features are often compact numerical indicators organized by hour, e.g. Number of bytes read or written to a certain file.

Step3: label the training data set. Determine a training horizon and a prediction horizon. Bin the number of access in prediction horizon and use the bin as basis of training data set labels.

Step4: train a classifier based on LSTM network to make predictions on a given file.

Step5: test the classifier. Slide the training horizon and prediction horizon forward for some time to collect a new training data set and test the trained classifier.

Step6: migrate a set of files. Use the classifier to predict file access heat changes and perform file migration (upward migration or downward migration) according to a predetermined migration strategy.

Step7: Evaluation. Calculate the classifier prediction accuracy. It would help the construction of file access features and model training method.

4   Experiments and Tests

Our first research is mainly for the high-energy physics experiment LHAASO located in Sichuan, China. The distributed storage system EOS provides storage services for the LHAASO experiment and LHAASO cooperation group user. The total number of user files currently in EOS has exceeded 70 million. According to statistics 5,842,207 files had been accessed during the month of 2018.4.1 to 2018.5.1. In the week of 2018.5.1 to 2018.5.7, the number of cold files accessed less than 3 times is 5,776,611. The number of hot files accessed more than 3 times is 66,196. In this paper, a sample data set consisting of 5,842,207 access feature vectors is obtained. These training feature vectors were divided into three groups, which are training data set (80%), evaluation data set (10%) and test data set (10%).

The LSTM prediction model is constructed based on tensorflow [19]. The model parameters are as follows:

A 4-layer fully connected artificial recurrent neural network with 64 LSTM nodes/cells per layer. The learning rate ranges between 0.001 and 0.0001. After each epoch, the learning rate was multiplied by an attenuation coefficient γ. In order to improve the convergence speed of the model, 256 samples are input into the network at the same time.

The trained LSTM model is tested using the test data set, and the evaluation data is used to verify the prediction results. The accuracy of the prediction and other evaluation indicators were calculated, as show below (Table 1).

Table 1. Key indicators of the classifier based on supervised machine learning

Evaluation indicators	Evaluation data set	Test data set
Predict accuracy	0.9060	0.8844
TPR	0.9211	0.8944
FPR	0.1076	0.1251
Precision	0.8358	0.8716
Recall	0.8712	0.8944
F-measure	0.9026	0.8829
AUC	0.7027	0.6611

5   Conclusion

Intelligent migration technology of tiered storage is the development trend of mass storage systems used in the field of high-energy physics. This paper proposes a method of supervising machine learning to predict data access heat. Based on the model prediction, we provided a migration strategy for tiered storage systems. We collected file access logs over a period of time from EOS, preprocessed them into a time series vector and divided files into different bins according to the number of accesses. Then we built a LSTM neural network for prediction. We created sample data sets from LHAASO users' real files in the production environment. The prediction accuracy is close to 91% for those files accessed within one month. At present, migration decisions are based on the predicted changes in data heat. But we didn't consider influences brought by data migration to the performance of the storage system. Next, we plan to introduce the concept of migration cost, consider the impact of migration on storage performance, study adaptive file migration strategy and seek to minimize migration costs.

Acknowledgments. This work was supported by the National key Research Program of China "Scientific Big Data Management System" (No. 2016YFB1000605).

References

1. Cheng, Z., Li, H., Huang, Q., et al.: Research on elastic resource management for multi-queue under cloud computing environment. In: Journal of Physics Conference Series, p. 092003 (2017)
2. Li, A., Yu, D., Shu, J., et al.: A tiered storage system for massive data: TH-TS. J. Comput. Res. Dev. **48**(6), 1089–1100 (2011)
3. Zhang, G., Chiu, L., Dickey, C., et al.: Automated lookahead data migration in SSD-enabled multi-tiered storage systems. In: IEEE Symposium on Mass Storage Systems & Technologies, pp. 1–6. IEEE Computer Society (2010)
4. Ari, I.: Using statistical correlation for dependency analysis of cache replacement policies. In: IEEE TOC, ACM TODS, ACM TOC, SIGMETRICS 2004, ICDCS 2003, FAST 2002, The Computer Journal (2004)
5. Jiang, S., Davis, K., Zhang, X.: Coordinated multilevel buffer cache management with consistent access locality quantification. IEEE Trans. Comput. **56**(1), 95–108 (2007)
6. 吴峰光, 奚宏生, 徐陈锋: 一种支持并发访问流的文件预取算法. 软件学报 **21**(8), 1820–1833(2010)
7. 刘爱贵, 陈刚: 一种基于用户的 LNS 文件预测模型. 计算机工程与应用 **43**(29), 14–16 (2007)
8. Palmer, M.L., Zdonik, S.B.: FIDO: a cache that learns to fetch. In: Proceedings of Conference on Very Large Data Bases (VLDB), Barcelona, pp. 255–264, September 1991
9. Griffioen, J., Appleton, R.: Reducing file system latency using a predictive approach. In: Proceedings of 1994 Summer USENIX Conference, pp. 197–207 (1994)
10. Lei, H., Duchamp, D.: An analytical approach to file prefetching. In: Proceedings of 1997 USENIX Annual Technical Conference, January 1997
11. Kang, S.J., Lee, S.W., Ko, Y.B.: A recent popularity based dynamic cache management for content centric networking. In: 2012 Fourth International Conference on Ubiquitous and Future Networks (ICUFN), pp. 219–224. IEEE (2012)
12. Catalina, T., Virgone, J., Blanco, E.: Development and validation of regression models to predict monthly heating demand for residential buildings. Energy Build. **40**(10), 1825–1832 (2008)
13. Lehmann, A., Overton, J.M.C., Leathwick, J.R.: GRASP: generalized regression analysis and spatial prediction. Ecol. Model. **157**(2–3), 189–207 (2002)
14. Braam, P.J.: The lustre storage architecture (2004)
15. Peters, A.J., Sindrilaru, E.A., Adde, G.: EOS as the present and future solution for data storage at CERN. In: Journal of Physics: Conference Series, vol. 664, no. 4, p. 042042. IOP Publishing (2015)

16. Dehdouh, K., Bentayeb, F., Boussaid, O., et al.: Using the column oriented NoSQL model for implementing big data warehouses. In: Proceedings of the International Conference on Parallel and Distributed Processing Techniques and Applications (PDPTA). The Steering Committee of the World Congress in Computer Science, Computer Engineering and Applied Computing (WorldComp), p. 469 (2015)
17. Graves, A.: Long Short-term Memory, pp. 1735–1780. Springer, Berlin (2012)
18. Wei, X.: From recurrent neural network to long short term memory architecture (2013)
19. Abadi, M., Barham, P., Chen, J., et al.: TensorFlow: a system for large-scale machine learning. In: OSDI, vol. 16, pp. 265–283 (2016)

Using Hadoop for High Energy Physics Data Analysis

Qiulan Huang[1(✉)], Zhanchen Wei[1,2], Gongxing Sun[1],
Yaodong Cheng[1], Zhenjing Cheng[1,2], and Qingbao Hu[1]

[1] IHEP Computing Center, P.O. Box 918-7 19B Yuquan Road,
Beijing 100049, China
huangql@ihep.ac.cn
[2] University of Chinese Academy of Sciences, Beijing, China

Abstract. With the development of the new generation of High Energy Physics (HEP) experiments, huge amounts of data are being generated. Efficient parallel algorithms/frameworks and High IO throughput are key to meet the scalability and performance requirements of HEP offline data analysis. Though Hadoop has gained a lot of attention from scientific community for its scalability and parallel computing framework for large data sets, it's still difficult to make HEP data processing tasks run directly on Hadoop. In this paper we investigate the application of Hadoop to make HEP jobs run on it transparently. Particularly, we discuss a new mechanism to support HEP software to random access data in HDFS. Because HDFS is streaming data stored only supporting sequential write and append. It cannot satisfy HEP jobs to random access data. This new feature allows the Map/Reduce tasks to random read/write on the local file system on data nodes instead of using Hadoop data streaming interface. This makes HEP jobs run on Hadoop possible. We also develop diverse MapReduce model for HEP jobs such as Corsika simulation, ARGO detector simulation and Medea++ reconstruction. And we develop a toolkit for users to submit/query/remove jobs. In addition, we provide cluster monitoring and account system to benefit to the system availability. This work has been in production for HEP experiment to gain about 40,000 CPU hours per month since September, 2016.

Keywords: Hadoop · HDFS · Mapreduce · HEP data analysis

1 Introduction

HEP experiments have been accumulated huge amount of datasets for decades. These datasets are produced by sophisticated detector systems that observe particle interactions. IHEP (Institute of High Energy Physics, Beijing, China) manages a number of China's major scientific experiments including BESIII [1], Dayabay [2], HXMT [3], LHAASO [4], JUNO [5] and so on. All of the experiments are typical data intensive applications, which require huge amount of computing power and storage. To meet the computing requirements, we have more than 15,000 CPU cores managed by HTCondor [6], 11 PB disk storage mainly deployed using Lustre [7] and EOS [8] file system and 5 PB tape storage. Figure 1 shows the computing system in IHEP. With the

© Springer Nature Switzerland AG 2019
J. Li et al. (Eds.): BigSDM 2018, LNCS 11473, pp. 146–153, 2019.
https://doi.org/10.1007/978-3-030-28061-1_16

development of a new generation of HEP experiments, more and more data are being generated. The traditional computing system has certain limitations in scalability, IO performance, fault tolerance and so on. The exploitation of new computing technologies has become an urgent practice to overcome a series of challenges in data analysis. Efficient parallel algorithms/frameworks and High IO throughput are key to meet the scalability and performance requirements of HEP offline data analysis.

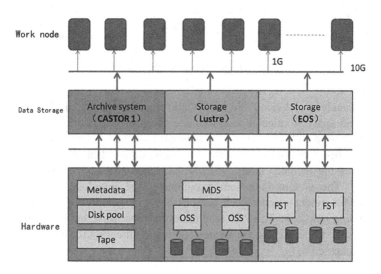

Fig. 1. Computing system in IHEP

Hadoop [9] is widely used to support data-intensive distributed application as it provides a software framework for distributed storage using HDFS [10] file system and processing big data using the MapReduce [11] programming model. Hadoop proposes the new idea of "Moving computation to data" instead of the traditional one "Moving data to computation". This fact becomes stronger while dealing with large data sets. The main advantage is that this increases the overall throughput of the system. It also minimizes network congestion. Though Hadoop has gained a lot of attention from scientific community for its scalability and parallel computing framework for large data sets, it is still difficult to make HEP data processing tasks run directly on Hadoop. In this paper we introduce the big data technologies into the HEP data analysis. The aim is to investigate the application of Hadoop to make HEP jobs run on it transparently.

2 HEP Data Analysis

The particular workflow of HEP data analysis [12] is illustrated in Fig. 2. Raw data is produced by sophisticated detector systems and simulation. The aim of the HEP data analysis is to mining the amount of experiment data to get the final results.

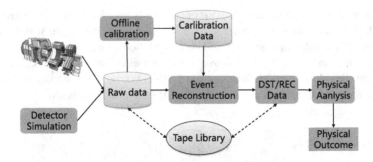

Fig. 2. The workflow of HEP analysis

HEP data analysis is divided into Mont Carlo simulation calculation, reconstruction calculation and analysis calculation. The three kinds of calculation behave as CPU-intensive computing and IO-intensive computing. CPU and IO are easily become the bottleneck of HEP data processing. Take LHAASO offline data analysis for example, the data processing includes three parts: system initialization, event processing loop and job end. The system initialization completes the initialization of application manager, service modular and object, and also the configuration of job attribute. The event processing loop contains event reading, event processing and event storage. The job end is to save the file, close the service and release the resource after all the events have been processed.

3   New Data Access in HDFS

Hadoop HDFS provides high throughput access to application data and is suitable for cases that handle large data sets. It releases a few POSIX requirements to enable streaming access to data. However, it cannot support random read and write operations. But physical software is designed and developed based on GAUDI [13] or SNIPER [14], and physical data is stored with ROOT format [15]. The I/O pattern of HEP software is random access. In order to make HEP jobs run on Hadoop, it's necessary to change the HDFS data access methods to support random read and write.

3.1   Implementation

HDFS consists of a Master/Slave architecture in which Master is NameNode that stores metadata and Slave is DataNode that stores the actual data. The NameNode contains all the information regarding which block is stored on which particular DataNode in HDFS, hence client needs to interact with the NameNode to get the address of the specific DataNode where the requested blocks are actually stored. The DataNode performs read and write operation from all clients' requests. Whatever clients read or write in HDFS, clients need to communicate to the NameNode to get number of blocks, block location, replicas and other details firstly. Based on the information, clients can communicate with the specific DataNodes to get block path to read/write by calling

HDFS streaming data access API. Additionally, NameNode is responsible to check whether clients are authorized to access that block. So it gives a security token to clients which they need to interact with the DataNode for authentication. After getting the address of DataNode that contains the specific blocks, clients will directly connect to the DataNode to read/write blocks with FSDataInputStream/FSDataOutputStream interfaces.

We implemented the new data access in HDFS by changing the data access methods of HDFS. Figure 3(a) shows the original HDFS data read flow. Figure 3(b) shows the new HDFS data read flow. From Fig. 3, we know this new data access allows the HEP software to read/write data directly in the local File System with the read interfaces of local File System instead of calling FSDataInputStream interfaces. The main advantage of new data read is that there is no data transmission, no network IO and low latency under one replica.

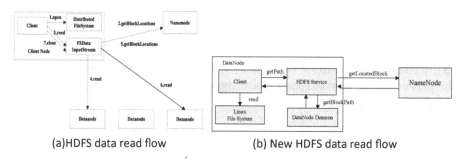

(a)HDFS data read flow   (b) New HDFS data read flow

Fig. 3. Read data flow

For write operation, the HDFS client sends a create request on Distributed FileSystem APIs to makes an RPC call to the NameNode to create a new file in the namespace. The Namenode performs various checks to make sure the file doesn't exist and the client has the permission to write the file. Then DistributedFileSystem call the FSDataOutputStream to write data in the specific DataNode. When the client has finished the file, it calls close() API on the stream. The write flow is more complicated than read. What we do is to change the part of writing data which is the process when client starts to write data in the block. When the DistributedFileSystem get the block location to know the address of the DataNode and block path, it lets HEP software to write data in the specific block location with the local file system API like ext3/ext4. Figure 4 shows the updated write data flow.

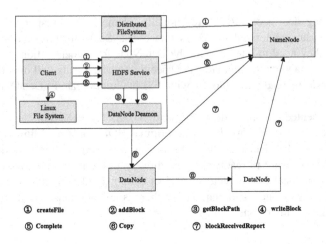

① createFile   ② addBlock   ③ getBlockPath   ④ writeBlock
⑤ Complete   ⑥ Copy   ⑦ blockReceivedReport

Fig. 4. New write data flow

3.2 Performance

We evaluated the performance of the new HDFS data access by submitting the real HEP jobs in Hadoop farm and traditional computing farm (Introduced in Paragraph 1).

Testbed Specification

HDFS: 1 NameNode, 5 DataNodes(6*6 TB, Raid 5), 1 Gigabit Ethernet
Lustre:1 Metadata server, 5 OSS servers with 2 Disk Arrays(24*3 TB, Raid 6), 10 Gigabit Ethernet

Test flow

(1) Prepare a large data set and copy into HDFS and Lustre;
(2) Submit the IO intensive jobs(Medea++) in Hadoop computing farm to access the prepared dataset, and record the job execution time of the job;
(3) Submit the same job in traditional computing farm to access the same dataset stored in Lustre, and record the job execution time of the job.

Results: Figure 5 displays comparison of medea++ job time consumption between HDFS and Lustre. It's clear to know the job access HDFS is one third of Lustre, which illustrate the job efficiency is higer than Lustre for IO intensive jobs. Because the IO intensive job requires large network IO as well as the Lustre client service consumes additional system overhead. Both of them affect the job operation efficiency.

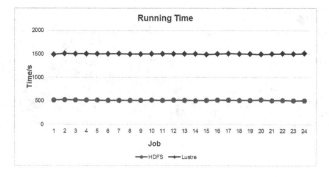

Fig. 5. The execution time comparison of medea++ job between HDFS and Lustre

4   Use Case in IHEP

4.1   User Interface and Monitoring

In order to provide a friendly interface, we develop a toolkit for users to submit/query/remove jobs. Diverse MapReduce models for HEP jobs are provided such as Corsika simulation, ARGO detector simulation and Medea++ reconstruction. For example, users can submit a job with the following command.

$$hsub + queue + jobType + jobOptionFile + jobname$$

queue: queue name
jobType: MC(simulation job), REC(Reconstruction job), DA(Analysis job)
jobOptionFile: Job option file, to describe the input files, output files, job execution environment settings, job execution commands and so on.
Jobname: job name

In addition, the cluster monitoring in terms of cluster healthy, number of running jobs, finished jobs and killed jobs is provided. We enable the ganglia to collect the specific metrics to monitor the CPU load, network IO and memory. And we write scripts to get the real-time job status including the number of running jobs, the number of finished jobs, the number of pending jobs and the killed jobs, and then make them visualized in ganglia and grafana. Figures 6 and 7 shows the real-time job monitoring.

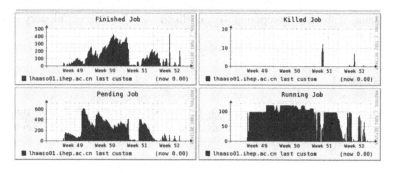

Fig. 6.   Real-time job monitoring

Fig. 7. Real-time job monitoring in Grafana

We also develop the accounting system to do job analysis in aspect of job numbers, task numbers, CPU time and memory usage group by username, jobId and job exit code. The system consists two parts: information collector and visualization. The information is collected by parsing the Hadoop logs and to get the items including username, group, CPU time, job create time, job finish time, memory usage and so on. All the information will be stored into Elasticsearch cluster and visualized with Grafana. Figure 8 shows the cpu time and memory used by user.

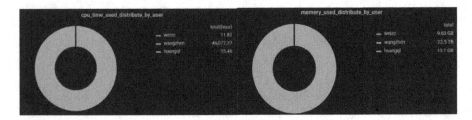

Fig. 8. The cpu time and memory used by user

4.2 Status in IHEP

This work has been in production for HEP experiment and the jobs use about 40,000 CPU hours per month since September, 2016. The production system has the storage 140 TB (88% used) and 120 CPU cores. It supports the HEP computing tasks like Corsika simulation, ARGO detector simulation and Medea++ reconstruction. The running status of the new computing system illustrates it's a successful story in HEP experiment in IHEP.

5 Conclusion

In this paper we present the application of Hadoop to make HEP jobs run on it transparently. This work has been in production for HEP experiment since September, 2016.

We discuss the new data access in HDFS to support HEP software to random access data in HDFS. This new feature allows the Map/Reduce tasks to random read/write on the local file system on data nodes instead of using Hadoop data streaming interface. This makes HEP jobs run on Hadoop possible. And diverse MapReduce model for HEP jobs are provided such as Corsika simulation, ARGO detector simulation and Medea++ reconstruction.

We also develop the friendly user interfaces and wrap them as a toolkit for users to submit/query/remove jobs. We provide the real-time cluster monitoring in terms of cluster healthy, number of running jobs, finished jobs and killed jobs and accounting system is included.

Acknowledgment. This work was supported by the National key Research Program of China "Scientific Big Data Management System" (No. 2016YFB1000605) and was supported by the National Natural Science Foundation of China (NSFC) "Research on the Key Technologies of Cloud Federation for High Energy Physics Experiments" under Contracts No. 11875283.

References

1. BESIII Collaboration: The construction of the BESIII experiment. Nucl. Instrum. Methods Phys. Res. Sect. A: Accel. Spectrom. Detect. Assoc. Equip. **598**(1), 7–11 (2009)
2. Cao, J., Luk, K.: An overview of the daya bay reactor neutrino experiment. High Energy Physics - Experiment (hep-ex) (2016)
3. HXMT-Hard X-Ray Modulation Telescope. http://spaceflight101.com/spacecraft/hxmt/
4. Sciascio, G., et al.: The LHAASO experiment: from gamma-ray astronomy to cosmic rays. In: CRIS 2015 Conference [hep-ex] (2015)
5. Jiangmen Underground Neutrino Observatory (JUNO). http://juno.ihep.cas.cn/
6. HTCondor. https://research.cs.wisc.edu/htcondor/
7. Wang, F., Oral. S., Shipman, G., et al.: Understanding Lustre filesystem internals
8. Peters, A.J., Sindrilaru, E.A., Adde, G.: EOS as the present and future solution for data storage at CERN. J. Phys: Conf. Ser. **664**, 042042 (2015)
9. Hadoop. http://hadoop.apache.org/
10. Shvachko, K., et al.: The hadoop distributed file system. In: Proceedings of IEEE 26th Symposium on Mass Storage Systems and Technologies (MSST), pp. 1–10 (2010)
11. Dean, J., Ghemawat, S.: MapReduce: simplified data processing on large clusters. Commun. ACM **51**(1), 107–113 (2008)
12. Li, W., Shi, J., et al.: Off-line computing of high energy physics experiments. Mod. Phys. **28**(3), 38–45 (2016)
13. Barrand, G., et al.: GAUDI-A software architecture and framework for building HEP data processing applications. Comput. Phys. Commun. **140**, 45–55 (2001)
14. Zou, J.H., et al.: SNiPER: an offline software framework for non-collider physics experiments. J. Phys: Conf. Ser. **664**, 072053 (2015)
15. Brun, R., Rademakers, F.: ROOT—an object oriented data analysis framework. Nucl. Instrum. Methods Phys. Res. Sect. A: Accel. Spectrom. Detect. Assoc. Equip. **389**(1), 81–86 (1997)

Cross-Domain Data Access System
for Distributed Sites in HEP

Qi Xu[1,2(✉)], Zhenjing Cheng[1,2], Yaodong Cheng[2], and Gang Chen[2]

[1] University of Chinese Academy of Sciences, Beijing, China
xuq@ihep.ac.cn
[2] Institute of High Energy Physics, Chinese Academy of Sciences,
Beijing, China

Abstract. A large amount of data is produced by large scale scientific facilities in high energy physics (HEP) field. And distributed computing technologies has been widely used to process these data. In traditional computing model such as grid computing, computing job is usually scheduled to the sites where the input data was pre-staged in. This model will lead to some problems including low CPU utilization, inflexibility, and difficulty in highly dynamic cloud environment. The paper proposed a cross-domain data access system (CDAS), which presents one same file system view at local and the remote sites, supporting directly data access on demand. Then the computing job can run everywhere no need to know where data is located. For the moment the system has been implemented including these functionalities such as native access for remote data, quick response, data transmission and management on demand based on HTTP, data block hash and store, uniform file view and so on. The test results showed the performance was much better than traditional file system on high-latency WAN.

Keywords: Experimental data in HEP · Cross-domain access ·
Remote file system · Cache · High performance

1 Introduction

Experiments in High Energy Physics (HEP) are mainly based on large-scale scientific devices, such as the European Large Hadron Collider (LHC), Beijing Electron Positron Collider (BEPC) and Daya Bay Neutrino Experiment. Lots of data is generated by the devices, every year 25 PB data is generated in by LHC and 100 TB data is generated by BEPC [1].

Computing in HEP is data-intensive and the essence of it is to mine rare events from massive data. The event is a unit of experiment data which is independent from others. So a data file is usually composed of series of events and different data files are scheduled to several nodes without communication. High throughput and concurrency make the computing mode of HEP different from others, cluster computing and isolation storage are main features of HEP computing. Based on the characteristics, data files will be scheduled to different nodes of different sites.

© Springer Nature Switzerland AG 2019
J. Li et al. (Eds.): BigSDM 2018, LNCS 11473, pp. 154–164, 2019.
https://doi.org/10.1007/978-3-030-28061-1_17

In traditional computing model, computing jobs and data files are usually scheduled to sites by grid computing for computing and analyzing. However, low resource utilization, inflexible scheduling and complicated maintenance restrict its development, and it cannot manage data files like distribute file systems which are not fit for applying in WAN [2]. While distributed computing is the common way in HEP. For data sharing between small sites with scare resources, a kind of efficient and flexible data management system is significant.

Focus on these, a new kind of cross-domain data access system (CDAS) for distribute sites in HEP is designed. With the technology of streaming and cache, it can access data of remote site as needed by an effective way. And the system provides a unified file view to manage files of remote site locally. Then it is tested and compared with other distributed file systems which are widely used in HEP to show its advantages in cross-domain data access between distrusted sites.

2   Streaming and Cache in File System

2.1   Traditional Distributed File System

In HEP, Lustre GlusterFS EOS are usually used as distributed file system to manage data files. Lustre is kind of parallel distributed file system and it is used in large computing cluster and supercomputer, it provides good underlying I/O performance but security and scalability are not available of it [3]. GlusterFS is mainly used in clusters and it is suitable for offline applications, it has good security and scalability but there is not metadata server of it which makes clients to be busy and perform badly for lookup [4]. EOS is an EB-level distributed file system developed by CERN. Comprehensive data management including master-slave, dynamic data migration, file replicas and so on make it suitable for HEP [5]. These distributed file systems are usually used to manage files in LAN and the comparison of their characteristics are shown in Table 1.

Table 1. Comparison of features between different distributed file systems.

Features	Lustre	GlusterFS	EOS
Metadata server	Double metadata servers	No metadata server Elastic hash	Double metadata servers
Data reliability	RAID	Data mirroring	Replica, stripe, etc.
System scalability	Storage scalable	Storage scalable	Storage scalable
Applying scene	Super computing	Multimedia application	HEP computing

Among these distributed file systems, only EOS takes cross-domain data access into consideration. But it just cluster IP addresses to access data from the closest file server. When clients access the data, communication layer of EOS identifies the IP address and redirect to the closest file server. If the file is not on the server, the server will broadcast to other file servers and the target file server will send the copy as response. Clients always communicate with the closest file server to access data file. During the process, data access is still based on files which is the same as grid computing and it cannot perform well.

Features of Lustre, GlusterFS and EOS are shown in Table 1, they are suitable for varies of application scenes but they all cannot work in WAN. Latency in WAN has a great impact on the performance of traditional distributed file systems. File system POSIX API cannot work in WAN with high latency. In 10000 Mbps bandwidth, the throughput of EOS with different latency is shown in Fig. 1. Histograms indicate read throughput with different latency and the curve shows the change of read throughput.

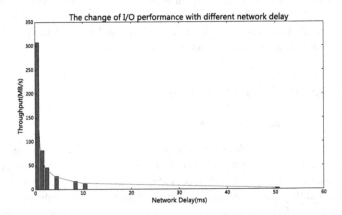

Fig. 1. The change of I/O performance with different network delay.

The interaction latency is usually about 1 ms of LAN, tens of milliseconds of WAN in China and hundreds of milliseconds of WAN between countries such as the latency is about 200 ms between China and Europe. As it is shown in Fig. 1, the performance of EOS with high latency decreases exponentially and the file system is unable to work with tens of milliseconds latency.

2.2 Streaming and Cache

Streaming transmission is mainly applied in streaming media (audio and video), it includes two types: sequential streaming and real-time streaming. For sequential streaming, media files are transmitted and downloaded in order, files cannot be read randomly. Real-time streaming makes the media bandwidth to match the WAN bandwidth, so the media can be watched in real time. But it needs particular streaming media server and protocol, it cannot ensure the quality of service faced with network congestion [6]. Steaming transmission is usually used with cache as system parts to provide an efficient and strong real-time way for data access.

Streaming and cache are used to solve the network latency and congestion. The latency and congestion destroy the real-time data access what makes users cannot enjoy the media data frequently. In HEP, cross-domain data access for distributed sites is an important feature. While there are great limitations of cross-domain data access between distributed sites with traditional computing mode such as grid computing, as it is shown below. Especially in small sites, short of kinds of computing resources makes it hard to develop grid computing.

1. Pre-allocation of storage and computing, low resource utilization
2. Transfer whole file, waste of resources and inflexible scheduling
3. No complete data management system
4. Based on grid computing, hard to develop and maintain

In experiments of HEP, a particle collision or interaction generates experimental data which constitutes an event and some events make up a physical data file – DST file [7]. Scientists are interested in some events of DST files, some data blocks of the file. So transferring whole DST files will waste the limited resources without any significance.

As is talked above, streaming transmission is suitable for cross-domain data access in HEP. It will transfer data blocks on demand to improve resource utilization and latency has little impact on it. In addition, streaming is based on HTTP protocol which is robust and portable and it is easy to develop and maintain without any operations of ports and firewall. Combined with cache, it will increase the throughput and speed up the response.

The cross-domain data access system is designed with streaming and cache. It is focus on data access between distrusted sites what is different from traditional distributed file systems. It has good performance of cross-domain data access, high resources utilization and network latency has little impact on its PSOXI API. In addition, it has complete data management functions.

In the system, events (data blocks) are transferred by streaming between distributed sites. Cache works as the core module to communicate with both the remote station and clients, all the data is handled by it. Requests from clients are sent to the cache and the cache will communicate with the remote station to pull the data by a long connection, as is shown in Fig. 2.

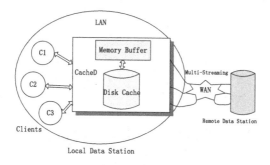

Fig. 2. The data flow of the cross-domain data access system.

3 System Design and Implementation

The system is consist of three loose coupling Unix Services: CacheD, TransferD and File Plugin and there are three parts in CacheD: MetaD, DataD and Daemon Process, the whole system structure is shown in Fig. 3.

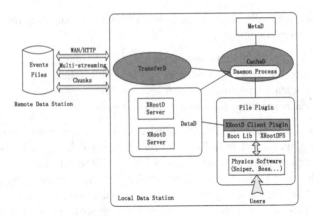

Fig. 3. The structure of the cross-domain data access system.

Building on loose coupling services, the system has good scalability and security. Loose coupling makes errors isolated and not spread. Unix Services run in the background and they do not communicate with the terminal to make systems security.

3.1 CacheD

CacheD is the core module (control module) of the system, data-flow and control-flow of the system are through it. Firstly, requests from clients are sent to Daemon Process then it establishes a long connection with remote site to pull data blocks. Data blocks will be cached and returned to clients. During the process clients are separated from data source site what ensures the security and increases the throughput of the system. CacheD makes cross-domain data access performs as well as accessing local data. CacheD is composed of three parts: Daemon Process, MetaD and DataD, as is shown in Fig. 4.

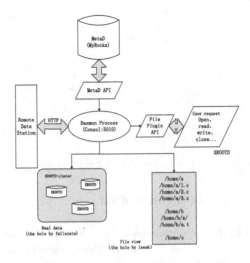

Fig. 4. The structure of the CacheD.

Each module of CacheD is loose coupling and the metadata is separated from data blocks. The metadata and data blocks are managed and stored independently, then their relationship is also stored in MetaD. In CacheD, hash map is token to keep load balance of storage nodes and enhance data security. Modules are described in detail as follows.

1. Daemon Process

   The message queue is consist of static thread pool and blocking queue, it is used for fast response to concurrent requests. Static thread pool avoids the cost of thread construction and destruction to improve the response speed. And it is designed as a service (Daemon) which runs in the background to listen the target port effectively. The daemon adopts stand-alone, signal-control mode and it does not depend on the super demon.

2. MetaD

   The MetaD stores the metadata of cache files and the information of data blocks based on RocksDB. RocksDB is a new key-value database based on memory. It can support large-scale storage and access effectively. It has better compressibility and low resource consumption. The data size in RocksDB is about 70% of InnoDB with compression, with the same size of memory the performance of it is about 500% of InnoDB. RocksDB is based on LSM-Tree (Log Structured Merge-Tree) which has better memory utilization and no write exploding. At last its raw data is always merged into log files, so it can provide good data security and quick disaster recovery.

3. DataD

   DataD provides data blocks redirection and load balancing based on HASH. Data blocks are hashed to different storage nodes and named by UUID (Universally Unique Identifier). The relationship between data blocks and target files is stored in MetaD by key-value format. By this way, the metadata of target files is separated from data blocks in order to simplify data migration and ensure data security. There are two kinds of file views reflected to target directories of remote sites. The first one is logical file view which is the structured output of metadata from MetaD and it do not need any storage resources. The other one is physical file view which is generated by function-lseek. The function builds the mirror of target directories with empty files, the empty files are as the same size as target files but they just take up a few space. The physical file view is optional.

When the first data block of a target file is cached, the system pre-allocates the space for the target file by function-fallocate. Then the subsequent data blocks of the target file are written into the space based on offset and size, by the way it accelerates the caching of data blocks.

Storage with Hash. The system converts absolute paths of target files to strings with fixed length by hash and the string is corresponding to an exclusive cache directory. By this way, the data set is divided into multiple cache directories for load balancing. The file in cache directories is named by a unique UUID, but the data is still accessed by the absolute path in remote sites, as is shown in Fig. 5.

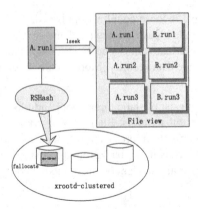

Fig. 5. The hash of data files.

RSHash is taken in the system, RSHash, KDRHash, APHash and DJBHash are commonly used for string hash. RSHash is good at the combination of letters and numbers and it is suitable for directories hash. Both the time and space complexity of RSHash is lower than others with lest computation cost [8].

Cache files are named by UUID (UNIX API – uuid_generate), the function is based on high quality random number generator (/dev/urandom). Then the random number will be converted to a string (36 Bytes) by function - uuid_uparse. High quality random numbers are unique, it avoids the collision between different cache files.

The absolute path of the target file in remote sites and the cache file is a pair of key-values which is stored by MetaD. So cache files can be accessed by absolute path as a result of MetaD which reflects the absolute path to cache files. The reflection makes cache files access implicit, so the data is more reliable scalable and flexible.

Cache Replacement. CacheD is the data warehouse for clients, all the data interaction is through it. In order to provide effective cache and improve cache hit ratios, CFLRU (Cache File Least Recently Used) based on LRU is designed. In the policy the reference count of cache files will increase when there is any data block of it being accessed, the least accessed file will be replaced recently according to the reference count.

In HEP, some events of a data file always have week correlation and physicists are commonly interested in several discrete events of a data file. So CFLRU is suitable for HEP computing, in which the data file is composed of weakly related data blocks.

3.2 TransferD

Tornado is adopted as the transfer service TransferD. Tornado is a Python web framework and asynchronous networking library, originally developed at FriendFeed. By using non-blocking network I/O, Tornado can scale to tens of thousands of open connections, making it ideal for long polling, WebSockets, and other applications that require a long-lived connection to each user [9].

Tornado is developed by Python. In order to communicate with CacheD efficiently, the C-API of Tornado clients is implemented by CPython and core interfaces are

Meta_rsync Data_get and Data_put. They support metadata synchronization, data blocks downloading and data blocks uploading.

Furthermore, multi-streaming, segmentation and retransmission are added into the Tornado server to provide an efficient and stable transmission service in WAN.

3.3   File Plugin

The File Plugin provides similar POSIX File System API (the Standard Interface of Portable UNIX File System), as is shown in Table 2. It supports UNIX local file systems such as FUSE, EXT4 and XFS and distributed file systems such as GlusterFS and EOS. All the file systems are mounted in CacheD and communicate with Daemon Process by the File Plugin API, the operation of the CacheD is transparent and changeless for clients.

For general applicability in HEP, XrootD framework is taken as the default file management system for clients. XrootD is a kind of high performance, scalable and fault tolerant data access framework. It is often applied in HEP computing and it is composed of an extensibility framework, a communication protocol and a set of plugin tools. High degree of freedom makes it suitable for almost data access and it supports ROOT framework which is applied in most HEP experiments [10].

When jobs are processed by physical analysis frameworks such as Sniper, BOSS and Gaudi, the data blocks can be accessed by XrootD from CacheD directly. Clients do not need copy the whole data, it makes data access effective and simple.

Table 2. File Plugin (XRootD) API.

API	Description
int xrd_open()	Open file (OW/OR)
int xrd_close()	Close file (Auto)
int xrd_getattr()	Get metadata from remote site
int xrd_read()	Read file (Transfer data block, if not cached)
int xrd_access()	Whether file is accessible
int xrd_opendir()	Open directory, get DIR_ID
int xrd_readdir()	Read directory (Metadata of files in it)
int xrd_rfsync()	Sync files to remote site
int xrd_refresh()	Update cache files
int xrd_unlink()	Delete cache files

4   Benchmarks

In order to ensure the security of HEP experiment data, the cross-domain data access system (CDAS) only supports read-only and write-only mode. Clients can read the data files or create private files but they do not have the permission to modify data files. In the section, comprehensive tests about read-only performance and stability of the system are carried out. For write-only mode, the new file from clients are cached by CacheD and it will synchronize to remote site regularly, it is a time-delay system.

In the test, two servers belong to different network segments work as the remote site and the local site. TC (Traffic Control) tool is used to simulate the transmission latency in WAN. The tool iperf is used to test the bandwidth. Iperf is a kind of network performance test tool, it can get the maximum TCP/UDP performance and it has multiple parameters and UDP characteristics [11]. By the test, the bandwidth is about 937 Mbit/sec.

4.1   Test Environment

1. Hardware
   CPU: Intel(R) Xeon(R) E5-2630 v3 @2.40 GHz * 32
   Memory: 8 GB DDR3 1600 MHz * 8
   Disk: SATA 7200 rpm 4.0 TB (Cache 128 MB)
2. Software
   OS: CentOS 6.0 x64 Linux
   Kernel: Linux version 2.6.32-696.1.1.el6.x86_64
   Test Tool: IOzone-3.424

IOzone is a kind of file system benchmark tool, it can test different R/W modes of the file system in different operating systems [12].

4.2   Test for Read Performance

In HEP computing, the size of most data files is about 600 MB. Therefore, the test file is selected with the size of 700 MB (Average Upper Limit of Data Files) and the performance of the system is compared with traditional distributed file systems Lustre and EOS, the result is shown is Table 3.

Table 3. Read performance (Throughput) of different file system (MB/s).

Network delay (ms)	Lustre	EOS	CDAS (file not cached)
0	79.85	30.12	77.36
1	45.93	7.26	72.71
10	13.87	1.03	64.02
100	1.49	0.12	51.38

Taking the throughput as the measure of the system I/O performance, it is shown that the cross-domain data access system has perfect I/O performance and stability. With the latency increasing, the I/O performance of Lustre and EOS decrease greatly as a result of that the file system semantics are broken by the latency and the file system cannot get the response in time. According to asynchronous and multi-streaming in HTTP protocol, the performance of the cross-domain data access system just decrease a little. In addition the test file is not cached, if it is cached the performance of the system is as the same as the local file system. Table 3 is shown that when the latency is about tens of milliseconds, traditional distributed file systems are almost crashing. While the I/O performance of the cross-domain data access system is about 51.38 MB/s with 100 ms network delays which shows that the system meet the requirements of cross-domain data access for distributed sites in HEP.

The speedup of CDAS with Lustre and EOS is shown in Fig. 6, it is obviously shown that the advantage of CDAS is more and more significant with the increasing of network delays, it is greatly suitable for cross-domain data access in WAN.

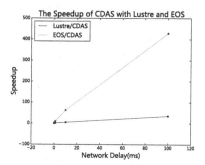

Fig. 6. The speedup of processing data between CDAS and Lustre, EOS.

4.3   Test for Stability with Network Delay Jitter

The I/O performance of the system with different network delay is tested and the result is shown in Fig. 7. The result shows that the system has perfect performance (50 MB/s–80 MB/s) in cross-domain data access and it is more stable. High network delay in WAN does not have a great impact on the system, especially compared with other distributed file systems. So the system meets the requirement of stability in cross-domain data access in WAN.

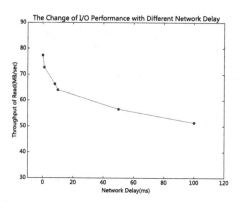

Fig. 7. The change of I/O performance with different network delay.

5 Conclusion

The cross-domain data access system (CDAS) for distributed sites in HEP is designed and implemented. It adopts streaming and cache to the system by which the system accesses data on demon and effective. It accesses cross-domain data by TransferD based on HTTP protocol, cache data blocks in CacheD and communicate with clients by File Plugin (XrootD). Cross-domain data access is localized and transparent for users in the system. The system is suitable for data access between distributed sites in HEP, it can make the most of resources in small sites to accelerate physical computing and it has perfect I/O performance and stability in WAN.

Acknowledgments. This work was supported by the National key Research Program of China "Scientific Big Data Management System" (No. 2016YFB1000604) and National Natural Science Foundation of China (No. 11675201 and 11575223).

References

1. WLCG - Worldwide LHC Computing Grid. http://lcg.web.cern.ch/LCG,2013.7
2. Berman, F., Fox, G., Hey, A.J.G.T.: Grid Computing: Making the Global Infrastructure a Reality, vol. 2, pp. 945–962. Wiley, Hoboken (2003)
3. Donovan, S., Symposium, L., Kleen, A., et al.: Lustre: building a file system for 1,000-node Clusters. In: Proceedings of the Linux Symposium, p. 9 (2003)
4. Davies, A., Orsaria, A.: Scale out with GlusterFS. Belltown Media (2013)
5. Peters, A.J., Janyst, L.: Exabyte scale storage at CERN. J. Phys. Conf. Ser. 241–244 (2011)
6. Zhang, L.: Technology of Streaming Media. China Youth, Beijing (2001)
7. Cheng, Y., Shi, J., Chen, G.: A survey of high energy physics computing system. e-Science Technol. Appl. **31**, 1189–1194 (2014)
8. Chen, C.: The principle of Hash algorithm and the application in rapid retrieval. Comput. Fujian (2009)
9. Byers, J.W., Luby, M., Mitzenmacher, M.: Accessing multiple mirror sites in parallel: using Tornado codes to speed up downloads. In: Proceedings of Eighteenth Joint Conference of the IEEE Computer and Communications Societies, INFOCOM 1999, vol. 1, pp. 275–283. IEEE (1999)
10. Dorigo, A., Elmer, P., Furano, F., et al.: XROOTD - a highly scalable architecture for data access. Wseas Trans. Comput. **4**(4), 348–353 (2005)
11. Tirumala, A., Cottrell, L., Dunigan, T.: Measuring end-to-end bandwidth with Iperf using Web100. In: Proceedings of Passive and Active Measurement Workshop (2003)
12. Norcutt, W.: The iozone filesystem benchmark. Software User Manual & Documentation (2003)

Multi-dimensional Index over a Key-Value Store for Semi-structured Data

Xin Gao$^{(\boxtimes)}$, Yong Qi, and Di Hou

The School of Electronic and Information Engineering,
Xi'an Jiaotong University, Xi'an, China
helloxiyue@foxmail.com, {qiy,houdi}@xjtu.edu.com

Abstract. The informal data structures and trillions of data volume are the challenges for databases to store and retrieve semi-structured data. Most researchers deal with the issues through R-Tree, KD-tree and space curves, but these structures are not suitable for default and discrete values of semi-structured data, and even require sampling before storage. We present MD-Index, a scalable multi-dimensional indexing system that supports high-throughput and real-time range queries. MD-Index builds bitmap index of sliced data over a range partitioned Key-value store. The underlying Key-value store guarantees high throughput, large data storage, high availability and fault tolerance of the system, and bitmap provides multi-dimensional index of data. Meanwhile, MD-Index encodes the discrete values as the hash code of a slice, and stores the data and the bitmap of a slice in the same region (a storage unit of the range partitioned Key-value store) to utilize distributed computing and data locality. Our prototype of MD-Index is built on HBase, the standard Key-value database. Experimental results reveal that MD-Index is capable of storing and retrieving trillions of semi-structured data and achieving a throughput of two million records per second.

Keywords: Multi-dimensional index · Key-value store ·
Bitmap index · Semi-structured data

1 Introduction

Server logs, user preferences collected by terminals, geographic location information and scientific experimental data are common semi-structured data, and these kinds of data do not conform to fixed models that can be illustrated as schemas in relation databases, which means that the relationships between data entities are difficult to described as ER [10]. The retrieved dimensions are usually more than ten or more. In addition, it is common that millions of semi-structured data records are generated per second by devices and terminals,

Supported by 2016YFB1000604.

J. Li et al. (Eds.): BigSDM 2018, LNCS 11473, pp. 165–175, 2019.
https://doi.org/10.1007/978-3-030-28061-1_18

and queries within several seconds are demanded, so semi-structured databases require to meet high throughput and real-time query. Relational databases such as MySQL support multi-dimensional queries, but trillions of data volumes and unfixed data schemas become performance and functional bottlenecks for relational databases; Key-value databases such as HBase [8] provide high-throughput and high-capacity storage, but a single primary key index makes it hard to support efficient multi-dimensional retrieval. At present, most of the research work focuses on R-Tree [1,6], KD-Tree [1,2], and space curves [3] to implement multidimensional index. However, these structures are not suitable for default and discrete values of semi-structured data, and even require sampling before storage [1]. MD-HBase leverages a multidimensional index structure layered key-value store. MD-HBase partitions entire data space into different subspaces by Z-order curve, and uses K-d tree or Quad-Tree to group these subspaces. However, Z-order curve is based on continuous values and K-d tree is time-consuming when split, so MD-HBase is only suitable for low dimensional data such as geographic location information. J K. et al. use space-filling curves for multi-dimensional indexing. Their paper proposes efficient algorithms to construct spatial curves, but isometric and segmenting dimensions are required. EDMI leverages Z-Order prefix R-Tree as the index layer. KD-Tree is used to partition data space into subspaces. In order to accelerate the construction of KD-Tree, EDMI will sample the data to be inserted. In addition, Z-Order prefix R-Tree is built through MapReduce which is a batch process. Therefore, EDMI cannot satisfy real-time insertion.

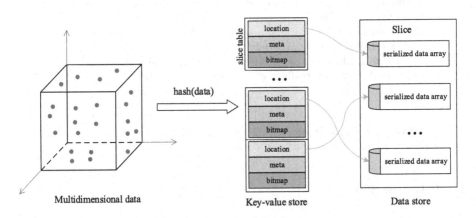

Fig. 1. Architecture of MD-Index.

In order to achieve multi-dimensional index of trillions of semi-structured data, we propose MD-Index. MD-Index takes each record of data as input to the system, which makes MD-Index support batch import as well as real-time insertion of streaming data. MD-Index encodes the discrete values as the hash code of a slice within which the bitmap index is built to realize multi-dimensional retrieve. Therefore, the global multi-dimensional query is the aggregation of results of each slice. Figure 1 illustrates MD-Index's top-level architecture showing that a slice stored in Key-value database consists of three parts: storage location pointing at the serialized semi-structured data arrays, meta information describing the states of the slice, and bitmap index. To ensure these three parts of a slice locate in the same region where network transmission does not exist when computing, the hash code is taken as the prefix of their key. Furthermore, the storage medium of semi-structured data can be a local file system as well as HDFS [9] whose storage strategies assure a copy located at the client host, or even directly stored in the key-value database if the slice if relatively small, either way the data locality is fulfilled when extracting data.

Contributions

- We propose the design of MD-Index that uses bitmap index to implement multi-dimensional index over a Key-value store for semi-structured data.
- MD-Index supports real-time insertion and retrieval, and meet the real-world demands of high-throughput, large-capacity storage, and real-time response.
- We implement a prototype of MD-Index based on HBase, and the experimental test is carried out on the trillion-level semi-structured data set, demonstrating the scalability and efficiency of MD-Index.

Organization. Section 2 provides related work on data slicing, bitmap index and the storage of serialized data. Section 3 describes the design and implementation of MD-Index. Section 4 presents a detailed evaluation of MD-Index on server and client side respectively. Section 5 concludes the paper.

2 Background

A. Data Slicing of Big Data. Data slicing is to take the fragmented data as a unit of storage and calculation, which can exert the abilities of parallel computing and scalability of distributed system. An appropriate slicing strategy is supposed to increase the parallelism of computing and reduce the amount of network data transmission required. There are two categories of conditions of the multi-dimensional query: equivalent query (EQ) and range query (RQ). A condition of the equivalent query is $p = v$ while that of the range query $p \in [v_1, v_2)$. Key-value store uses B+Tree as the index layer that has the advantage of range query, so it is better to carry out range query within a slice. Semi-structured data generally has discrete attributes such as machine number and event type etc. If there are no discrete attributes, a slice attribute is required to fragment data into buckets, usually the self-incrementing number or timestamp of the data. A record of data can be denoted by $r = (\tau, d, c)$, where τ is the slice attribute and can be default, $d = (d_1, d_2, \ldots, d_m)$ the set of discrete attributes, $c = (c_1, c_2, \ldots, c_n)$ the set of continuous attributes. Assuming tz is the bucket size of the slice attribute, the hash code of a record of data is as below.

$$hashCode(\boldsymbol{r}) = hash(\lceil \tau/tz \rceil, \boldsymbol{d}) \tag{1}$$

The hashCode is taken as the prefix of the key of a row in Key-value store, then data with identical prefix will be stored in the same region. Hence, hook functions added to the region can manipulate the data within a slice and create bitmap index dynamically.

B. Bitmap Index. The bitmap index [4] identifies whether the attribute of a record of data is equal to a certain value by a boolean known as bit. Because the discrete attributes are used to generate hash code, MD-Index's bitmap is established on the continuous attributes set $\boldsymbol{c}$. Suppose the data within a slice is $\boldsymbol{R} = (r_1, r_2, \ldots, r_k)$ and an attribute $c \in \boldsymbol{c}$ has l values denoted as $\boldsymbol{v} = (v_1, v_2, \ldots, v_l)$, the bitmap index for the attribute c is a $k \times l$ matrix $\boldsymbol{M}_c$ of booleans.

$$\boldsymbol{M}_c[x][y] = \begin{cases} 1, & \text{if } \boldsymbol{r}_x.c = v_y \\ 0, & \text{else} \end{cases} \tag{2}$$

Using $ebmi_c(y)$ to represent a column of matrix $\boldsymbol{M}_c$, it demonstrates the data whose value of attribute c equals to y, which is similar to the idea of inverted index. In actual use, in addition to the equivalent bitmap index (EBMI) above, the range bitmap index (RBMI) as well as the interval bitmap index (IBMI) can be established. RBMI will be more efficient and space-efficient to demonstrate the data whose value of an attribute is less than a specific value, and so is IBMI to show the data whose value of an attribute is between an interval. $rbmi_c(y)$ denotes the data $\boldsymbol{r}$ that $\boldsymbol{r}.c < y$, and $ibmi_c(z, y)$ denotes the data $\boldsymbol{r}$ that $z \leq \boldsymbol{r}.c < y$. Both $rbmi_c(y)$ and $ibmi_c(z, y)$ can be obtained by the or-operation of $ebmi_c(y)$ as below.

$$\begin{cases} rbmi_c(y) = \bigcup_{t<y} ebmi_c(t) \\ ibmi_c(z, y) = \bigcup_{t=z}^{t<y} ebmi_c(t) \end{cases} \tag{3}$$

C. The Storage of Serialized Data. Serialization is the process of transferring data into a format that can be stored persistently. JSON and Protobuf are the most common methods to serialize data structures, or you can even leverage the writable interface of java to implement the work of serialization. Whatever method you take, you will get a byte array $s = serialize(data)$, whose length is $l = length(s)$. Assuming the data stored within a slice is $\boldsymbol{s} = (s_1, s_2, \ldots, s_n)$, MD-Index concatenates $\boldsymbol{s}$ to realize the process of dynamical insert, so the meta information $\boldsymbol{l} = (l_1, l_2, \ldots, l_n)$ needs to be updated after s inserted in order to split the concatenated serialized data. Figure 2 illustrates the process of inserting data and updating meta information.

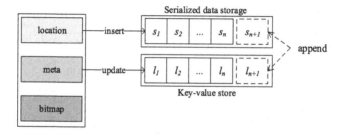

Fig. 2. The process of inserting data and updating meta information.

3 Multi-dimensional Index Design

Key-value store requires a well-designed schema to implement bitmap storage and multi-dimensional index of MD-Index. The combination of and-or operations of bitmap calculates the result of multi-dimensional index of a slice by hook functions attached to the region, aggregations of which are the final result of MD-Index. B+Tree [7] has efficient performance of range query, so schema supporting range query will increase performance of bitmap index. In addition, EBMI is applicable to the case where the number of distinct values of each attribute is modest, for EBMI is to increase in proportion to the number of values. When the number of distinct values is remarkably large, merging EBMI into RBMI or IBMI is leveraged by MD-Index. Specially, data skew is unavoidable and a slice needs to be split when over the threshold. Thus, the merging of bitmap and the splitting of slice are the keys of MD-Index's scalability and efficiency.

A. Schema Design of Key-Value Store. Table 1 demonstrates the schema design of Key-value store. MD-Index is made up of three parts: storage location pointing at the serialized semi-structured data arrays, meta information describing the states of the slice, and bitmap index. hashCode#splitCount is a self-incremental number, showing the order of current split of a slice and incrementing 1 when a slice split. MD-Index has three categories of bitmap index. In initial state, only EBMI is built since few values are inserted into MD-Index. With the expansion of distinct values, EBMI needs to be merged into RBMI or IBMI to save storage space and speed up retrieval. In the schema of key, a special function named *pad* is leveraged to encode value to make sure that the lexicographic order is consistent to the numerical order, which can take advantages of B+Tree's efficiency of range query. For example, 8 is smaller than 13 while "8" is greater than "13" in the format of string. The easiest method to encode a number is to pad it with zero. Such as "008" is smaller than "013" which is consistent to the numerical order. If we update bitmap index every time a record of data inserted, a lot of rows need to be appended by a single bit, which is unefficient of any Key-value store system. Therefore, MD-Index launches daemon threads to update bitmap index in batch mode when the data needs to be retrieved is accumulated to a specific amount. This is actually a kind of cache or buffer operation.

Table 1. Schema design of Key-value store

	Key	Value
Location	hashCode#splitId#data	Serialized data array
Bitmap	hashCode#splitId#ebmi#attribute#pad(value)	$ebmi_{attribute}(value)$
	hashCode#splitId#rbmi#attribute#pad(value)	$rbmi_{attribute}(value)$
	hashCode#splitId#ibmi#attribute#pad(v_1)#pad(v_2)	$ibmi_{attribute}(v_1, v_2)$
Meta	hashCode#splitId#attribute#count	The number of distinct values of this attribute
	hashCode#splitId#serialized#format	Format information of serialized data array
	hashCode#splitId#size	The size of bitmap index
	hashCode#splitId#count	The number of data stored in this slice
	hashCode#splitCount	Self-incremental order

Algorithm 1. Insert a record of data

Input: r be the data to be inserted
Output: how to insert into slice and update bitmap
$hashCode = hash(r)$;
$region = getRegionByPrefix(hashCode)$;
// Invoke insert function of the region by rpc
if *region.rpc.invoke("insert", r)* **then**
    $split = regoin.getCurrentSplit(hashCode)$;
    if *split.isFull()* **then**
        $split = region.createNewSplit(hashCode)$;
    end
    $s = serialize(r); l = length(s)$;
    $split.concat(s)$;
    $meta.concat(l)$;
    for *ebmi in slice's bitmap* **do**
        if *ebmi's (attribute, value) in r* **then**
            $ebmi_{attribute}(value)$ appends 1;
        else
            $ebmi_{attribute}(value)$ appends 0;
        end
    end
end

B. Process of Inserting Data. Determining the slice to which a record of data belongs is to calculate the hash code. Recall that $hashCode = hash(\lceil\tau/tz\rceil, d)$, so maybe hundreds of thousands of data will get the same hash code. Hook functions attached to the region receive those data with the same hash code prefix and append them in the row whose key equals $hashCode\#splitId\#data$. The initial value of $splitId$ is 0, and the suffix $data$ is just a identification whose value may the "data" itself to increase readability. The steps for inserting data are shown in Algorithm 1. Since there is a threshold to the number of data a slice can hold, the insertion algorithm check the size of current split of the slice and create a new split for the slice when the split is full. Creating a split is to increase the $splitId$ by 1 and append data to the current split.

C. Merge EBMI into RBMI. Bitmap index is considered to work well for low-cardinality columns. The exploding of distinct values will waste storage space and reduce computing speed. In the real world, the Pareto and Locality principle universally exist. Merging EBMI into RBMI can reduce the storage space and make it even more efficient when searching for ranges. Of course, missing bitmaps of certain values after merging results in positive cases with errors, which can be deleted by adding a filter function.

Algorithm 2. Merge EBMI into RBMI

Input: $ebmi_c$ be the EBMI of attribute c
Output: merge EBMI into RBMI when the number of distinct values
        over threshold

$v = ebmi_c$;
if $\|v\| > threshold$ **then**
  | `// cnt be the number of RBMI after merging`
  | `// get the most frequently visited values of top cnt`
  | $v^* = mostFrequentlyVisited(v, cnt)$;
  | **for** $v \in v^*$ **do**
    | `// get previous value of v in v*`
    | $v^- = getPreValue(v)$;
    | $rbmi_c(v) = rbmi_c(v^-) \cup (\bigcup_{t=v^-}^{t=v} ebmi_c(t))$;
    | $insertOrUpdate(rbmi_c(v))$;
  | **end**
end

Algorithm 2 shows the process of merging EBMI into RBMI. MD-Index launches daemon threads to merge RBMI when over the threshold. cnt be the number of RBMI after merging, we can remain the most frequently visited values of top cnt. Equation 3 can calculate the RBMI, but a large number of repeated computations are carried out. Figure 3 shows how to calculate $rbmi_c(v)$ by its previous RBMI.

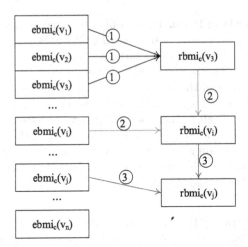

Fig. 3. The process of calculating $rbmi_c(v)$ by its previous RBMI.

Algorithm 3. Process of retrieving data

Input: $q = (\delta, \xi, \mu)$ be the conditions of a query
Output: the data satisfied the searching conditions
$result = \emptyset$;
for $\delta \in \delta$, $\xi \in \xi$ **do**
    $hashCode = hash(\delta, \xi)$;
    $region = getRegionByPrefix(hashCode)$;
    $bitmap = 11...11...1111$;
    for $\mu \in \mu$, $(attribute, value) \in \mu$ **do**
        if μ *is EBMI* **then**
            $bitmap \cap = ebmi_{attribute}(value)$;
        else if μ *is RBMI* **then**
            $bitmap \cap = rbmi_{attribute}(value)$;
        else if μ *is IBMI* **then**
            split $value$ into (v_1, v_2);
            $bitmap \cap = (rbmi_{attribute}(v_2) \oplus rbmi_{attribute}(v_1))$
    end
    $result \cup = region.getDataByBitmap(bitmap)$;
end
$result = filter(result)$;

D. Process of Retrieving Data. Before retrieving data in MD-Index, we should abstract the query conditions. Assume that query conditions are denoted by $q = (\delta, \xi, \mu)$, where $\delta = (\lceil \tau_1/tz \rceil, \lceil \tau_1/tz \rceil + 1, \ldots, \lceil \tau_2/tz \rceil)$ is a list of buckets describing the search scope of the slice attribute $\tau \in [\tau_1, \tau_2)$, and $\xi = (d_x = v_x, d_y = v_y, \ldots, d_z = v_z)$ is a set of key-value pairs of discrete attributes describing equivalent queries, and $\mu = (c_x \in [vx_1, vx_2), c_y \in [vy_1, vy_2), \ldots, c_z \in [vz_1, vz_2))$ is the set of range queries for continuous attributes. Algorithm 3 provides the pseudo code for the process of retrieving data. It seems to be a serial process architecturally, but in fact, the regions are distributed across the hosts, and the client sends asynchronous requests to regions through rpc and handles response by the callback functions, so this is a distributed parallel process, similar to a fork/join framework. The region a bitmap index belongs to is found by the hash code, and hash codes calculated from all permutations of $\delta \in \boldsymbol{\delta}$ and $\xi \in \boldsymbol{\xi}$ are actually the indexes of the first two query conditions of q. Therefore, the combination of hash code and bitmap index realizes the multi-dimensional index of semi-structured data.

4   Experimental Evaluation

We implemented our prototype using HBase [8] 1.2.0 and Hadoop [9] 2.6.0 as the underlying system. Our experiments were performed on astronomical satellite data which amounts to trillion with 7 dimensions. The data source can be accessed and downloaded at http://hsuc.ihep.ac.cn/web/hxmtdata/, and its formats are described on https://fits.gsfc.nasa.gov/ in detail. The size of our cluster was varied from 4 to 16 nodes. Each nodes consists of 24 cores and 48 threads, 64 GB memory, 256 GB SSD, 2TB HDD, and 64 bit Linux(v2.6.32). We used throughput per second to evaluate the performance of insertion speed, equivalent query speed, and range query speed. In order to test the concurrency performance, the number of clients increased from 1 to 50. Each client synchronously and continuously sent requests to MD-Index.

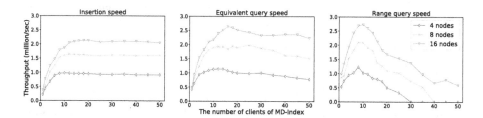

Fig. 4. Throughput of insertion, equivalent equery and range query change with the number of clients. Each client synchronously and continuously send requests to MD-Index.

Figure 4 illustrates the performance of MD-Index. It should be noted that the vertical coordinates are the throughput of the MD-Index as a whole, rather than the throughput of a single client. The performance of range query drops sharply when the number of clients increases to a certain number, even causing MD-Index to crash, because the result set of range query is too huge. MD-Index does not paginate, and the result set accumulated by the client explodes the physical memory.

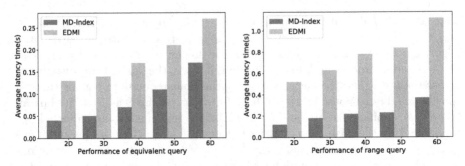

Fig. 5. Experimental results on MD-Index and EDMI.

We completed a series of experiments to compare the performance of MD-Index with EDMI [1], another multi-dimensional index system over Key-value store with lowest latency time as far as we know. We used the average latency time of requests to evaluate the query performances. Figure 5 shows that MD-Index performs better than EDMI on equivalent query and range query.

5   Conclusion

In this paper, we propose an efficient multi-dimensional index framework—MD-Index. MD-Index encodes the discrete values as the hash code of a slice, and builds bitmap index on the continuous values. Moreover, MD-Index takes each record of data as input to the system, which makes MD-Index support batch import as well as streaming data of real-time insertion. MD-Index is capable of storing and retrieving trillions of semi-structured data and achieving a throughput of two million records per second.

References

1. Zhou, X., Zhang, X., Wang, Y., Li, R., Wang, S.: Efficient distributed multidimensional index for big data management. In: Wang, J., Xiong, H., Ishikawa, Y., Xu, J., Zhou, J. (eds.) WAIM 2013. LNCS, vol. 7923, pp. 130–141. Springer, Heidelberg (2013). https://doi.org/10.1007/978-3-642-38562-9_14
2. Nishimura, S., et al.: MD-HBase: a scalable multi-dimensional data infrastructure for location aware services. In: 2011 12th IEEE International Conference on Mobile Data Management (MDM), vol. 1. IEEE (2011)
3. Lawder, J.K., King, P.J.H.: Querying multi-dimensional data indexed using the Hilbert space-filling curve. ACM Sigmod Rec. **30**(1), 19–24 (2001)
4. Chan, C.-Y., Ioannidis, Y.E.: Bitmap index design and evaluation. ACM SIGMOD Rec. **27**(2), 355–366 (1998)
5. Bentley, J.L.: Multidimensional binary search trees used for associative searching. Commun. ACM **18**(9), 509–517 (1975)
6. Guttman, A.: R-trees: a dynamic index structure for spatial searching, vol. 14, no. 2. ACM (1984)

7. Jensen, C.S., Lin, D., Ooi, B.C.: Query and update efficient B+-tree based indexing of moving objects. In: Proceedings of the Thirtieth International Conference on Very Large Data Bases, vol. 30. VLDB Endowment (2004)
8. Apache HBase - Apache HBase^{TM} Home. base.apache.org/
9. Apache Hadoop. hadoop.apache.org/
10. Entity-Relationship Model Wikipedia, Wikimedia Foundation, 4 October 2018. en.wikipedia.org/wiki/Entity

Technology Dependency Graph (TDG): A Scientific Literature Mining Based Method for Technology Insight

Hui Gao[1,2(✉)], Wei Luo[2], Lin Gui[1], and Ting Wang[1]

[1] School of Computer, National University of Defense Technology,
Changsha, China
gaohui_baixiang@163.com
[2] Information Research Center of Military Science, Beijing, China

Abstract. The increasingly massive amount and open access of literature provide a data foundation for technology insight based on big data analysis. This paper proposes a new technology insight framework based on the text mining-Technology Dependency Graph (TDG). Firstly, an adversarial multitask learning model and distantly-supervised learning model are applied to extract the technology entities and dependency relations with a little labeled sample. Then, a weighted directed graph, i.e., a TDG, is constructed with the technology entities as vertices and the dependency relations as edges. A TDG contains rich and valuable semantic information which represents the support, contribution or relying on relationship between technologies. At the same time, the social network properties of TDG allow researchers to analyze and mine hot topics, key technologies, and technology architecture by using network theories, methods and tools. In the case study, the TDG of DSSC (dye-sensitized solar cell) is constructed. Furthermore, the technology dependency architecture for DSSC is constructed according to a spanning tree out of the TDG, which provides a global perspective for the research of DSSC.

Keywords: Information extraction · Text mining · Scientific literature · Technology insight · Dependency graph · DSSC

1 Introduction

Currently, researchers have access to a huge and rapidly growing amount of scientific literature available on-line [1]. Recent estimates reported that a new paper is published every 20 s [2]. At the same time, during the last few years the number of scientific papers that are freely accessible on-line considerably grew. Sometimes between 2017 and 2021, more than half of the global papers are expected to be published as Open Access articles [3]. At the same time, the scientific literature not only contains the authors' original research, but also a large number of reviews and summaries of previous research works. From the beginning of the 20th century, researchers began to use the bibliometrics method for technology insight. However, bibliometrics generally focuses on the metadata of literature to find co-citation and co-occurrence information, lacking deep mining to the content.

© Springer Nature Switzerland AG 2019
J. Li et al. (Eds.): BigSDM 2018, LNCS 11473, pp. 176–185, 2019.
https://doi.org/10.1007/978-3-030-28061-1_19

The achievement of big data analysis and artificial intelligence technologies has provided new routes for literature-based technical insight, which lie in two aspects: the data focused on is not only the metadata, but also the unstructured full text of the literature; the representation of knowledge is no longer co-citation or co-occurrence information, but the knowledge graph containing various technology entities and semantic relationships between technology entities. The framework in the related researches generally consist of two tasks: extracting structural information from the literature content and analysis for technology insight based on the extracted information. However, there was rarely a research in which both of the two tasks could be accomplished sufficiently.

In this paper, we applied deep learning approaches to solve the problem of text mining with only a small number of labeled samples, and implemented the extraction for technology entity and dependency relation between entities. A weighted directed graph was constructed with the technology entities as vertices and the dependency relations as edges, which is called as Technology Dependency Graph (TDG). A TDG contains rich and valuable semantic information which represents the support, contribution or relying on relationships between technologies. At the same time, the social network properties of TDG allow researchers to analyze technologies by using network theories, methods and tools. The main contributions of this paper are: (1) we constructed a comprehensive architecture for building and defining TDG; (2) we presented an adversarial multitask learning model and distantly-supervised learning model for the information extraction with limited labeled sample. (3) we discussed the potential applications of TDG for technology insight, and applied TDG into technology architecture generation.

The whole paper is organized as follows. Section 2 introduces the related work. We describe our methods in Sect. 3. In Sect. 4, we present the case study on DSSC (dye-sensitized solar cell). Finally, in Sect. 5 we conclude this paper and explore the future work.

2 Related Work

Related works are mainly from two research communities: management and computer science. The former often uses simple text mining methods, such as: clustering [4], rules-based methods [5, 6], etc., to extract keywords [7] or technical concepts; but its advantage is that it can analyze text mining results with mature technical insight methods, such as technology roadmap [5], network theory [7], technology life cycle theory [8], TRIZ [4], etc., to achieve professional conclusions and discovery. The most similar study with this paper is the SAO-based (Subject – Action – Object) network analysis method with subject and object as technological key concepts and action as relations [9]. Examples include constructing technology trees [10], analyzing technological trends [11] and detecting signals of new technological opportunities [12]. However, the SAO network is generally built by heuristic rules or existing software, which requires a lot of manual work and contains a lot of noise. The researches from computer science community have strong information extraction ability, and can extract richer semantic information based on various methods such as LDA [13], CRF

[14], and machine learning [15]. However, its shortcoming is the lack of technical analysis theory, so only some preliminary conclusions can be drawn in technical analysis, such as the distribution of key-phrase or topic over time [13–15].

In summary, there was rarely a research in which both of extracting structural information from the literature content and analysis for technology insight were accomplished sufficiently. In this paper, we applied deep learning approaches to improve text mining, and presented a network of TDG which can be used for technology insight with existing network theories, methods and tools.

3   System Architecture

3.1   Basic Concepts

Technology Entity: There are many different expressions for technology in previous researches, such as technology concept, technology entity, technical terms, and key phrase. Because the TDG is defined as a special knowledge graph, the expression of technology entity was adopted to represent the technology in this paper.

Dependency Relation: The dependency relations refer to the semantic relations of promotion, dependence, contribution, use, etc. between two different technology entities. If the technology entity v_1 has dependency relation with the technical entity v_2, it can be expressed as $<v_1, v_2>$.

Technology Dependency Graph: The TDG is a weighted directed graph with the technology entities as vertices and the dependency relations as edges. The TDG can be expressed as $G = <V, E>$, where the vertices set is represented as $V = \{v_1, v_2, v_3, \ldots\}$ and edges set is represented as $E = \{e_1, e_2, e_3, \ldots\}$. The edge e_k from v_i to the v_j represents that technology of v_j depends on technology of v_i. And the weight of e_k is expressed as $W(e_k)$, which equals the number of times the corresponding dependency relationship e_k appears in the literature.

3.2   Architecture

The architecture of TDG consists of three modules – information extraction module, TDG construction module and technology analysis module. The information extraction module is developed for extracting technology entities and dependency relations from a massive amount of academic papers or technical patents. In the TDG construction, a weighted directed graph is generated by connecting the technology entities according to dependency relations. In the technology analysis module, with the help of network theories, methods and tools, we can obtain insight into the technology architecture, key technologies or hot topic, etc. for a certain domain.

3.3   Information Extraction Module

Entity recognition and relation extraction are both sub-tasks of information extraction [16]. The biggest challenge for information extraction from scientific literature is the lack of labeled corpus, and large-scale manual annotation is very costly. This paper proposed semi-supervised, weak supervision and transfer learning to realize the extraction of technology entities and dependency relations based on a small amount of labeled corpus.

Entity Extraction

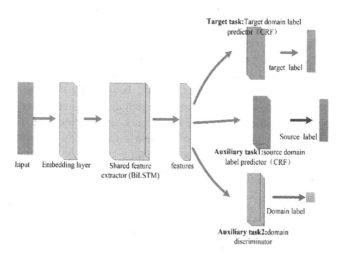

Fig. 1. The adversarial multitask learning model. The model includes a shared feature extractor (green), a target domain label predictor (blue), a source domain label predictor (purple) and a domain discriminator (yellow). (Color figure online)

A corpus for task ScienceIE [17] at Semeval2017 was built which consists of 500 journal articles distributed among the domains Computer Science, Material Sciences and Physics. Based on ScienceIE corpus, we developed a Neural Network of transfer learning to apply the knowledge gained in the three domains above to recognize entities in new domains (e.g., solar-energy). This model is a multi-tasks framework which consists of a target task and two auxiliary tasks:

Target Task: Technology entity recognition for target domain (e.g., solar-energy) based on a small number of labeled samples.

Auxiliary Task1: Science keyphrase recognition for the source domain, i.e. Computer Science, Material Sciences and Physics.

Auxiliary Task2: Inspired by the work on domain adaptation [18], an adversarial task is introduced as a domain discriminator to recognize whether source domain or target domain the input sample is belong to. When the domain discriminator cannot distinguish the domains of the input sample, it means that the model learns the sharable features, and ensuring better transfer learning ability.

This adversarial multitask learning model is shown in Fig. 1. An adversarial multitask learning model (BiLSTM) is used to extract sharable features of input text, two conditional random fields models (CRF) are used to predict technology entities through sequence tagging, and a domain discriminator is used to recognize whether source domain or target domain the input sentence is belong to.

Relationship Extraction

Distantly-supervised relation extraction has been proven to be effective in many researches when the labeled data set is small. A knowledge base (KB) of relation or concept instances is used to train a distantly-supervised learning system [19], but in many cases the knowledge base is lacking or incomplete: e.g., there is no release of a dependency relation knowledge base. Bing [20] extracted seeds from the well-structured corpus, and then extended to a complete instance-base through the semi-supervised method of label extension, combined with the distantly-supervised method to extract the medical concepts and relationships. Inspired by the work of Bing, we present a method which combines Bootstrapping and distantly-supervised to extracting technology dependency relation. Firstly, based on a small amount of well annotated corpus, bootstrapping method is used to generate a technology dependency relation instance-base that is sufficient to support distantly-supervised learning. Then, the distantly-supervised relation extraction model is trained to achieve the generalization of Bootstrapping.

3.4   TDG Construction Module

The construction for TDG consists of two tasks: entity linking and weight assignment. Entity linking is the task to link entity mentions in text with their corresponding entities in a knowledge base [21], which will be used to filling the extracted technology entities and dependency relations to the correct positions in the TDG. Because the expression in the scientific literature is rigorous, there is little ambiguity about the full name of the technology entity. The main challenge of entity linking is to find the correspondence between the abbreviation and the full name of the same entity, and we use a simple pattern matching method [22] to achieve this correspondence. The weight of the directed edge in TDG is assigned by the number of scientific literatures that contains the corresponding dependency relation.

3.5   Technology Insight Based on TDG

As a weighted directed graph, the TDG has rich properties of social networks that allow researchers to accomplish technology insight by using theories, methods and tools of graph, social network and even traditional bibliometrics. This section will explore some technical insight scenarios of TDG based on social network analysis theory.

Path Based Technology Architecture Construction: In graph theory, a directed path in a graph is a directed sequence of edges which connect a sequence of vertices which are all distinct from one another. The terminal vertex in a path is pointed to by all other vertices, which means that the technology of the terminal vertex depends on the technologies of all the other vertices in a path of TDG. All the paths with a certain technology as the terminal construct a complete technical dependency architecture which provides a global perspective for the certain technology.

Centrality Based Evaluation of Technical Importance and Maturity: In social network analysis, indicators of degree centrality and betweenness centrality can identify the importance of vertices within a graph. The degree increases as the frequency of occurrences of corresponding technology increases in the literature, which means that we can obtain the hottest technology by finding the vertex of the highest degree. As the maturity of a technology increases, its application becomes more extensive, and the outdegree of corresponding vertex will increase. Therefore, the ratio of the outdegree and the intdegree can reflect the maturity of a technology. Betweenness centrality quantifies the number of times a node acts as a bridge along the shortest path between two other nodes. In a TDG, betweenness can be used to find the key technologies.

Structural Equivalence Based Discovery for Alternative Technology: In a network, if two actors do not change the structure of the entire network after replacing each other, the two actors are structurally equivalent. The structural equivalence can measure the similarity of relational patterns or network positions between two vertices. In a TDG, alternative technologies can be discovered according to the structural equivalence between technologies, which help researchers to bypass insurmountable technical difficulties in technological innovation activities.

Community Evolution based technology trend prediction: In the TDG, technologies similar in disciplines, fields, applications, etc. are grouped into communities. By studying the birth, merger, decomposition, and extinction of these communities in the time dimension, we can predict the technology trend. For example, it is indicated the emergence of a new interdisciplinary when two different communities are merging together.

4   Case Study

This empirical study applied the proposed framework to literatures of related to DSSC (dye-sensitized solar cell) which is a low-cost solar cell. We retrieved 473935 related articles with keyword of 'dye-sensitized solar cell' from an aggregation digital Library (consisting of INSPEC, WPI, EI, Elsevier, Springer, etc.). All were published in between 1999 and 2015.

4.1   Construction for TDG of DSSC

We employed a framework of adversarial multitask learning introduced in 3.3.1 to extract technology entity. The corpus of source domain was released by Semeval2017 [17] which consists of 500 journal articles distributed among the domains Computer Science, Material Sciences and Physics. The corpus of target domain is about DSSC built by manual annotation from 20 articles. We implemented the extraction through modifying the source code shared by Chen[1] [23]. For the dependency relation

---

[1] https://github.com/FudanNLP/adversarial-multi-criteria-learning-for-CWS.

extraction, we also labeled 38 relations instances from 20 articles, and expanded to 573 instances by bootstrapping. After that, we applied the distantly-supervised learning to extract the dependency relations between technologies. At last, we obtained a TDG of DSSC which consists of 472 vertices and 1027 edges. Being rendered by Gephi[2], the TDG is shown in Fig. 2.

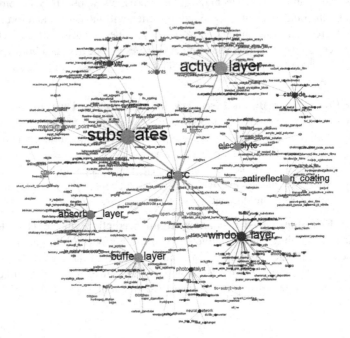

Fig. 2. The TDG of DSSC. Every node represents a technology, and the directed edge represents the dependency relation between technologies.

4.2   Technology Insight

In this section, we applied the TDG to analyze the technology dependency architecture of DSSC. Firstly, we got a spanning tree with DSSC as the terminal out of the TDG using Breadth First Search. In this spanning tree, all other technologies point to DSSC, which means the technology of DSSC is directly or indirectly depend on the other technologies. So, the spanning tree represents the technology dependency architecture of DSSC (Fig. 3). In order to make a clear description, we converted the spanning tree into a table (Table 1), which contains the dependency relations whose weight bigger than 10. Based on expert knowledge and literature verification, it is confirmed that the technology architecture of DSSC obtained by this analysis method is close to the actual situation.

---

[2] https://gephi.org/.

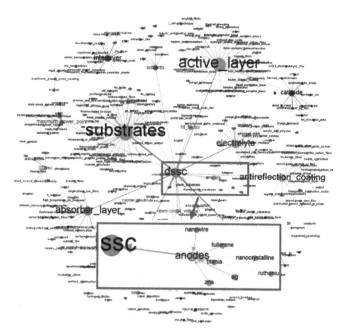

Fig. 3. The dependency architecture of DSSC. For example, DSSC depends on anodes which in turn depends on nanocrystalline technology (shown in the red wireframes). (Color figure online)

Table 1. Table of the dependency architecture for DSSC. The DSSC is dependent on technologies in the second column, and technologies in the second column are dependent on technologies in the third column.

DSSC	Window layer	thin-film cds, flexible plastic substrates, gallium arsenide, microcrystalline silicon
	Electrolyte	potassium iodide, electron donor mediator, polysulfide, ionic liquid, sodium polysulfide solution
	Photocatalyst	nanoparticles, titanium oxide, optical radiation, quantum dots, titanium dioxide, znin2s4
	Buffer layer	znse, zn-doped cigs layer, znte, porous silicon, znpc, zno particles, azo films, bathocuproine, cds nano-layer
	Interlayer	copolymer encapsulant, transparent multiwalled carbon nanotube sheets, carrier transportation process
	Antireflection coating	pecvd-grown sinx film, silicon nitride films, thin nanoporous layer, moth-eye antireflection nanostructures
	Absorber layer	ternary chalcopyrite semiconductor cu, single-phase sns films, colloidal nanocrystals, evaporated sns
	Anodes	titania, ruthenium, zns, nanocrystalline, nanowire, fullerene
	Active layer	novel cyanine-fullerene dyad, organic films, vhf-gd technique, bulk heterojunction structure

5   Conclusion

This paper introduced Technology Dependency Graph (TDG), a big data analysis framework will facilitate the technology insight from scientific literature. We applied deep learning methods to implement the construction of TDG, and discussed the potential applications of TDG for technology insight. Several extensions of the framework we presented can be performed in future works. The types of technology (such as PROCESS, ATTRIBUTE, and MATERIAL) should be identified further to enrich the semantic information in TDG. And we will do more case studies to evaluate the application of TDG for technical insights.

Acknowledgement. This research was financially supported by the National Natural Science Foundation of China (Grant NO. 61602490).

References

1. Ronzano, F., Saggion, H.: Knowledge extraction and modeling from scientific publications. In: González-Beltrán, A., Osborne, F., Peroni, S. (eds.) SAVE-SD 2016. LNCS, vol. 9792, pp. 11–25. Springer, Cham (2016). https://doi.org/10.1007/978-3-319-53637-8_2
2. Munroe, R.: News: the rise of open access. Science **342**(6154), 58–59 (2013)
3. Lewis, D.W.: The inevitability of open access. Coll. Res. Libr. **73**(5), 493–506 (2012)
4. Zhang, Y., Zhou, X., Porter, A.L., et al.: Triple Helix innovation in China's dye-sensitized solar cell industry: hybrid methods with semantic TRIZ and technology roadmapping. Scientometrics **99**(1), 55–75 (2014)
5. Yang, C., Zhu, D., Zhang, G.: Semantic-based technology trend analysis. In: International Conference on Intelligent Systems and Knowledge Engineering, pp. 222–228. IEEE (2015)
6. Guo, J., Wang, X., Li, Q., et al.: Subject–action–object-based morphology analysis for determining the direction of technological change. Technol. Forecast. Soc. Chang. **105**(4), 27–40 (2016)
7. Silva, F.N., Amancio, D.R., Bardosova, M., et al.: Using network science and text analytics to produce surveys in a scientific topic. J. Informetr. **10**(2), 487–502 (2016)
8. Durmuşoğlu, A.: A pre-assessment of past research on the topic of environmental-friendly electronics. J. Clean. Prod. **129**, 305–314 (2016)
9. Yoon, J., Kim, K.: Detecting signals of new technological opportunities using semantic patent analysis and outlier detection. Scientometrics **90**, 1–17 (2012)
10. Choi, S., Park, H., Kang, D., et al.: An SAO-based text mining approach to building a technology tree for technology planning. Expert Syst. Appl. **39**(13), 11443–11455 (2012)
11. Yang, C., Cui, H., Su, J.: An improved SAO network-based method for technology trend analysis: a case study of graphene. J. Informetr. **12**(1), 1577–1751 (2018)
12. Yoon, J.: Detecting signals of new technological opportunities using semantic patent analysis and outlier detection. Scientometrics **90**(2), 445–461 (2012)
13. Cano-Basave, A.E., Osborne, F., Salatino, A.A.: Ontology forecasting in scientific literature: semantic concepts prediction based on innovation-adoption priors. In: Blomqvist, E., Ciancarini, P., Poggi, F., Vitali, F. (eds.) EKAW 2016. LNCS (LNAI), vol. 10024, pp. 51–67. Springer, Cham (2016). https://doi.org/10.1007/978-3-319-49004-5_4

14. Prabhakaran, V., Hamilton, W.L., Dan, M.F., et al.: Predicting the rise and fall of scientific topics from trends in their rhetorical framing. In: Proceedings of the 54th Annual Meeting of the Association for Computational Linguistics, pp. 1170–1180. ACL (2016)
15. Mckeown, K., Daume, H., Chaturvedi, S., et al.: Predicting the impact of scientific concepts using full-text features. J. Assoc. Inf. Sci. Technol. 67(11), 2684–2696 (2016)
16. Li, S., Lin, C.Y., Song, Y.I., et al.: Comparable entity mining from comparable questions. IEEE Trans. Knowl. Data Eng. 25(7), 1498–1509 (2014)
17. Augenstein, I., Das, M., Riedel, S., et al.: SemEval 2017 task 10: ScienceIE - extracting keyphrases and relations from scientific publications. In: Proceedings of the 11th International Workshop on Semantic Evaluations, pp. 546–555. ACL, Vancouver (2017)
18. Ganin, Y., Ustinova, E., Ajakan, H., et al.: Domain-adversarial training of neural networks. J. Mach. Learn. Res. 17(1), 2030–2096 (2017)
19. Ru, C., Tang, J., Li, S., et al.: Using semantic similarity to reduce wrong labels in distant supervision for relation extraction. Inf. Process. Manag. 54(4), 593–608 (2018)
20. Bing, L., Ling, M., Wang, R.C., et al.: Distant IE by bootstrapping using lists and document structure. In: AAAI 2016. AAAI (2016)
21. Shen, W., Wang, J., Han, J.: Entity linking with a knowledge base: issues, techniques, and solutions. IEEE Trans. Knowl. Data Eng. 27(2), 443–460 (2015)
22. Adar, E., Datta, S.: Building a scientific concept hierarchy database (SCHBASE). In: Proceedings of the 53rd Annual Meeting of the Association for Computational Linguistics, pp. 606–615. ACL, Beijing (2015)
23. Chen, X., Shi, Z., Qiu, X., et al.: Adversarial multi-criteria learning for Chinese word segmentation. In: Proceedings of the 55th Annual Meeting of the Association for Computational Linguistics, pp. 1193–1203. ACL, Vancouver (2017)

PGCAS: A Parallelized Graph Clustering Algorithm Based on Spark

Dongjiang Liu[1,2(✉)] and Jianhui Li[1]

[1] Computer Network Information Center, Chinese Academy of Science,
No. 4, Zhongguancun South 4th Street, Haidian District, Beijing 100190, China
ldongjiang@cnic.cn
[2] University of Chinese Academy of Sciences,
No. 80, Zhongguancun East Road, Haidian District, Beijing 100190, China

Abstract. Nowadays plenty of data are in graph format. For example, knowledge graph use vertices to represent entities and use edges to represent relations between entities; graph data in microbiology contain microorganisms and relations between them etc. So information can be obtained by graph mining from these data. Graph clustering is a part of graph mining. Recent years, many graph clustering algorithms have been proposed. But most of them are Sequential Algorithms. So they cannot run in distributed environment. In this case the volume of data that can be processed by the algorithms is limited. In this paper we propose a new parallelized graph clustering algorithm based on Spark. And some methods have been adopted in the algorithm to improve its running speed. From the experimental results we can find that the proposed algorithm is better than the parallelized graph clustering algorithm for comparison.

Keywords: Parallelized graph clustering · Community detection · Spark · Graph mining · Graph data

1 Introduction

Graph contains many nodes and edges. So it can be used to model the data of many fields. Generally, in graph data, nodes are used to represent data items of the field. Edges are used to represent the relationship between data items. For example, in microbiology, a node represents a specific microorganism or the property of a microorganism. The edge between the microorganisms represents the relationship between the microorganisms. In World Wide Web, a node represents a specific webpage. An edge specifies whether a hyperlink exits between two page or not. In social networks, a node of the graph represents a user and an edge represents the relationship between two users which can be subscription, friendship etc.

Nowadays as graph is used to describe data in many fields, an efficient graph mining algorithm can help to find out the valuable information from the graph data related to these fields. In graph mining, graph clustering is a very important

© Springer Nature Switzerland AG 2019
J. Li et al. (Eds.): BigSDM 2018, LNCS 11473, pp. 186–198, 2019.
https://doi.org/10.1007/978-3-030-28061-1_20

mining task. It is also called community detection. Main task of graph clustering is dividing the vertices of graph data into different clusters. In the result of clustering, the connection between the vertices in the same cluster should be as close as possible and the connection between the vertices in different clusters should be as remote as possible. The connection can be measured by the value calculated based on the number of links between the vertices. Graph clustering has been adopted in many different fields. For example, in microbiology [1], graph clustering has been used to detect the clusters of proteins and the network of diseases. In graph visualization [2], graph clustering is used to put the similar vertices of a graph into same cluster. A cluster is shown as a vertex on the screen. In this way, the graph can be shown more clearly. In computer networks [3], graph clustering can also be performed on the graph consisted by network users. By doing so, the users that are close with each other in geographical distance or share same interests can be found. With the information, the service quality of the network can be improved.

In recent years, the volume of graph data increased very fast. So graph clustering algorithm should perform on big graph data. The most efficient way to deal with big graph data is performing parallelized graph clustering on distributed system. Now the number of parallelized graph clustering algorithms is still relatively small. And most of them are based on a distributed computing framework of Hadoop called MapReduce or parallelized framework called MPI [4–8]. The parallelized graph clustering algorithm based on Spark is relatively scarce. Though MapReduce and MPI are very famous, they still have some shortage. For MapReduce, during computation it needs to read and write data stored on disk repeatedly. So performing computing based on MapReduce will cost plenty of time. For MPI, the reliability of this computing framework is very poor and it also has the problems of communication delay and load unbalance. The performances in all these aspects are improved greatly by Spark [20]. So we choose to propose a parallelized graph clustering algorithm based on Spark. Nowadays many graph clustering algorithms have been proposed [9–13]. Among them, one part is the graph clustering algorithm based on spectrum [14–16]. A very famous and representative algorithm in this kind of algorithms is the one proposed by Donetti et al. [14]. This algorithm tries to model the graph data by using Laplacian matrix and describes the vertices of the graph data by using eigenvectors of the Laplacian matrix. In this way a vertex of graph data will become a point in high dimensional space. So the vertices will be more distinguishable. And then the clustering result will be greatly improved. In this paper a new parallelized graph clustering algorithm based on the algorithm described in [14] is proposed. Moreover, some strategies are adopted to promote execution speed of the proposed algorithm.

The rest of this paper is organized as follows. In Sect. 2, the newly proposed parallelized graph clustering algorithm based on spectrum will be described. The experimental evaluation is presented in Sect. 3. Finally, the paper will be concluded in Sect. 4.

2   PGCAS Algorithm

In this section, we will propose a new parallelized graph clustering algorithm based on spectrum. In this algorithm, firstly all the 1 degree vertices in graph data should be filtered out. As the connected neighbors of 1 degree vertices are closer to them than other vertices in graph data, all the 1 degree vertices should be put to the cluster that their neighbor belongs to. So during the clustering process, 1 degree vertices need not to be judged. In this case, 1 degree vertices can be filtered out. After filtering, the volume of graph data will decrease greatly and at the same time a new edge set E_{new} will be obtained. The elements in set E_{new} are in the form of $e_j = (k_j, v_j)$. k_j and v_j are the indexes of vertices. So these two values are integer. Then a Laplacian matrix must be obtained. Each row and each column in the matrix correspond to a vertex in graph data. If a value of the matrix is -1, it means that there is an edge between the vertices represented by the row and column the value belongs to. The values of main diagonal elements of Laplacian matrix are the degrees of the vertices represented by the rows or columns the values belong to. While the Laplacian matrix is obtained, D eigenvectors of the matrix should be calculated. We can use Lanczos algorithm [17] to calculate all the eigenvectors. Then a matrix L can be built based on the D eigenvectors we have obtained, as shown in (1):

$$L = [l_1, l_2, ..., l_D] \tag{1}$$

Columns l_1, l_2, ..., l_D of matrix L is D eigenvectors of the Laplacian matrix. Now we can use the rows of matrix L to represent the vertices of graph data. Each row corresponds to a single vertex. In this way, all vertices of graph data will be map to D dimensional space. Then we can perform clustering based on bottom-up hierarchical clustering algorithm. During the clustering we can use cosine similarity to calculate the similarity of two vertices. The pseudo code of PGCAS algorithm is shown in Algorithm 1:

In the third step, a Laplacian matrix is built based on set E_{new}. The Laplacian matrix is represented by a set named $matrix$. This set is a triple set. Elements of this set are in the form of $(index_0, index_1, value)$. $index_0$ is row index of the element. $index_1$ is column index of the element. $value$ is the value of the element which located in row $index_0$ and column $index_1$. In order to reduce the storage space the triple set consumes, the values of the elements in matrix that are equal to 0 will not be stored in the triple set. Moreover, as Laplacian matrix is a symmetric matrix, if the triple set that represents Laplacian matrix contains element $(index_0, index_1, value)$, it must contain $(index_1, index_0, value)$. In this case, $index_0$ of each element in the triple set could also be treated as column index and $index_1$ could be treated as row index. The pseudo code of Laplacian matrix building process is shown in Algorithm 2:

In Algorithm 2, set E_{new} is updated at 2. Based on the updated set E_{new} a new set $matrixA$ is obtained at 3. Set $matrixA$ represents a matrix. Rows and columns of the matrix represent the vertices of graph data. It is obvious that set $matrixA$ is a symmetric matrix. Set $matrixB$ described in step 4 also represents

---

Algorithm 1. PGCAS Algorithm

Intput: edge set E_{new}, dimension value D
Output: clustering result set $result$
1: $\beta_0 = 0$
2: Filter out the vertices whose degree is 1.
3: Build a Laplacian matrix based on set E_{new}, the laplacian matrix is represented by a set called $matrix$.
4: Build a zero vector q_0 and a random vector q_1, these two vectors are also represented by set.
5: **for** $i = 1 \rightarrow D$ **do**
6:     $z = matrix * q_i$, $\alpha_i = q_i^T z$, $z = z - \alpha_i q_i - \beta_{i-1} q_{i-1}$, $\beta_i = ||z||_2$
7:     **if** $\beta_i = 0$ **then**
8:         Break
9:     **else**
10:         $q_{i+1} = \frac{z}{\beta_i}$
11:     **end if**
12: **end for**
13: Calculate the tri-diagonal matrix T based on all the values of α_i and β_i obtained above and compute the matrix Q based on all vector q obtained.
14: Performing eigenvalue decomposition based on matrix T, $T = PUP^T$. This task can be fulfilled by using QR decomposition algorithm, and then matrix P will be obtained.
15: $L = QP$.
16: Build a graph model $G(V(id, vector), E(srcid, dstid))$.
17: Build a result set $result$.
18: Perform hierarchical clustering based on graph model $G(V, E)$. During the clustering process, set $result$ will be updated repeatedly. When the clustering process is finished, set $result$ will be returned as final result of clustering.

Algorithm 2. building Laplacian matrix

Intput: edge set E_{new}
Output: Laplacian matrix $matrix$
1: Exchanging position of key and value of each element in set E_{new} by map operation. Then we will obtain a new set E_1. The elements of set E_1 are in the form of $e_j = (v_j, k_j)$.
2: $E_{new} = E_{new} \cup E_1$. Then cluster the elements of set E_{new} with the same key by groupbykey operation. After that a new set named N is built. The elements of set N are in the form of $e_i = (k_j, D_j)$. k_j is the key of the element and D_j is a set of values that share the same key k_j. Each element of set N corresponds to a vertex in graph data.
3: Change the form of elements in set E_{new} by map operation. All the changed elements form a new set $matrixA$. Elements of set $matrixA$ are in the form of $e_j = (k_j, (v_j, -1))$.
4: Then the form of elements in set N should also be changed. All the changed elements form a new set $matrixB$. The elements of set $matrixB$ are in the form of $e_j = (k_j, (k_j, deg_j))$. Value deg_j is the number of elements in set D_j. So deg_j is the degree of vertex indexed by k_j.
5: $matrix = matrixA \cup matrixB$
6: return set $matrix$ as result.

---

a matrix. This matrix is a diagonal matrix. Rows and columns of this matrix represent the vertices of graph data. The values of diagonal elements of $matrix B$ are the degree values of the vertices of graph data. Set $matrix$ obtained in step 5 also represents a matrix. This matrix is a Laplacian matrix. It is calculated by summing up matrix represented by set $matrix A$ and matrix represented by set $matrix B$. As both matrices represented by set $matrix A$ and set $matrix B$ are symmetric matrix, matrix represented by set $matrix$ is also a symmetric matrix.

In the 4-th step of Algorithm 1, vector q_0 and q_1 should be built. q_0 is a zero vector and q_1 is a random vector. Both of these two vectors are represented by set. Vector q_0 is represented by set q_0. Vector q_1 is represented by set q_1. Both set q_0 and set q_1 are built by map operation based on set N which is obtained in second step of Algorithm 2. Elements of set q_0 are in the form of $e_j = (k_j, 0)$. Elements of set q_1 are in the form of $e_j = (k_j, rand)$. $rand$ represents a random value.

In the for loop of 5-th step of Algorithm 1, $\alpha_1, \alpha_2, ..., \alpha_D, \beta_1, \beta_2, ..., \beta_{D-1}$ and $q_1, q_2, ..., q_D$ are calculated. All these values are used to form the tri-diagonal matrix T and matrix Q. The form of matrix T is shown in Eq. (2):

$$T = \begin{bmatrix} \alpha_1 & \beta_1 & \cdots 0 & 0 \\ \beta_1 & \alpha_2 & \cdots 0 & 0 \\ \vdots & \vdots & \ddots & \vdots & \vdots \\ 0 & 0 & \cdots \alpha_{D-1} & \beta_{D-1} \\ 0 & 0 & \cdots \beta_{D-1} & \alpha_D \end{bmatrix} \tag{2}$$

The form of matrix Q is shown in Eq. (3):

$$Q = [q_1, q_2, ..., q_D] \tag{3}$$

In step 5–12 of Algorithm 1, the sum of two vectors, the product of two vectors and the product of a vector and a matrix should be calculated. As the volume of graph data is big, the sum and product values should be calculated in parallelized way. So we will try to fulfill the calculation tasks by set operations. In step 5, vector z is calculated firstly. During the calculation process, set $matrix$ and set q_i should be joined by join operation. Then a new set E_2 is created. The elements of set E_2 is in the form of $e_j = (k_j, ((v_j, value_{1j}), value_{2j}))$. $(v_j, value_{1j})$ is a value of an element with key k_j in set $matrix$. $value_{2j}$ is the value of an element with key k_j in set q_i. While we get set E_2, we need to process elements of the set by map operation. Then we get a new set E_3. The elements of set E_3 are in the form of $e_j = (v_j, value'_j)$. $value'_j$ is the product of $value_{1j}$ and $value_{2j}$. Finally in set E_3 the values with same key should be summed up by reducebykey operation. All these processed key value pairs in set E_3 form a new set z. z represents a vector. While vector z is obtained, α_i should be calculated. During the process, set q_i and z are joined by join operation. Then a new set qiz is obtained. The elements of set qiz are in the form of $e_j = (k_j, (ele_{qj}, ele_{zj}))$. ele_{qj} is the value of an element of set q_i whose key is k_j. ele_{zj} is the value of an element of set z whose key is k_j. After that, process all the elements of set qiz by map operation.

All the processed elements form a new set qiz_2. Each element of set qiz_2 can be computed by summing up ele_{qj} and ele_{zj}. Finally all the values in set qiz_2 will be summed up by using reduce operation. The result of summing is the value of α_i. Now vector z should be updated. During updating, firstly, join set z, q_i and q_{i-1} by join operation. Then a new set $middleVec$ is created. All the elements of set $middleVec$ are in the form of $e_j = (k_j, (v_{zj}, v_{qij}, v_{qij-1}))$. v_{zj}, v_{qij} and v_{qij-1} are the values of elements of set z, q_i and q_{i-1}. These elements share the same key k_j. After that set $middleVec$ should be updated by map operation. Set z is set to the updated set $middleVec$. Elements of set z are in the form of $e_j = (k_j, (v_{zj} - v_{qij} * \alpha_i - \beta_{i-1} * v_{qij-1}))$. When the updating process is finishing, value β_i should be calculated based on vector z. β_i can be calculated by summing up all the values of set z. This task can be fulfilled by map operation and reduce operation.

In step 7, the value of β_i should be checked. If it is equal to 0, break out of the loop. Otherwise, form a new set q_{i+1} based on set z by map operation. Elements of set q_{i+1} are in the form of $e_j = (k_j, v_j/\beta_i)$.

While all the α_i, β_i and q_i are obtained, tri-diagonal matrix T and matrix Q should be computed. In step 13, matrix T is composed by all the values of α_i and β_i and matrix Q is formed by all the vectors $q_1, q_2, ..., q_D$.

Now the tri-diagonal matrix T and matrix Q have been obtained. The dimension of the rows and columns of the matrix T is D. We will calculate a D dimensional vector for each vertex in graph data. Obviously, value D is not very large. So we can perform QR decomposition for the tri-diagonal matrix T based on the sequential algorithm presented in Algorithm 3. Then matrix P will be obtained. When both matrix P and matrix Q are obtained, matrix L can be calculated by multiplying matrix P and matrix Q. Each row vector of matrix L represents a vertex of graph data.

Algorithm 3

Intput: tri-diagonal matrix T
Output: matrix P
1: $T_1 = T$
2: **for** $k = 1 \rightarrow n$ **do**
3:      Performing QR decomposition based on matrix T_k, $T_k = Q_k * R_k$
4:      $T_{k+1} = R_k * Q_k$
5: **end for**
6: $P = Q_1 * Q_2 * ... * Q_n$

In step 17 of Algorithm 1, graph model $G(V(id, vector), E(srcid, dstid))$ is built. Set V contains all the vertices of graph model. Each vertex of set V is a cluster of vertices in graph data. It may contain only one vertex of graph data or multiple vertices of graph data. id is the index of the element in set V. $vector$ is a D dimensional vector corresponding to the element indexed by id. Initially, each element in set V only contains one vertex in graph data. So at this time

id is the index of vertex of graph data and *vector* is the vector corresponding to the vertex specified by id. Set E contains all the edges of the graph model. Each element of set E represents an edge which connects two elements in set V. *srcid* is the index of source vertex of the edge; *dstid* is the index of destination vertex of the edge.

In step 18 of Algorithm 1, a result set *result* is created. This set is used to store clustering result. Each element of set *result* corresponds to a vertex in graph data. So the set can be built based on set V by map operation. Each element of set *result* represents a vertex of graph data and all the elements are in the form of $e_j = (clusterid_j, id_j)$. id_j is the index of the vertex; $clusterid_j$ is index of the cluster that the vertex belongs to. obviously, During the process of creating set *result*, each vertex in set V will be mapped to a new element which contains id_j and $clusterid_j$. Initially the value of id_j equals to the value of $clusterid_j$.

In step 19 of Algorithm 1, the bottom-up clustering algorithm is performed based on the graph model $G(V(id, vector), E(srcid, dstid))$. The parallelized clustering algorithm is shown in Algorithm 4:

Algorithm 4

Intput: graph model $G(V(id, vector), E(srcid, dstid))$, the number of vertices need to be clustered *clustnum*, set *result*

Output: set *result*

1: **while** the number of vertices in set V is greater than 1 **do**

2:    Calculate the similarity values for all the pairs of connected vertices based on triples of graph model. Then a new element can be created based on each triple. All the newly created elements are in the form of $e_j = (srcid_j, dstid_j, sim_j)$. $srcid_j$ is the index of source vertex of the edge which belongs to the triple. $dstid_j$ is the index of destination vertex of the edge which belongs to the triple. sim_j is the similarity value between source vertex and destination vertex. All the new elements form a new set E_1.

3:    Collect *clustnum* elements from set E_1 that have the largest sim values. All these elements form a new set E_2. sim values of all the elements in set E_2 should be deleted. The form of elements in set E_2 will be changed to $e_j = (dstid_j, srcid_j)$.

4:    Find all clusters in set E_2.

5:    Update the *clusterid* value of all the elements in set *result* based on the cluster information obtained from set E_2.

6:    Update set V and set E based on the cluster information, then a new vertex set V_{new} and a new edge set E_{new} will be got.

7:    Create a new graph model $G(V_{new}, E_{new})$ based on set V_{new} and E_{new}.

8: **end while**

In step 2 of Algorithm 4, the similarity values of all the connected vertices are calculated. Each pair of connected vertices is contained in a triple. Each triple of graph model contains two vertices and an edge. The triples are in the form of $(srcnode, edge, dstnode)$. *srcnode* and *dstnode* represent two vertices. They are connected by an edge represented by *edge*. *srcnode* represents the source vertex of the edge. *dstnode* represents the destination vertex of the edge. Both *srcnode* and *dstnode* belong to vertex set V of graph model G. Element *edge* belongs to edge set E of graph model. In this case, the similarity value can be calculated based on each triple. We can calculate the similarity value of two connected vertices of an triple by mapTriplets operation. Then a new element is created based on a triple. Each element contains index of the source vertex, index of the destination vertex and the similarity value of the two vertices. Set E_1 consists of all the new elements.

In step 3 of Algorithm 4, *clustnum* elements with largest similarity values should be found. The process can be fulfilled by the parallelized algorithm described in Algorithm 5:

Algorithm 5

Intput: set E_1
Output: set E_2
1: Change the form of the elements in set E_1. After that the elements in set E_1 are in the form of $e_j = (sim_j, (srcid_j, dstid_j))$.
2: The elements of set E_1 should be sorted based on keys of the elements by sortbykey operation.
3: Add an index to each element in set E_1 in order. Now the elements of set E_1 are in the form of $e_j = ((sim_j, (srcid_j, dstid_j)), id_j)$. id_j is index of the element.
4: Delete all elements whose index is greater than *clustnum* by filter operation.
5: All the elements of set E_1 should be changed by using map operation. The form of all the elements in set E_1 are changed to $e_j = (srcid_j, dstid_j)$. All these changed elements form a new set E_2.

In step 4 of Algorithm 4, all clusters of vertices in set E_2 should be found. In set E_2, each element is a pair of vertices. So each element of E_2 represents an edge. In this situation, a vertex set V_{part} and an edge set E_{part} could be built based on set E_2. Then a graph model $G_1(V_{part}, E_{part})$ could be built based on set V_{part} and E_{part}. Finally all the connected subgraphs of graph model G_1 can be found by connectedComponents operation. Each connected subgraph represents a vertex cluster. The minimum vertex index of each connected subgraph is treated as the index of the cluster. Now the elements of set V_{part} are in the form of $v_j = (id_j, clustid_j)$. id_j is index of the j-th vertex. $clustid_j$ is index of the cluster which contains the j-th vertex. At last, all the elements of set V_{part} should be added into a new set V_1.

In step 5 of Algorithm 4, *clusterid* values of some elements in set *result* should be updated. Now set V_1 obtained in step 4 is used. Firstly set *result* and set V_1 should be joined by leftoutjoin operation. The elements in set *result* are in the form of $e_j = (clusterid_j, (id_{j1}, id_{j2}))$. id_{j1} is index of j-th vertex of set *result*. id_{j2} is a cluster index of an element in set V_1. After that the elements of set *result* should be updated by map operation. During the updating process, each element in set *result* should be checked. If an element of set *result* contains id_{j2}, the value of id_{j2} will be used to replace the value of $clusterid_j$. Then id_{j2} will be deleted. Now all the elements of set $result$ are in the form of $e_j = (clusterid_j, id_{j1})$.

During clustering process, some vertices of graph model G are put into same cluster. So in step 6 of Algorithm 4, the vertex set V and edge set E of graph model G must be updated. The updating process is shown in Algorithm 6:

Algorithm 6

Intput: graph model $G(V(id, vector), E(srcid, dstid))$, set V_1

Output: vertex set V_{new}, edge set E_{new}

1: Join vertex set V and V_1 by using leftoutjoin operation. After that the form of elements of set V are changed to $v_j = (id_{j1}, (vector_j, id_{j2}))$. id_{j1} is index of the elements of set V and V_1. $vector_j$ is a vector that corresponds to the element of set V whose index is id_{j1}. id_{j2} is cluster index of the vertex contained in set V_1 whose index is id_{j1}.

2: The elements of set V should be updated by using map operation. Check whether an element contains id_{j2}. If so, use the value of id_{j2} to replace the value of id_{j1}. Otherwise, keep the value of id_{j1} unchanged. While the updating is finished, all id_{j2} should be deleted. Now the elements of set V are in the form of $v_j = (id_j, vector_j)$.

3: Cluster the elements that share the same index in set V by groupbykey operation. Then a new set V_2 is created. All the elements in set V_2 are in the form of $v_j = (id_j, vecset_j)$. id_j is the element index. $vecset_j$ is a vector set.

4: Update the elements in set V_2 by using map operation. Calculate the mean vector vec_j of all the vectors of $vecset_j$ of each element and use this vector to replace $vecset_j$. Then a new set V_{new} is created. All the elements of set V_{new} are in the form of $v_j = (id_j, vec_j)$. id_j is index of the element and vec_j is the mean vector.

5: Join the edge set E and set V_1 by leftoutjoin operation. Then the form of elements in set E are changed to $e_j = (srcid_j, (dstid_j, id_{j2}))$.

6: Update the elements in set E by map operation. Check whether an element of set E contains id_{j2}. If so, the value of id_{j2} is used to replace the value of $srcid_j$. Otherwise, keep $srcid_j$ of the element unchanged.

7: Continue to update all the elements in set E. After that the form of all elements in set E are changed to $e_j = (dstid_j, srcid_j)$.

8: Repeat all the operation described in step 5 and 6. Then change the form of elements of set E into $(srcid_j, dstid_j)$ by map operation. All the updated elements form the new set E_{new}.

---

3   Experiment Descriptions

In this section, experiment will be described in detail. In the experiment the proposed parallelized graph clustering algorithm based on spectrum will be tested. And the proposed algorithm will be compared with the parallelized algorithm described in [18]. The experiment is performed on a cluster that contains ten virtual machines. CPU of each virtual machine has 24 cores. The memory of each virtual machine is 64G and the disk capacity is 1T. Version of Hadoop used in the experiment is 2.7.3. Version of spark used is 2.3.0.

In this experiment, the proposed algorithm will be tested based on two datasets. One is WDCM dataset; the other is YAGO dataset. As all the data in these two dataset are in RDF format, both of these two datasets need to be preprocessed. During the preprocessing, each vertex gets a unique index value. Then each RDF triple is replaced by a pair of index values. The first index value is the index of source vertex and the second index value is the index of destination vertex.

3.1 Datasets and Measurement Metrics Introduction

WDCM dataset is a microbiological data set. Data volume of WDCM dataset is very big. It contains plenty of information related to germ, bacterial and fungus etc. All the microorganisms appeared in this dataset are connected by edges. All these connected microorganisms form a large graph model. WDCM is a data center. This organization tries to collect all the microbiological data, integrates them and provides management so that user can access the whole data easily.

YAGO dataset is also a graph dataset released by Max Planck Institute. The data of YAGO is collected from Wikipedia, WordNet and GeoNames. The entities collected from Wikipedia are connected with the entities collected from WordNet. Otherwise, the time attribute and location attribute are added to many entities of YAGO dataset.

Table 1 shows the number of vertices and number of edges in WDCM dataset and YAGO dataset:

Table 1. Numbers of nodes and edges

Dataset	Number of vertices	Number of edges
WDCM	92720410	286439613
YAGO	72136513	244795904

The experiment will measure and compare the running time of the proposed algorithm and the algorithm described in [18] based on WDCM dataset and YAGO dataset. And modularity values [19] of the clustering results of the two algorithms are also measured and compared. Modularity value is a measurement metrics for graph clustering result and this metrics has been widely used. The way to calculate the modularity value is shown below:

$$modularity = \sum_i (e_{ii} - a_i^2)$$
$$a_i = \sum_j e_{ij} \tag{4}$$

e_{ii} is the ratio of the edge number in cluster i to the total edge number in graph data. e_{ij} is the ratio of the edge number between cluster i and cluster j to the total edge number. In graph clustering, better clustering result always has larger modularity value.

3.2 Experimental Results

In this section, the experimental results will be presented. Figure 1 shows the experimental results based on WDCM dataset. Figure 2 shows the experimental results based on YAGO dataset. PGCAS shown in Figs. 1 and 2 is the parallelized graph clustering algorithm proposed in this paper. Compared_Algorithm represents the parallelized algorithm described in [18]. As the proposed PGCAS algorithm contains the parameter *clustnum* which should be set by users, results based on different *clustnum* values will be presented in this experiment. PGCAS(500000) column represents the experimental results of the proposed algorithm in which parameter *clustnum* is set to 500000. And PGCAS(1000000) column represents the experimental results of the proposed algorithm in which parameter *clustnum* is set to 1000000. Moreover, two parallelized algorithms will be tested based on two measurement metrics. One is the running time of the algorithms; the other is modularity value of the clustering results. In Figs. 1 and 2, part (a) shows the running time of the two algorithms; part (b) shows the modularity values of the clustering results.

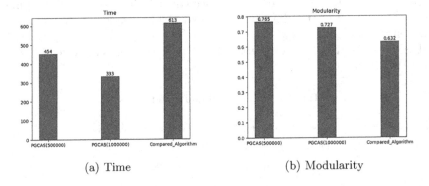

(a) Time   (b) Modularity

Fig. 1. Experimental result based on WDCM

In Figs. 1 and 2, it is obvious that no matter which value the parameter *clustnum* is set to, 500000 or 1000000, the running time of PGCAS algorithm is always less than the Compared_Algorithm. It means that the time-reducing methods adopted by the proposed PGCAS algorithm, including filtering 1 degree vertices and clustering multiple vertices at one time, can reduce the running time of the proposed algorithm efficiently. Besides, modularity values of the clustering results of PGCAS algorithm based on different parameter values are greater than modularity value of the clustering result of Compared_Algorithm. It means that clustering result of the proposed algorithm is better than the clustering result of Compared_Algorithm.

At the same time, it is obvious that PGCAS algorithm gets different results by setting different values to parameter *clustnum*. The running time of PGCAS(500000) is longer than PGCAS(1000000). But the modularity value

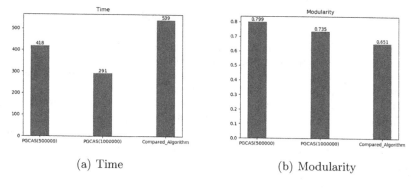

(a) Time   (b) Modularity

Fig. 2. Experimental result based on YAGO

of PGCAS(500000) is higher than PGCAS(1000000). So the clustering result of PGCAS(500000) is better than PGCAS(1000000). It means that, although when parameter *clustnum* is set to 1000000, the running time of PGCAS algorithm can be shortened efficiently, the clustering result becomes coarser.

4 Conclusions

A new parallelized graph clustering algorithm is proposed based on Spark in this paper. In the proposed algorithm some new methods are adopted to shorten the execution time of the algorithm. In the experimental result, it is obvious that the algorithm proposed in this paper runs faster and can get better clustering result.

Acknowledgements. The work is supported by the National Key Research and Development Plan under grant No. 2016YFB 1000600.

References

1. Huttlin, E.L., Bruckner, R.J., Paulo, J.A., et al.: Architecture of the human inter-actome defines protein communities and disease networks. Nature **545**(7655), 505 (2017)
2. Vehlow, C., Beck, F., Auwärter, P., et al.: Visualizing the evolution of communities in dynamic graphs. Comput. Graphics Forum **34**(1), 277–288 (2015)
3. Krishnamurthy, B., Wang, J.: On network-aware clustering of web clients. ACM SIGCOMM Comput. Commun. Rev. **30**(4), 97–110 (2000)
4. Wickramaarachchi, C., Frincu, M., Small, P., et al.: Fast parallel algorithm for unfolding of communities in large graphs. In: High Performance Extreme Comput-ing Conference (HPEC), pp. 1–6. IEEE (2014)
5. Lu, H., Halappanavar, M., Kalyanaraman, A.: Parallel heuristics for scalable com-munity detection. Parallel Comput. **47**, 19–37 (2015)
6. Moon, S., Lee, J.G., Kang, M.: Scalable community detection from networks by computing edge betweenness on MapReduce. In: International Conference on Big Data and Smart Computing (BIGCOMP), pp. 145–148. IEEE (2014)

7. Shi, J., Xue, W., Wang, W., et al.: Scalable community detection in massive social networks using MapReduce. IBM J. Res. Dev. **57**(3/4), 1–12 (2013)
8. Chen, Y., Huang, C., Zhai, K.: Scalable community detection algorithm with MapReduce. Commun. ACM **53**, 359–366 (2009)
9. Girvan, M., Newman, M.E.J.: Community structure in social and biological networks. Proc. Nat. Acad. Sci. **99**(12), 7821–7826 (2002)
10. Peel, L., Larremore, D.B., Clauset, A.: The ground truth about metadata and community detection in networks. Sci. Adv. **3**(5), e1602548 (2017)
11. Lancichinetti, A., Fortunato, S., Kertsz, J.: Detecting the overlapping and hierarchical community structure in complex networks. New J. Phys. **11**(3), 033015 (2009)
12. Abdelbary, H., El-Korany, A.: Semantic topics modeling approach for community detection. Int. J. Comput. Appl. **81**(6), 50–58 (2013)
13. Nguyen, T., Phung, D., Adams, B., et al.: Hyper-community detection in the blogosphere. In: Proceedings of Second ACM SIGMM Workshop on Social Media, pp. 21–26. ACM (2010)
14. Donetti, L., Munoz, M.A.: Detecting network communities: a new systematic and efficient algorithm. J. Stat. Mech: Theory Exp. **2004**(10), P10012 (2004)
15. Gulikers, L., Lelarge, M., Massoulié, L.: A spectral method for community detection in moderately sparse degree-corrected stochastic block models. Adv. Appl. Probab. **49**(3), 686–721 (2017)
16. Zhang, X., Newman, M.E.J.: Multiway spectral community detection in networks. Phys. Rev. E **92**(5), 052808 (2015)
17. Golub, G.H., Van Loan, C.F.: Matrix Computations. Johns Hopkins University Press, Baltimore (1996)
18. Zhang, Q., Qiu, Q., Guo, W., et al.: A social community detection algorithm based on parallel grey label propagation. Comput. Netw. **107**(P1), 133–143 (2016)
19. Newman, M.E.J., Girvan, M.: Finding and evaluating community structure in networks. Phys. Rev. E **69**(2), 026113 (2004)
20. Zaharia, M., Chowdhury, M., Franklin, M.J., et al.: Spark: cluster computing with working sets. HotCloud **10**(10–10), 95 (2010)

Query Associations Over Big Financial Knowledge Graph

Xiaofeng Ouyang, Liang Hong[(⊠)], and Lujia Zhang

School of Information Management, Wuhan University, Wuhan 430072, China
hong@whu.edu.cn

Abstract. Knowledge graph, as the core technology of artificial intelligence, is playing a more and more important role in the financial field. In this paper, we study the problem of querying associations over big financial knowledge graph formed by equity network. This type of queries is building block for many financial services, which reveal the complex equity structure that covers several vertices of interest including enterprises, shareholders and so on. Specifically, we propose efficient algorithms to find Top-k path with largest control power between two vertices, namely Dual-node Association Query (DAQ). Differently from typical path queries, first, there are heterogeneous edges in financial knowledge graph, including shareholding and holding. Second, DAQ calculates path weights based on product rather than sum of edge weights on the path. Further, we propose an efficient algorithm for Multi-node Association Query (MAQ) that generalizes DAQ. Experimental evaluation and extensive case study on a real financial knowledge graph demonstrate the efficiency and effectiveness of the proposed algorithms.

Keywords: Financial knowledge graph · Equity association · Top-k path

1 Introduction

The rise of big data and artificial intelligence has prompted the financial industry to transform into intelligent finance. However, in Chinese financial sector, the accumulation of big data is not enough, and the labeled financial data corpus is in small number. In the face of massive multi-source heterogeneous data, knowledge graph is a powerful representation with semantic processing and open data organization capabilities. Financial knowledge graph can support multiple smart applications in large-scale financial fields, especially, we focus on equity risk control using equity-networks-based financial knowledge graph.

Equity structure is an important feature of the financial field. The primary significance of studying equity structure in enterprise governance is to provide relevant empirical evidence for the decision-making and management of relevant departments, especially for the optimization of equity structure and the improvement of enterprise governance structure of Listed Companies in China [14]. Equity ratio means the size of control power, if we find the connection of financial equity, we can grasp the basis for the formation and transmission of financial risks. From the perspective of finance, such as relation, equity and control, to the perspective of knowledge graph, we focus on the

© Springer Nature Switzerland AG 2019
J. Li et al. (Eds.): BigSDM 2018, LNCS 11473, pp. 199–211, 2019.
https://doi.org/10.1007/978-3-030-28061-1_21

query of equity structure and propose knowledge association query, it refers to the equity association between two or more entities, which include banks, enterprises and natural persons. In this way, we can find the Top-k largest equity control power chain between entities for association analysis of equity.

In terms of knowledge association query, we define the problem of finding the Top-k path with heterogeneous edges and largest product of weights between two vertices. The main challenge lies in two aspects. On the one hand, there are heterogeneous edges in the knowledge graph with different labels of shareholding or holding. The controlling power of shareholding is higher than holding. On the other hand, Literature [12, 13, 15–17] and so on all proposed their own query algorithms for find k shortest paths in directed graph, similar but different, the edge weight in financial knowledge graph does not represent the length of the path but the equity ratio. In the sense of financial equity, we need to calculation the product of all the edge weights on the path to get the weight of the path, and finally find Top-k equity chains with larger weight, where loops are allowed.

We present knowledge association query based on the financial knowledge graph and focus on equity structure. In particular, we modeled the problem as finding the Top-k path with largest equity control power path between two vertices. From the perspective of equity association between the two entities, we can generalize the equity association to multiple entities. We describe related query algorithms, the use of heuristic lead us to find the Top-k path with less time and memory. We test our algorithm on a real big financial knowledge graph, where it is indeed effective to find Top-k equity control power path.

The remainder of the paper is organized as follows. Section 2 reviews the literature and presents some k shortest path query algorithms. Section 3 defines the query method and describes query framework. Section 4 presents our algorithm. Section 5 discusses the empirical results and provides some robustness test results. Section 6 concludes this study.

2 Related Works

Financial Knowledge Graph

With the development of artificial intelligence, Semantic technologies in financial have recently received increasing attention from both the research and industrial side [6]. The concept of knowledge graph was originally proposed by Google Knowledge Graph project on 16 May 2012 [7]. DBpedia, Wikidata, Freebase are community-driven efforts to integrate hundreds of datasets into a general-purpose knowledge base [8]. There are some researches in the financial field. Song D et al. built and queried an enterprise knowledge graph, and proposed TR Discover, a natural language interface [4], Gao et al. studies the enterprise knowledge composition of business process and their application [5].

Financial Application

The earliest application of knowledge graph in financial field Garlik is the representative, the main business is online personal information monitoring. Besides, Dataminr, a real-time risk analysis company based on Twitter and other public information. Wen Yin Interconnection [1] is a company founded in Beijing in 2013, Through financial knowledge graph and financial semantic search, to help investors to target A shares, obtain new third board information, mine the information behind transaction data (such as abnormal transactions) and discover investment opportunities.

Graph Path Query

Recently, the research and application of large-scale knowledge graph has attracted enough attention in academia and industry. Meanwhile, graph path querying is a primary operation in the network graph space, for both real time querying and inferential analysis [9]. Nardelli E et al. presented a fast solution to the problem of finding a shortest path between two nodes in a graph [15], but query efficiency and accuracy need to be improved to meet the increasing query requests of data volume.

There are some algorithms to solve the k simple shortest path problem. Yen et al. proposes Yen's algorithm, to find k shortest paths without loop from the origin to the sink [13], which is the earliest algorithm to use migration path to solve k simple shortest path problem. Martins et al. improved Yen algorithm and proposed MPS algorithm [18]. Hershberger et al. present an algorithm to enumerate the k shortest simple paths in a directed graph [12], the difference between Hershberger and Yen algorithm is that Hershberger divides candidate paths into several equivalent classes, uses alternative path algorithm to find the shortest path of each equivalent class and stores it in the heap.

Hart et al. put forward a heuristic search algorithm A* algorithm, which laid a foundation for heuristic search [17]. Eppstein gives algorithms called EA algorithm for finding the k shortest paths (not required to be simple) connecting a pair of vertices in a digraph [16], But the time cost of building $\mathcal{P}(G)$ is expensive. Jimenez et al. noticed the shortcomings of EA algorithm and proposed a lazy version of EA algorithm called LVEA algorithm [19]. Aljazzar present K* algorithm which was inspired by EA, differently, it apply A* to graph rather Dijkstra. It can operate on-the-fly with less time and memory than EA [10]. All of the above methods are based on the edge weight as the path length to calculate the sum of edges. However, our edge right represents the equity ratio, the edge weights should be the multiplied rather than sum, so, their algorithm can't address our problem directly.

3   Preliminaries

In this section, we start with the underlying RDF graph, and then map the RDF graph into an abstract model to facilitate the definition of our subsequent problems. Next, we define the corresponding query methods for financial problems, then show the correlation between algorithms. In the following, all our queries are based on the big financial knowledge graph. First, we introduce query mapping.

3.1   Query Mapping

Equity structure is an important feature of the financial industry. Therefore, we constructed a big financial knowledge graph based on equity structure. For example, we inquired about the equity association between China Merchants Bank Company and China Merchants Steamship Company. As shown in Fig. 1 below, entities are transformed into vertices in the abstract models. Edges are the relationships in RDF, and edge weights are labels on edges, p_{ij} is the equity ratios and c_{ij} is the semantic label. So the edge weights can be expressed as:

$$w_{ij}(p_{ij}, c_{ij})$$

The red circles in Fig. 1(a) correspond to start node v_0 and end node v_5 in Fig. 1(b) respectively. There is a loop in the graph. because of the phenomenon of cross-shareholding, that is, two or more entities holding equity each other.

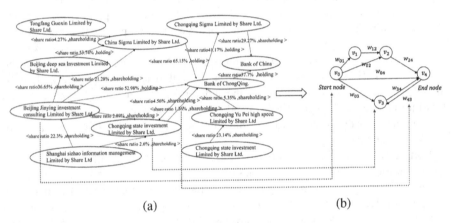

(a)   (b)

Fig. 1. Mapping RDF graph to abstract models (Color figure online)

3.2   Problem Definition

Knowledge association query mainly refers to the equity relationship among entities including enterprises, financial institutions, or natural person (shareholders, supervisors) and the relationship between shareholding and holding; identifying the relationship between the two enterprises' control and being controlled, and even further judging the degree of control. In the financial knowledge graph, the algorithm of graph traversal can be used to find the k largest equity paths between entities conveniently, and measure the intimacy between entities. The proportion of equity is larger, the capital flow in the equity chain between the two entities is more, therefore, it has stronger control power.

Aiming at the problem of financial knowledge association, we propose two algorithms: Dual-nodes association query (DAQ), Multi-nodes association query (MAQ), The following are their definitions respectively.

Definition 1. Dual-nodes association query (DAQ): In an edge-weighted graph $G(V, E)$, its edge weights w_{ij} include p_{ij} and c_{ij}, p_{ij} equals the equity ratio from v_i to v_j, $p_{ij} \in [0,1]$. c_{ij} represents the label on the edge, give dual nodes, a path from $srt = v_s \in V$ to $end = v_t \in V$. is a node sequence $(v_s, v_1, \ldots, v_t)$ for which $(v_i, v_j) \in E$. The length of a path P is the product of its edge weights ∂P.. so, We can find the k largest equity paths $EPG(v_s, v_t)$. As shown in Fig. 2, v_0 is start node, v_4 is end node.

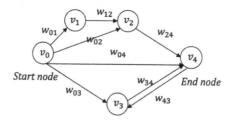

Fig. 2. Dual-nodes association query (DAQ)

Definition 2. Multi-node association query (MAQ): in order to query the number of associated nodes. In an edge-weighted graph $G = (V, E)$, give n nodes, assumed n = 3, query node set V_S (v_s, v_m, v_t), Select a pair of nodes from the set, a path from $start = v_s \in V$ to middle $= v_m \in V$ and then to $end = v_t \in V$ is a node . sequence $(v_s, v_1, \ldots, v_m, \ldots, v_t)$ for which $(v_i, v_j) \in E$.we can find $EPG(v_s, v_m)$ and $EPG(v_m, v_t)$, join all path. Then sort Top-k shareholding paths $CEPG(v_s, v_m, v_t)$. As shown in Fig. 3, the start node is v_0, the middle node is v_4, the end node is v_6.

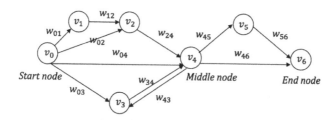

Fig. 3. Multi-nodes association query (MAQ)

We define two queries, namely DAQ, MAQ. There is actually an interrelated process between them. MAQ is a fusion operation of DAQ. The relationship between multiple entities can be queried through the association path between two entities, then union path, finally the association paths of multiple enterprises or banks are sorted out. The algorithm supports the upper financial knowledge service, DAQ and MAQ support knowledge association query.

4   Query Processing

4.1   Dual-Nodes Association Query (DAQ)

In order to find out more equity relationships between two enterprises or banks, we look for Top-k equity chains between two entities. Unlike the algorithms of finding k shortest paths, there are two challenges in our algorithms. Firstly, in the knowledge graph, there are heterogeneous edges, each side has label, that is, the difference between shareholding and holding. The controlling power of shareholding is higher than holding.

So, the label on edge, i.e., c_{ij} can be quantified as follows.

$$c_{ij} = \{shareholding : 0,\ holding : 1\}$$

Secondly, our edge weight does not represent the length of the path, but represents the equity ratio. Therefore, in the sense of financial equity, we need to calculate the product of all the edge weights on the path, and finally Top-k equity chains with larger weight are output for users to carry out equity analysis. The formula for weight is as follows.

$$\partial P = \prod_{(v_i,v_j \in P)} p_{ij}$$

In order to solve the problem of the product of edge weights, we carried out relevant research. Considering that multiplying edge weights will be very costly, so we changed our mind, that is, to transform the problem of the product of edge weights into the problem of the sum of edge weights. We can do a transformation of p_{ij}, that is,

$$E_{ij} = -\ln p_{ij}$$

So, the sum of T_{ij} is as follows.

$$E(P) = \sum T_{ij} = -\ln(\partial(P))$$

The maximum value of $\partial(P)$ corresponds to the minimum value of the sum of T_{ij}. Thus, we can successfully convert the path with the Top-k largest product into the problem with the Top-k shortest path.

The design of DAQ algorithm was inspired by A* algorithm. Thus, we use the heuristic evaluation function $f(v)$ in our algorithm, which is computed as the sum of two functions $g(v)$ and $h(v)$:

$$f(v) = g(v) + h(v)$$

The function $f(v)$ gives the estimated length from v_s to end node v_t through v, Where $g(v)$ is the actual length from start node v_s to v, and $h(v)$ is the estimated length from current node v to v_t.

Algorithm 1: Find Top-k dual-nodes equity control power chain

Input: a knowledge graph $G = (V, E)$, start node v_s, end node v_t, natural number k

Output: Top-k ranking path $EPG(v_s, v_t), \partial(P), \alpha$

1: $open_p \leftarrow$ empty priority queue.
2: $closed_p \leftarrow$ empty priority queue.
3: P $\leftarrow$ empty list.
4: Reverse running Dijkstra on G until $h(v)$ was calculated for all vertex
5: **if** v_t was not reached
6:     **then** exit without a path between v_s and v_t
7: Add v_s to $open_p$.
8: **while** $open_p$ is not empty **do**
9:         **foreach** find the all contact vertex v_{next} of v **do**
10:            Let v be the head of search queue $open_p$
11:            Find through the contact vertex v_{next} of v
12:            delete v in $open_p$ and add v in $closed_p$
13:            Add v_{next} to $open_p$.
14:            $e(v, v_{next}) = -\ln(p(v, v_{next}))$
15:            $g(v_{next}) = g(v) + e(v, v_{next})$
16:            $f(v_{next}) = g(v_{next}) + h(v_{next})$
17:            m is the number of the vertex in P
18:            $\alpha \leftarrow \sum c_{ij}/m$
19:        **if** $v == v_t$
20:            **then** count++.
21:        **if** count $==$ k **then** go to line 22
22: **Return** Top-k $EPG(v_s, v_t), \partial(P), \alpha$

In algorithm 1, we first reverse running Dijkstra algorithm to calculate the $h(v)$ value of each node v to v_t, create a new priority queue and add the source point v_s to the queue. Use the head of the $open_p$ as an extended node to find all its connection points v_{next}, calculate the value of $g(v_{next}), f(v_{next})$ and α. the α is the ratio of holding edge to total number of paths m. The point v pops up from the $open_p$, and the priority rule $(c > f > g)$, if c of v_1 is 1, c of v_2 is 0, we will choose v_1. Because the control power of holding is stronger than shareholding. For f and g, we choose smaller ones. If the current node v is v_t, the length of the current path is the path from v_s to v_t, which will add to P. when the count of path equals to k or $open_p$ is empty, output Top-k EPG (v_s, v_t) which are according to α and $\partial(P)$, $\partial(P)$ is equals to $\exp(-E(p))$.

According to the designed algorithm, Top-k paths can be obtained. The real path is reserved, while the dashed path is discarded. As shown in the following Fig. 4, we can find Top-k equity chain from Beijing Jinying investment consulting company to Chongqing state investment company to Chongqing Bank, when k = 3, the path is discarded.

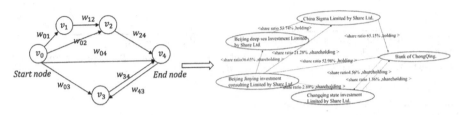

Fig. 4. The Top-k equity chain between the two companies

4.2   Multi-nodes Association Query (MAQ)

On the basis of algorithm 1, we can find the relationship between Top-k shareholding chains among multiple nodes. Firstly, we can deduce the results from local optimum and global optimum, It is proved as follows.

Theorem 1. In any case, the Top-k path between v_s, v_m and v_t must be the union of the Top-k $EPG(v_s, v_m)$ and Top-k $EPG(v_m, v_t)$, and the maximum K × K path can be formed from which Top-k paths of v_s, v_m and v_t can be selected.

Proof: We assumed that there's a Top-k paths of v_s, v_m and v_t not in the union of Top-k $EPG(v_s, v_m)$ and Top-k $EPG(v_m, v_t)$, $\partial(P_k) < \partial(P_{k-1}) < \ldots < \partial(P_1)$, so top-k paths of v_s, v_m and v_t exits a path $P_{ij} = \{P_i(v_0, v_m), P_j(v_m, v_h)\}$. $\partial(P_i) < \partial(P_k)$, $\partial(P_{ij}) = \partial(P_i) *\partial(P_j)$, so $\partial(P_i) * \partial(P_j) < \partial(P_k) * \partial(P_j) < \partial(P_{k-1}) * \partial(P_j) < \ldots < \partial(P_1) * \partial(P_j)$, so P_{ij} is not top-k path, so the assumption is not true, the Top-k path between v_s, v_m and v_t must be the union of the Top-k $EPG(v_s, v_m)$ and Top-k $EPG(v_m, v_t)$

The equity chains between multiple nodes are dual nodes of joint equity chains. Therefore, we can use algorithm 1 many times to get $EPG(v_s, v_t)$, $\partial(P)$, then we can get the chain of Top-k among multiple nodes by judging the combination.

Algorithm 2: Find Top-k multi-nodes equity control power chain

Input: a knowledge graph $G = (V, E)$, v_s, v_m, v_t, natural number k
Output: Top-k ranking path $CEPG(v_s, v_m, v_t)$, $\partial(P)$

1: Initialize candidate path list P_1 for path from v_0 to v_m in Q
2:   call algorithm1 (v_s, v_m)
3:   **Return** Top-k $EPG(v_s, v_m)$, $\partial(P)$
4. Initialize candidate path list P_2 for path from v_m to v_t in Q
5:   call algorithm1 (v_m, v_t)
6:   **Return** Top-k $EPG(v_m, v_t)$, $\partial(P)$
7: $EPG(v_s, v_m)$ union $EPG(v_m, v_t)$
8:   find Top-k path
9: **Return** Top-k $CEPG(v_s, v_m, v_t)$, $\partial(P)$, α

According to the designed algorithm, Top-k paths can be obtained. As shown in the following Fig. 5, we can find Top-k equity chain among companies. The path from Beijing Jinying investment consulting company to Chongqing Bank to China Bank, when k = 5, the Top-k path is as follows.

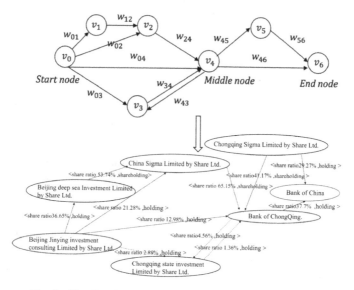

Fig. 5. The Top-k equity chain between the multi companies

5   Experiments

5.1   Dataset

We used a large dataset to test the different times of our query algorithms and analyze the effectiveness of the algorithms in different situations. In the process of experiments, the entities can be randomly selected from the graph to generate the query, and we also selected specific entities to analyze the application of algorithm in finance (Table 1).

Table 1. Statistics of the data sets in graph database.

Dataset	Number
Triple	293580726
Entity	52309823
Subject	35641336

5.2   Query Efficiency

We test the query efficiency of our algorithm. The first to test is a relational query for two vertices. We randomly selected 10 companies from the dataset in graph database, and then tested the efficiency of query using our DAQ algorithm. As shown in the Fig. 6 below, using the DAQ algorithm is faster in time, the slowest time do not exceed 0.6 s. For we have a huge number of data and randomly selected data, the performance of the algorithm is pretty good.

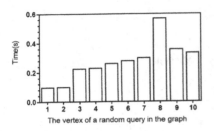
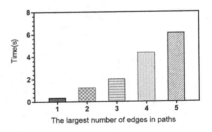

Fig. 6. Two vertex relationship query times

Fig. 7. The average time of the longest edge number query in the path

As shown in Fig. 7, after the overall query, we randomly query 100 times in the graph and calculate the average response time according to the largest number of edges in the two companies' paths. It can be clearly seen that the response time grows gradually with the increase of number of edges. Because the more edges in the path, the longer it takes to traverse.

We not only analyzed the query time, but also tested the number of vertices randomly in different cases. Randomly select 100 pairs of vertices, we counted the number of paths produced and then count the number of vertices with different paths. As shown in the figure below, in the selected 100 pairs of vertices, the most number is without shareholding chain relationship, the next is three-path relationship. We did not find four-path and five-path vertex pairs (Fig. 8).

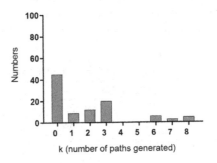

Fig. 8. The numbers of different k paths

5.3 Case Study

We illustrate how DAQ algorithm works using the following case. Through the query of China Merchants Bank Co., Ltd. and China Merchants Steamship Co., Ltd., the relationship between the two enterprises' control and controlled can be clearly identified, and the degree of control can be judged by the equity calculation on the equity chain. First, we get the abstract model by mapping as shown in Fig. 9.

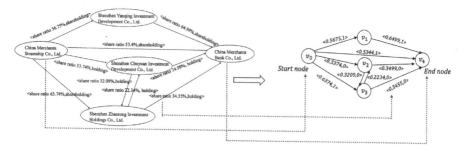

Fig. 9. Two companies equity chain graph

Then, we do the edge weight conversion and reverse running Dijkstra to calculate $h(v)$ value from each point to the end point v_4 (Fig. 10).

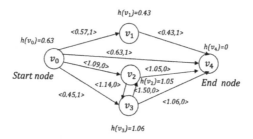

Fig. 10. Graph with h(v) after transformation

After that, we extended nodes are based on v_0, so, the $v_{next}(v,g,f,c)$ are $<v_4,0.63,0.63,1>$, $<v_1,0.57,1,1>$, $<v_2,1.14,2.19,0>$, $<v_3,0.45,1.51,1>$, the v_4 is equal to v_t, so, the top-1 is found. Next, We compute v_1 as the next extended node, and so on, until K paths are found. We can easily see the Top-k control paths enumerated in Table 2.

Table 2. Top-5 equity chain

Path	$v_0 - v_4$ Path	α	$E(P)$	$\partial(P)$
p_1	v_0, v_4	1	0.63	0.534
p_2	v_0, v_1, v_4	1	1	0.369
p_3	v_0, v_3, v_4	2/3	1.51	0.220
p_4	v_0, v_3, v_2, v_4	1/2	3	0.050
p_5	v_0, v_3, v_2, v_3, v_4	2/5	4.15	0.016

6   Conclusion

In this paper, we propose query associations based a big financial knowledge graph, define knowledge association query to analyze the financial equity structure. In the proposed two algorithms are divided into DAQ, MAQ, which find out the Top-k largest equity chains in two or more nodes, we combined with the semantic label, adds a semantic priority sequence based on the edge weight ratio, which can effectively find the k shortest paths. The control power of the first k paths is weakened in turn, providing strong support for financial risks. The proposed approach is generalized for a family of path-based metrics, allowing it to support a very large graph.

Acknowledgments. The research is supported in part by National Science Foundation, China No. 91646206, and National Science Foundation, Hubei Province, No. 682, 2018.

References

1. Bao, J.: How can knowledge graph help to realize intelligent finance. Gold Card Proj. (7), 45–49 (2016)
2. Chao, Y., Wang, H.: Developed Dijkstra shortest path search algorithm and simulation. In: International Conference on Computer Design and Applications, pp. V1-116–V1-119. IEEE (2010)
3. Myoupo, J.F., Fabret, A.C.: A modular systolic linearization of the Warshall-Floyd algorithm. IEEE Trans. Parallel Distrib. Syst. 7(5), 449–455 (2002)
4. Song, D., Schilder, F., Hertz, S., et al.: Building and Querying an Enterprise Knowledge Graph. IEEE Trans. Serv. Comput. **PP**(99), 1 (1939)
5. Gao, J., Fu, X., Liang, Y., et al.: The construction of enterprise knowledge map based on business process. International Conference on Business Intelligence & Financial Engineering, pp. 285–289. IEEE (2012)
6. Galkin, M., Auer, S., Scerri, S.: Enterprise knowledge graphs: a backbone of linked enterprise data. In: IEEE/WIC/ACM International Conference on Web Intelligence, pp. 497–502. IEEE (2017)
7. Wu, T., Qi, G., Li, C., et al.: A survey of techniques for constructing Chinese knowledge graphs and their applications. Sustainability **10**(9), 3245 (2018)
8. Bizer, C., et al.: Dbpedia-a crystallization point for the web of data. Web Seman.: Sci. Serv. Agents World Wide Web 7(3), 154–165 (2009)
9. Balaji, J., Sunderraman, R.: Distributed graph path queries using spark. In: Computer Software and Applications Conference, pp. 326–331. IEEE (2016)
10. Aljazzar, H., Leue, S.: K*: a heuristic search algorithm for finding the k shortest paths. Artif. Intell. **175**(18), 2129–2154 (2011)
11. Yang, S., Han, F., Wu, Y., et al.: Fast top-k search in knowledge graphs. In: IEEE International Conference on Data Engineering, pp. 990–1001. IEEE (2016)
12. Hershberger, J., Maxel, M., Suri, S.: Finding the k shortest simple paths: a new algorithm and its implementation. ACM (2007)
13. Yen, J.Y.: Finding the K shortest loopless paths in a network. Manag. Sci. **17**(11), 712–716 (1971)

14. Wang, H., Guoliu, H.: The role and efficiency of ownership structure in corporate governance-literature review and future research direction based on Chinese listed companies. J. Hunan Univ. (Soc. Sci. Ed.) **18**(3), 42–47 (2004)
15. Nardelli, E., Proietti, G., Widmayer, P.: Finding the most vital node of a shortest path. Theor. Comput. Sci. **296**(1), 278–287 (2001)
16. Eppstein, D.: Finding the k shortest paths. In: Symposium on Foundations of Computer Science, pp. 154–165. IEEE Computer Society (1994)
17. Hart, P.E., Nilsson, N.J., Raphael, B.: A formal basis for the heuristic determination of minimum cost paths. IEEE Trans. Syst. Sci. Cybern. **4**(37), 28–29 (1972)
18. Martins, E.Q.V., Pascoal, M.M.B.: A new implementation of Yens ranking loopless paths algorithm. Q. J. Belgian French Italian Oper. Res. Soc. **1**(2), 121–133 (2003)
19. Jiménez, V.M., Marzal, A.: A lazy version of Eppstein's K shortest paths algorithm. In: Jansen, K., Margraf, M., Mastrolilli, M., Rolim, J.D.P. (eds.) WEA 2003. LNCS, vol. 2647, pp. 179–191. Springer, Heidelberg (2003). https://doi.org/10.1007/3-540-44867-5_14

CMCloud: An Autonomous Monitoring System for Big Data Processing Platform

Zhonglei Fan$^{(\boxtimes)}$, Mengru Xu, Jialin Xi, and Donghai Li

Chang'An University, Xi'an 710064, China
zlfan@chd.edu.cn

Abstract. Big data analytic needs a reliable data processing platform, which usually consists of large amount of distributed monitored objects, sometimes geographically dispersed ones. The rapidly increasing scale and complexity of a big data processing platform are making autonomous monitoring and management become much more crucial than before. In this paper, we design and implement an autonomous monitoring system - CMCloud to deal with these challenges faced by current big data processing platform. By introducing sequential flow control for multi-step operations of an action, CMCloud implements autonomous interaction between monitoring server and monitored objects as well as automatic fault diagnosis and recovery. CMCloud can be deployed into a big data processing platform to find, locate and process potential system faults timely and precisely, and then enhance the reliability of it.

Keywords: Big data analytic · Big data processing platform · Reliability · Autonomous monitoring · Fault diagnosis

1 Introduction

A big data processing platform usually includes large amount of distributed and changing objects such as physical servers, physical storage, physical network, virtual machines, virtual volumes, virtual network and so forth [1]. Monitoring and managing these objects to make them work reliably is an essential requirement for a big data processing platform. However, with the rapidly increasing scale and complexity in a big data processing platform, current monitoring systems that mostly employ manual intervention or weakly automatic processing will no longer meet this requirement. To deal with it, autonomous monitoring and management system with proactive, automatic and rapid fault processing is proposed as a promising solution and becoming much more crucial than before [2].

There are lots of researches on autonomous monitoring and management from different perspectives. MON [3] builds an on-demand management overlay network on the PlanetLab [4] that allows users to execute instant management commands, such as querying the current status of the application, pushing software updates to all the nodes and so forth. By this method, MON can achieve automatic fault elimination to some degree, but it does not provide a general approach, and its functionalities are not powerful enough to achieve automatic fault elimination.

© Springer Nature Switzerland AG 2019
J. Li et al. (Eds.): BigSDM 2018, LNCS 11473, pp. 212–223, 2019.
https://doi.org/10.1007/978-3-030-28061-1_22

Also, there are many excellent open-source monitoring systems to emphasize on these challenges faced by large distributed systems such as big data processing platforms. Zabbix [5, 6], ganglia and Nagios [7–9] are typical and widely adopted by lots of applications. However, in terms of automatic fault elimination, they employ a weak mechanism so that they are not able to meet requirements of complex scenarios of a big data processing platform. For example, Zabbix allows only single-way operations (commands or scripts) triggered by event to be executed on a monitored node; monitoring server has no message whether some operation is executed correctly and no control of the executing order of operations on a monitored node. An active management framework (AMF) with two-way information channel was presented in [10] to solve the issue of single-way operation mentioned above, but it did not present an effectively sequential flow control for operations, and thus limits its application in big data processing platform.

In this paper, we design and implement an autonomous monitoring system - CMCloud to deal with these challenges faced by current big data processing platform. By introducing sequential flow control for multi-step operations of an action, CMCloud implements autonomous interaction between monitoring server and monitored objects and automatic fault diagnosis and recovery.

The rest of this paper is organized as follows. Section 2 introduces the autonomous monitoring architecture of CMCloud and its distinctive features. Section 3 discusses the implementation of CMCloud. Section 4 evaluates CMCloud by deploying it into our self-developed CSCloud storage system for testing via representative experiments. Section 5 concludes.

2 Autonomous Monitoring Architecture

The autonomous monitoring architecture of CMCloud is shown in Fig. 1 consisting of three major subsystems: Frontend subsystem, Backend subsystem and Client subsystem (see Fig. 1).

The Frontend subsystem provides web-based interface for configuration and display of monitoring and control information. Backend subsystem includes Monitoring Information Manager (MIM), Flow Control Manager (FCM) and Artificial Intelligence Libraries Manager (AILM). Based on configuration of the Frontend subsystem, MIM collects and saves data from local or remote monitored objects, and then create triggers for monitored objects by defined threshold. By providing sequential flow control for multi-step operations of an action defined by the Frontend subsystem and executed by triggers created in MIM, FCM realizes autonomous interaction between monitoring server and monitored objects as well as automatic fault diagnosis and recovery. AILM is responsible for managing handlers of fault diagnosis and recovery called by FCM during execution procedure of actions, which enables proactive fault processing possible. The Client subsystem contains monitored objects and communicates with monitoring server via agent.

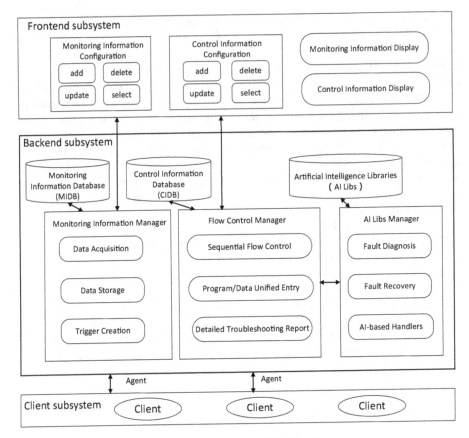

Fig. 1. Architecture of CMCloud

2.1 Sequential Flow Control

The sequential flow control of an action is defined by the Frontend subsystem per specific fault and executed in case the preset triggering conditions is satisfied. Each action consists of multi-step operations, and each operation owns a specific handle in AI Libs and is executed sequentially based on its priority and execution time. Each operation has return values after execution and the next operation is executed based on the result of the previous operation. All operations are configured in sequence by software-defined control flow. This flow is iterative until all operations are executed. If a fault is resolved after some operation is executed, next operation will abort and operation with lower priority is no longer executed. Sequential processing flow of an action is shown in Fig. 2.

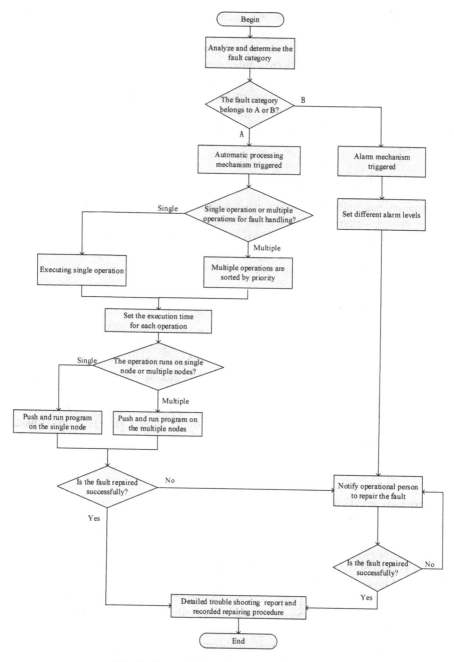

Fig. 2. Sequential processing flow of an action

In order to combine automatic and manual fault processing complementarily, fault processing are categorized into two categories: (A) Pre-defined automatic processing flow and (B) Unknown processing flow. According to analysis and diagnosis result of monitored objects, FCM can decide which category should be adopted. If a fault falls into category A, an automatic fault handling mechanism will be triggered. The failure recovery may require one or multi-step operations to be executed in some sequence and each operation needs to be set priority, programming language, execution time, objects location, etc. If a fault fall into category B, a fault alarm mechanism is triggered to ask for manual repairing by administration and maintenance persons and then the fault repair procedure will be recorded. If similar faults occur again, automatic recovery mechanism can be used to recover from faults without notification for manual intervention.

2.2   Program/Data Unified Entry

The program/data unified entry module is a manager for program and data communication between monitoring server and monitored objects. Monitoring server sends processing programs to monitored objects and gets return value from monitored objects after executing these processing programs through this unified entry. FCM encapsulates the entire process and make it transparent to users.

2.3   Artificial Intelligence Libraries Manager

The Artificial Intelligence Libraries (AI Libs) Manager is responsible for managing and providing handlers in AI Libs called by FCM for fault diagnosis and recovery during the processing procedure of an action.

CMCloud is designed for large distributed systems, so causes of each fault are complex as well as processing procedures. By employing AI-based handlers, root causes of faults can be found precisely and quickly, and faults can be eliminated timely and effectively. The prediction capability offered by AI gives a great help on AI-based handlers to detect potential faults and eliminate them proactively, and thus increase reliability and robustness of systems.

3   Implementation

CMCloud is implemented under Linux environment, using Python [11] as the development language and LAMP [12] and Django [13] as the frameworks.

Zabbix [10] provides efficient data collection, flexible triggers, highly customizable alarms, real-time drawing capabilities, multiple visual presentations, storage of historical data, and fast third-party API based on HTTP protocol and JSON-RPC. CMCloud utilizes some components of Zabbix to reduce the development burden, e.g. MIM uses API to monitor underlying items.

As shown in Fig. 3, FCM receives configuration information of an action from the Frontend subsystem including action name, host that triggers this action, monitoring item, trigger condition, and operation type (remotely executing or sending message). As shown in Fig. 4, if remote execution is chosen, FCM will present configuration interface for remote execution, including step number, step duration, host on which remote operation is executed, and methods of providing operation program/handler.

FCM supports two methods for users to provide customized programs/handles: file uploading and online editing.

After the above configuration is completed, FCM will control defined operations in an action to be executed step-by-step autonomously. Monitoring server delivers operation programs/handles to monitored nodes in two ways: uploading compiled processing programs at local storage to monitored nodes or editing programs online and uploading them directly to monitored hosts.

Autonomous interaction mainly depends on how to gather return value of each operation to guide the following operations. CMCloud sets up a YAML [14] manager to read and write results of each operation. Return value of each operation of monitored objects is stored in a specified YAML file, and then return back. Flow of sequential operation consists of a set of steps with strict priorities. The highest priority is A1, and the descending order of priority is automatically named A2, A3... An. YAML manager writes the return value of the previous operation A1 to the YAML file, and then the next operation A2 automatically reads the YAML file to obtain return value of A1. Return value of monitored object is passed to monitoring server via real-time delivery.

Fig. 3. Action configuration

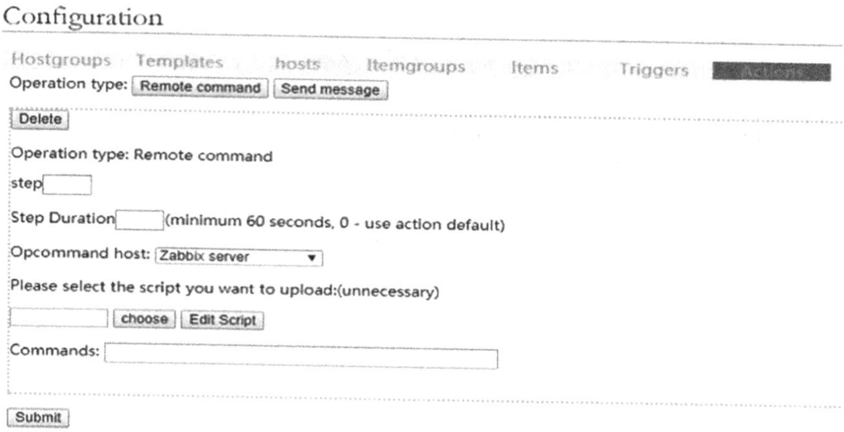

Fig. 4. Configuration interface of remote execution

4 Evaluation

In order to evaluate whether CMCloud is able to satisfy the expected targets for using in big data processing platforms, we deployed it into our self-developed cloud storage system – CSCloud for testing. CSCloud adopts scalable distributed storage architecture and consists of a lot of storage servers which work together to provide storage service. Dynamic extension by adding new storage servers and load balancing among existing storage servers are crucial features for CSCloud. Considering these two features are also crucial to other alike systems (e.g. big data processing platforms, cloud computing platforms), we choose them as our experimental scenarios to evaluate whether CMCloud can work well as predicted.

4.1 Dynamic Extension of Storage Server

To test dynamic extension, we build a system with four in-use storage servers and three standby servers that will be joined dynamically by CMCloud later. The information of each storage servers is shown in Table 1.

Table 1. Information of storage server

IP	Status	User data path
192.168.1.128	in-use	/data/
192.168.1.129	in-use	/data/
192.168.1.130	in-use	/data/
192.168.1.133	in-use	/data/
192.168.1.101	Standby	/data/
192.168.1.102	Standby	/data/
192.168.1.103	Standby	/data/

At the beginning, each storage server in CSCloud is in a light load status, around 2% in our experiment. The trigger threshold is configured as shown in Fig. 5. The threshold connection relationship among these four in-use storage servers is "and", which means the action is triggered when each threshold is higher than 50%. The flow configuration of sequential operation is shown in Fig. 6.

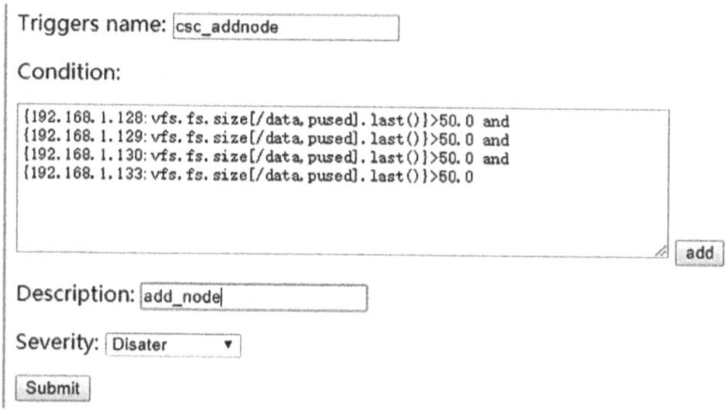

Triggers name: csc_addnode

Condition:

```
{192.168.1.128:vfs.fs.size[/data.pused].last()}>50.0 and
{192.168.1.129:vfs.fs.size[/data.pused].last()}>50.0 and
{192.168.1.130:vfs.fs.size[/data.pused].last()}>50.0 and
{192.168.1.133:vfs.fs.size[/data.pused].last()}>50.0
```

add

Description: add_node

Severity: Disater ▾

Submit

Fig. 5. Configuration of trigger threshold

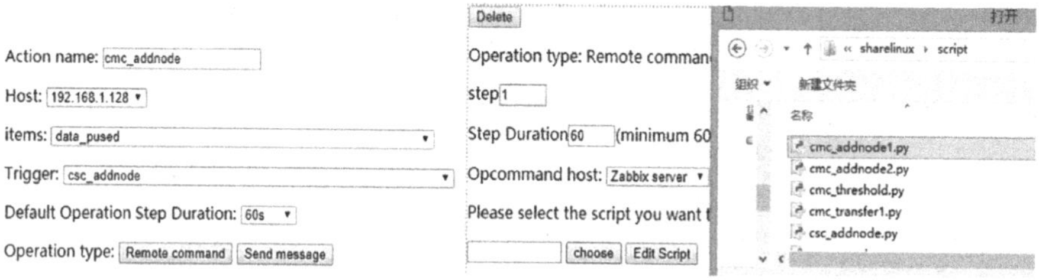

Action name: cmc_addnode

Host: 192.168.1.128 ▾

items: data_pused ▾

Trigger: csc_addnode ▾

Default Operation Step Duration: 60s ▾

Operation type: Remote command / Send message

Delete

Operation type: Remote command

step 1

Step Duration 60 (minimum 60

Opcommand host: Zabbix server ▾

Please select the script you want t

choose   Edit Script

打开

◄◄ sharelinux › script

组织 ▾   新建文件夹

名称

cmc_addnode1.py
cmc_addnode2.py
cmc_threshold.py
cmc_transfer1.py
csc_addnode.py

Fig. 6. Configuration of sequential operation flow

In order to simulate real application case, we register 100 different users in batches and let them ingest data concurrently to increase the overall load of the existing storage servers for simulating the overloaded status of storage servers. Once the overall load of the in-use storage servers is higher than the preset threshold, the predefined flow of sequential operation for automatic fault processing will be triggered. CMCloud will select a storage server from standby server group, install and run necessary software, add corresponding monitoring items and thresholds, and finally add this server to in-use server group. In our experiment, the standby server 192.168.1.101 is selected for extension, which can be found from the return value in Recovery Log.

The return value of the first action step is shown in the "Monitoring" column (shown in Fig. 7(a)). The first step is executed at 15:03:51. As the sequential operations have not been completed at this time, the trigger status is still "PROBLEM". But the return value shows that the 192.168.1.101 has been selected as the newly added storage server. The return value of the second action step is shown in the "Processed" column (shown in Fig. 7(b)). The processing is completed at 15:06:05, and the sequential operation is performed in multiple steps so the trigger status is changed to "OK". The return value shows that the csc and cmc-agent services have been installed on 192.168.1.101 and the process has been started.

Recover   Processed   Monitoring

Time	Action Name	Trigger	Action Descriptions	returnValue
2018.09.09 15:03:51	cmc_addnode	Trigger status: PROBLEM Trigger name: csc_addnode	Trigger severity: Disaster Action name: cmc_addnode Event ID: 2838 Event value: 1 Event status: PROBLEM Recovery time: 15:03:51 Recovery date: 2018.09.09 Item values: data_pused (192.168.1.128:vfs.fs.size[/data,pused]): 50.55 % Original event ID: 2838	Action Step: 1 ReturnValue:192.168.1.101 has been selected as the newly added storage server.

(a). Return value of the first action step

Recover   Processed   Monitoring

Time	Action Name	Trigger	Action Descriptions	returnValue
2018.09.09 15:06:05	cmc_addnode	Trigger status: OK Trigger name: csc_addnode	Trigger severity: Disaster Action name: cmc_addnode Event ID: 2838 Event value: 0 Event status: OK Recovery time: 15:06:05 Recovery date: 2018.09.09 Item values: data_pused (192.168.1.128:vfs.fs.size[/data,pused]): 50.55 % Original event ID: 2838	Action Step: 1 ReturnValue:192.168.1.101 has been selected as the newly added storage server. Action Step: 2 ReturnValue:192.168.1.101 has installed csc and cmc-agent services and started the process

(b). Return value of the second action step

Fig. 7. Return values of sequential action operations for dynamic extension

4.2   Load Migration

The dynamic extension of storage server is implemented above. However, the actual load has not been migrated to the new storage server. Load migration is employed to migrate data from the server with heavier load to the new extended storage server.

In practice, it takes a certain amount of time to install and configure services after the new storage server is added. In order to improve the system responsiveness, the migration threshold is set a little higher than the dynamic extension threshold, 55% in our experiment.

The trigger threshold for migration is configured as shown in Fig. 8. The threshold connection relationship among these five servers (containing the newly added server 192.168.1.101) is "or", which means the action is triggered as long as one storage server threshold is higher than 55%. The flow configuration of sequential operations is shown in Fig. 9.

Triggers name: `csc_transfer`

Condition:

```
{192.168.1.128: vfs.fs.size[/data,pused].last()}>55.0 or
{192.168.1.129: vfs.fs.size[/data,pused].last()}>55.0 or
{192.168.1.130: vfs.fs.size[/data,pused].last()}>55.0 or
{192.168.1.133: vfs.fs.size[/data,pused].last()}>55.0 or
{192.168.1.101: vfs.fs.size[/data,pused].last()}>55.0
```

add

Description: transfer

Severity: Disater ▼

Submit

Fig. 8. Configuration of migration threshold

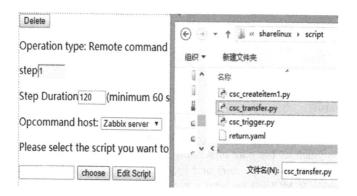

Fig. 9. Configuration of migration sequential operations

In our experiment, we ingest data into the 192.168.1.130 server to trigger the load migration operation by causing its load to be higher than 55%. The return value of sequential operation of load migration in Fig. 10 shows that transferring from 192.168.1.130 to 192.168.1.101 is successful.

Recover   Processed   Monitoring

Time	Action Name	Trigger	Action Descriptions	returnValue
2018.09.10 18:28:07	cmc_transfer	Trigger status: OK Trigger name: csc_transfer	Trigger severity: Disaster Action name: cmc_transfer Event ID: 2839 Event value: 0 Event status: OK Recovery time: 18:28:07 Recovery date: 2018.09.10 Item values: data_pused (192.168.1.130:vfs.fs.size[/data,pused]): 57.01 % Original event ID: 2839	Action Step: 1 ReturnValue:from 192.168.1.130 to 192.168.1.101 transfer successfully!

Fig. 10. Return value of sequential operation of load migration

Through dynamic extension of storage server and following load migration, the overall load of the CSCloud reaches a balanced distribution. The load values of the storage servers during the entire experimental scenarios are shown in Table 2, which prove CMCloud can autonomously realize dynamic extension and load migration of CSCloud based on real-time monitoring of the overall load information of CSCloud. When the load is unbalanced in CSCloud, the dynamic extension and load migration mechanism is automatically triggered by CMCloud to balance the load. Practice has proved that using CMCloud system, load balance can be reached with the shortest migration time, the minimum number of migrations, and the least labor cost.

Table 2. Load values of storage servers during entire experimental scenarios

Storage server	Beginning	After simulated overloaded status	After dynamic extension	Before load migration	After load migration
192.168.1.128	2.10%	50.05%	50.05%	50.05%	50.05%
192.168.1.129	2.45%	51.05%	50.05%	50.05%	50.05%
192.168.1.130	1.75%	55.05%	55.05%	57.06%	29.05%
192.168.1.133	1.95%	56.00%	50.05%	50.05%	50.05%
192.168.1.101	None	None	0.90%	0.90%	24.13%

5 Conclusion

CMCloud monitoring system is designed and implemented towards large-scale distributed system such as big data processing platforms, Cloud services, etc. Based on autonomous interaction architecture and sequential flow control for multi-step operations of an action, CMCloud presents a powerful distributed monitoring system with automatic fault detection and elimination. Fault processing in CMCloud is completely triggered by faults and is handled automatically without manual intervention, which make it become an efficient solution for big data processing platforms.

CMCloud proposes a new framework and idea for research, design and implementation of intelligent operation and maintenance platform. As deployed in existing systems, CMCloud can integrate with them seamlessly and endow them new abilities of autonomous and intelligent fault processing and enhance the reliability of them.

CMCloud is still under development and improvement such as evaluating accuracy and cost quantitatively, adding more fault types, extending AI Libs and so forth.

Acknowledgment. This work is partially supported by the Special Fund for Basic Scientific Research of Central Colleges, Chang'An University (CHD2011TD009). The authors also gratefully acknowledge the helpful comments and suggestions of the reviewers, which have improved the presentation.

References

1. EMC Education Services: Cloud Infrastructure and Services. EMC Corporation, Hopkinton Massachusetts (2014)
2. Kutare, M., Eisenhauer, G., Wand, C., Schwan, K., Talwar, V., Wolf, M.: Online monitoring and analytics for managing large scale data centers. In: Proceedings of the 7th International Conference on Autonomic Computing, pp. 141–150. ACM, US (2010)
3. Liang, J., Ko, S.Y., Gupta, I., Nahrstedt, K.: MON: on-demand overlays for distributed system management. In: WORLDS 2005: Second Workshop on Real, Large Distributed Systems, pp. 13–18. USENIX, San Francisco (2005)
4. PlanetLab Consortium. https://www.planet-lab.org/. Accessed 2017
5. Zabbix LLC. https://www.zabbix.com/. Accessed 2018
6. Wu, Z.: Zabbix Enterprise Distributed Monitoring System. Mechanical Industry Press, Beijing (2014)
7. Ganglia Community. http://ganglia.info/. Accessed 07 Mar 2018
8. Nagios Enterprises: https://www.nagios.org/. Accessed 2018
9. Zhang, X.Y., Chen, G.S.: Intelligent monitoring system on cloud computing platform based on Ganglia and Nagios. J. Anhui Univ. Sci. Technol. (Nat. Sci.) **36**(4), 69–74 (2016)
10. Fan, Z.: An active management framework for automatic fault detection and elimination in distributed systems. ICIC Express Lett. **2**(1), 31–35 (2011)
11. Python Software Foundation. https://www.python.org/. Accessed 2018
12. Wikipedia. https://en.wikipedia.org/wiki/LAMP (software_bundle). Accessed 2018
13. Django Software Foundation. https://www.djangoproject.com/. Accessed 2018
14. Ben-Kiki, O., Evans, C., döt Net, I.: http://www.yaml.org/. Accessed 2018

Short-Timescale Gravitational Microlensing Events Prediction with ARIMA-LSTM and ARIMA-GRU Hybrid Model

Ying Sun, Zijun Zhao, Xiaobin Ma, and Zhihui Du[✉]

Department of Computer Science and Technology, Tsinghua University, Beijing, China
duzh@tsinghua.edu.cn

Abstract. Astronomers hope to give early warnings based on light-detection data when some celestial bodies may behave abnormal in the near future, which provides a new method to detect low-mass, free-floating planets. In particular, to search short-timescale microlensing (ML) events from high-cadence and wide-field survey in real time, we combined ARIMA with LSTM and GRU recurrent neural networks (RNN) to monitor all the observed light curves and to alert before abnormal deviation. Using the good linear fitting ability of ARIMA and the strong nonlinear mapping ability of LSTM and GRU, we can form an efficient method better than single RNN network on accuracy, time consuming and computing complexity. ARIMA can reach smaller alerting time and operating time, yet costing high false prediction rate. By sacrificing 15% operating time, hybrid models of ARIMA and LSTM or GRU can achieve improved 14.5% and 13.2% accuracy, respectively. Our work also provide contrast on LSTM and GRU, while the first type is commonly used for time series predicting systems, the latter is more novel. We proved that in the case of abnormal detection of light curves, GRU can be more suitable to apply to as it is less time consuming by 8% while yielding similar results as LSTM. We can draw a conclusion that in the case for short-timescale gravitational microlensing events prediction, hybrid models of ARIMA-LSTM and ARIMA-GRU perform better than separate models. If we concentrate more on accuracy, ARIMA-LSTM is the best option; on the other hand, if we concentrate more on time consuming, ARIMA-GRU can save more time.

Keywords: Gravitational lensing · Recurrent neural networks · ARIMA · Time series prediction and alarming

1 Introduction

Astronomy is the origin of information explosion, and it is the first field to meet the challenge of big data [1]. In the 21st century, astronomical data is growing

© Springer Nature Switzerland AG 2019
J. Li et al. (Eds.): BigSDM 2018, LNCS 11473, pp. 224–238, 2019.
https://doi.org/10.1007/978-3-030-28061-1_23

at a rate of terabytes or even PB. By 2010, the information file had been as high as 1.4×242 bytes. The ground-based Wide-angle Camera array (GWAC) [15] is part of the SVOM space project, which searches for various types of optical transient sources by continuously imaging the 5000 square-degree field of view (FOV) every 15 s. Each exposure contains $36 \times 4k \times 4k$ pixels, usually 36×175, 600 sources can be extracted. A GWAC camera produces a 32-megapixel map every 15 s. According to the limit star 16.5, which produces 2,400 images per night, each night the GWAC project will generate 1010 recorded star catalog data. The data rate per camera is approximately 12,000 points per second (2.4 MB), which means the total data rate for the entire GWAC camera array system is 85 MB/s. The image data has about $1.7*105$ records per image. The entire camera array produces $6.12*106$ records in 15 s. Each night is observed for 10 h, and each record has 22 columns of attributes. GWAC will generate about every day. With 2.5 TB of photometric star table data, the 10-year design cycle will form a very large-scale database with a total number of stars with a scale of 3 PB 6 PB, which requires the database management system to have a very strong processing capacity for massive data. The China GWAC Observation Core Station is located in the Xinglong Observation Base of Hebei Province. GWAC's international station in Chile is under discussion. Xinglong Station will have 9 GWAC turntables and 36 CCD cameras of 18 cm [15]. In addition, the core station has a set of mini-GWAC system, which is the leading project of the ground-based wide-angle camera array. It consists of 12 sets of 8 cm large field of view cameras and has been built and placed in the Xinglong Observatory of the National Astronomical Observatory. In this paper, we rely on the mini-GWAC data as samples of training and testing. We investigate the problem of a real-time search for short-timescale gravitational ML events from a huge number of light curves by applying hybrid model of ARIMA-LSTM and ARIMA-GRU. Experiments are conducted on mini-GWAC dataset to evaluate the performance of these two models.

2   Related Work

Impressive works have been conducted in the area of short-timescale gravitational microlensing events prediction. Early Warning System (EWS) was first developed by Udalski in 1994, which is a real-time search system that has succeeded in operation in OGLE project for more than 20 years [7,8]. In 2006, Wyrzykowski et al. [6] have succeeded in systematic search for short-timescale gravitational microlensing events from the variable baselines in OGLE-III data. Recently, an approach of ARIMA that increases efficiency with a multi pro- cess parallel approach was developed by Bi [12]. The ARIMA model is improved with dynamic and parallel processing to detect anomaly. Also, a vari- ant form of ARIMA called D-ARIMA is developed to adjust the parameters of ARIMA for real-time anomaly detection on light curves [9]. Additionally, we look into algorithms of time series detection based on neural networks. In 1987, Farmer and Sidorowich [11] have yield robust results by applying neural network to predict chaotic time series, reaching high effi- ciency in Mackey-Glass delay differential equation, Rayleigh-Benard convection, and Taylor-Couette flow. Later, the advantages of recurrent neural networks are discovered and one of its variant, LSTM, becomes most widely applied [10]. In addition to LSTM, A gated recurrent unit (GRU) was proposed by Cho et al. [13] to make each recurrent unit to adaptively capture dependencies of different time scales, yielding similar results as LSTM yet less computing complexity.

3   ARIMA-LSTM and ARIMA-GRU Hybrid Models

Light-detection data from mini-GWAC implies over 900 time series files for each planet. It is obvious that data is assumed to be consisted of both linear part and the nonlinear part [2]. Thus, we can express the data as follows.

$$x_t = L_t + N_t + \epsilon_t$$

In the function of data x_t, L_t represents the linearity of data at time t, while N_t signifies nonlinearity. The ϵ_t value is the error term. In previous work [3], the Autoregressive Integrated Moving Average (ARIMA) has a great performance on linear problems of over 85% accuracy in general. It is a traditional method in time series prediction. On the other hand, the Long Short-Term Memory (LSTM) model can capture nonlinear trends in the dataset. As a result, we develop one hybrid ARIMA-LSTM to encompass both linear and nonlinear tendencies in the model.

3.1   ARIMA+LSTM Model

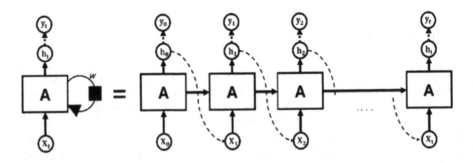

Fig. 1. Structure of recurrent neural network

As we know, ARIMA is a kind of traditional model used to predict the results and LSTM is a kind of neueal network. Our hybrid model which combines ARIMA and LSTM model to improve precision (Fig. 1).

The meaning of ARIMA is the one-product autoregressive moving average process. Assuming that a stochastic process contains d unit roots, which can be transformed into a stationary autoregressive moving average process after d difference, the stochastic process is called integral autoregressive moving average process. In this hybrid model, ARIMA is used for linear part of prediction.

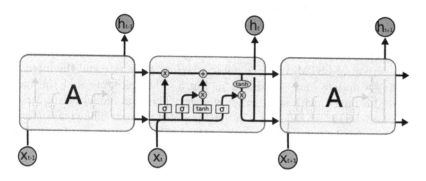

Fig. 2. Structure of LSTM

LSTM is a special type of RNN that can learn long-term dependency information, as shown in Fig. 2. LSTM was proposed by Hochreiter and Schmidhuber [4] and has recently been improved and promoted by Graves [5]. In many problems, LSTM has achieved considerable success and has been widely used. In this hybrid model, LSTM is used to predict nonlinear part.

In terms of final predictions, the first half of each file from the dataset is used as testing sets and the second half of each file from the dataset is used as training sets.

(1) In the linear predictions, training sets can be used directly to predict the future. In each file, the first 20% of the series is used as the basis and this is the size of the moving window. When a new result is predicted, the new result will add in and the whole window will move one step forward until the end. At the same time, the residuals, which means the training sets for LSTM, are also collected.

(2) In the nonlinear predictions, testing sets are from the residuals produced by the linear predictions. And the way how to train the neural network is of significance. Each training set is divided into two parts. The first part which is the first 20% of this series is as the training sets for LSTM and the rest is the test sets to study in order to help LSTM improve precision. As a result, the nonliear predictions are produced.

(3) Finally, predictions consist of linear part and nonliear part.

There are a lot of advantages of this hybrid model. Compared to pure ARIMA model, this model can produce preciser results with sacrificing little time. And compared with pure neural network, it is easier for user to adjust the parameters in a quicker and proper way. However, there are some challenges this hybrid model needs to face, such as reducing time cost and so on.

3.2   ARIMA+GRU Model

This model is designed to compare the efficiency of hybrid model of ARIMA and LSTM. We replace LSTM with GRU as GRU was widely applied since 2015 and it can yield better results in some situation than LSTM. Datasets of Short-timescale Gravitational Microlensing Event are used to evaluate whether hybrid models of GRU is more efficient in this background.

As a variant of LSTM, GRU combines the forgotten gate and the input gate into a single update gate, as shown in Fig. 3. The cell state and the hidden state are also mixed, with some other changes. The final model is simpler than the standard LSTM model and is a very popular variant. The construction of the GRU is simpler: one gate less than the LSTM, so there are fewer matrix multiplications. As a result, GRU can save a lot of time when the training data is large, which is also proved to be true in our case.

And the difference between this model and the previous hybrid model combined with LSTM is only the neural network. All the steps are identical.

And the parameters used in neural network are listed below.

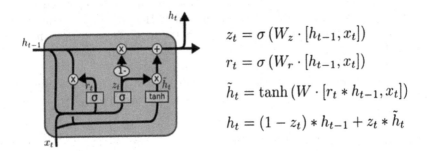

$$z_t = \sigma\left(W_z \cdot [h_{t-1}, x_t]\right)$$

$$r_t = \sigma\left(W_r \cdot [h_{t-1}, x_t]\right)$$

$$\tilde{h}_t = \tanh\left(W \cdot [r_t * h_{t-1}, x_t]\right)$$

$$h_t = (1 - z_t) * h_{t-1} + z_t * \tilde{h}_t$$

Fig. 3. Structure of GRU

Table 1. Sizes of the models tested in the experiments

Algorithms	Units	Parameters
LSTM	128	70400
GRU	128	56192

4   Experiment Method

4.1   Dataset

The data of GWAC has not been open till recently, so our algorithms are tested on mini-GWAC dataset. Mini-GWAC system is consisted of 12 sets of 8 cm large field of view cameras and has been built and placed in the Xinglong Observatory of the National Astronomical Observatory. In this paper, we rely on the mini-GWAC data as samples of training and testing. For each planet, mini-GWAC dataset contains average of 980 txt files, each indicating a thorough observation of one single approach. In each file, we can receive 900 data consisting a part of time series. Contrast of main station of GWAC and mini-GWAC is presented in Table 2.

The experimental environment is based on CentOS Linux and python 3.6.

Table 2. Comparison of GWAC and mini-GWAC.

GWAC station	Main station	Mini-GWAC
Numbers of turrets	9	6
Numbers of cameras	36 cameras in total and 4/turret	12 cameras in total and 2/turret
Diameter	18 cm	7 cm
Focal length	213 mm	85 mm
Length of wave	500–850 nm	500–850 nm
Limiting magnitude	16.5 V	13.0 V
Binocular FOV	4k*4k in pixel;12.8*12.8 = 160	12.8*12.8 = 16
Monocular FOV	5000°	5000°
Pixels	12	2

4.2 Model Fitting

Before fitting the ARIMA model, the order of the model must be specified. The ACF plot and the PACF plot can be used to aid the decision process. ACF/PACF plots of the light-curve data for star11 is plotted in Fig. 4.

In our ARIMA algorithm, we set $(p,d,q) = (0,1,1)$ because it can yield the best outcome of Akaike Information Criterion (AIC) [14]. We compute the log likelihood function for the AIC metric by using the maximum likelihood estimator. We fit the ARIMA model with data from one planet at a time. For each file, we assume that the first 50 data is normal and learn 51th out of them. Then, we move forward at step = 1 at a time until we go through the whole file. After fitting the ARIMA model, we generate predictions of the linear part of the data. Then, we feed the residence of prediction results and reality value into LSTM and GRU networks for training and testing. For each individual planet, we train all previous samples and test light-curve data of the latest day to measure whether there is short-timescale gravitational microlensing events. LSTM and GRU are both imported from keras in python, using 128 units. Final results are

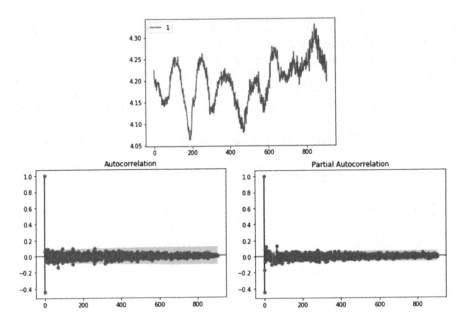

Fig. 4. Notable trends in the data and its ACF/PACF

the sum of the outputs of ARIMA and LSTM or GRU, forming ARIMA-LSTM and ARIMA-GRU hybrid model. An alarm will be set out if the difference of prediction and reality reaches a threshold. To determine the value of threshold, we tested several value shown in Table 3. The threshold is 0.2 eventually.

Table 3. Evaluation results with different thresholds

Thresholds	Loss of epoch	Alerting time
0.1	e−5	−37.5%
0.15	e−5	42.6%
0.2	e−5	42.2%
0.25	e−5	45.4%
0.35	e−5	47.5%

4.3   The Algorithm

Algorithm 1. ARIMA-LSTM/ARIMA-GRU Hybrid Model

1: $Result = []$
2: $Resid = emptylist$
3: **for** $dataset$ in $datasets$ do **do**
4:    $Predict = []$
5:    $Models = emptylist$
6:    $Order = arima.aicminorder$
7:    **if** unstable: **then**
8:       $Model = diff(model)$
9:    **end if**
10:    $Model = fitarima(dataset, order)$
11:    $Addresidual[0]toResid$
12:    $Addpredict[0]toPredict$
13:    $AddpredicttoResult$
14: **end for**
15: $SaveResult[-1], Resid$

---

5   Results and Evaluation

5.1   Testing Results on Simulation Dataset

Before testing on mini-GWAC dataset, we first test on a simulation dataset for evaluation under GWAC-like environment was generated, including 12,960 constant sample light curves and 3240 variable sample light curves. We first combine the microlensing magnification and baseline by three steps: (1) zero-pad the microlensing magnification at the right to expand its duration from T' days to 3+T' days; (2) element-wisely add IA and Is0; and (3) horizontally reverse the combined signal.

To meet the convention term of signal processing, source brightness lensed part in light curve (time length is T) will be called the signal part, while the other part is defined by the background (no lensing effect) in our paper. In our simulation, we will generate a background light curve with a time span of 3 + T' days.

We measured time consuming and accuracy of our hybrid model on simulated dataset. The dataset includes 12,960 constant sample light curves and 3240 variable sample light curves, divided into 636 different files. In this section, the ratio of training set to testing set is 5:1.

In each testing sample, each file is consisted of roughly 1,200 data. After training LSTM or GRU model with training set, we further use the first quarter of each sample, about 300 data in advance, for testing. The line of reality as well as prediction are drawn in Figs. 5, 6, 7, 8 and Fig. 9.

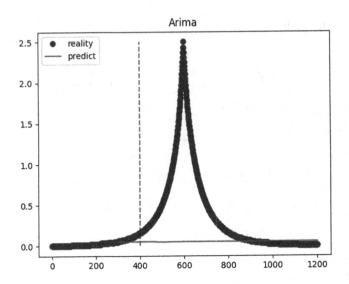

Fig. 5. ARIMA test on simulated dataset

The total time of evaluating these 16,200 sample stars by hybrid model is about 8 h and 24 min, which means that the model is capable to handle at least 16,200 stars in one day and can be applied to online daily renewable systems.

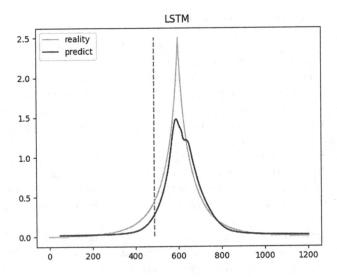

Fig. 6. LSTM test on simulated dataset

Results are drawn as follows:

(1) The average time consuming of one star is less than 2 s.
(2) Accuracy of hybrid models can reach over 90%, yet the time of alerting differs from one another.

Table 4. Evaluation results of different algorithms

Algorithms	Accuracy/Alerting time	Execution time
ARIMA	84.375%/40.0%	1.84 s/star
LSTM	88.67%/51.2%	0.56 s/star
GRU	87.33%/52.3%	0.54 s/star
ARIMA-LSTM	98.31%/40.2%	1.85 s/star
ARIMA-GRU	98.30%/41.4%	1.82 s/star

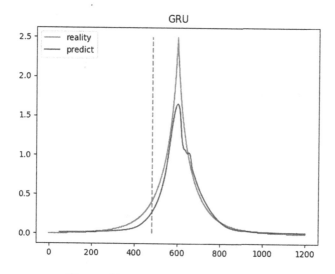

Fig. 7. GRU test on simulated dataset

5.2 Testing Results on Mini-GWAC

In this section, we measured three aspects of the efficiency of the model - accuracy, time consuming and computing complexity. Accuracy is determined by the percentage of alerting point and false prediction rate. That is, we hope the model should be precise on predicting and sensitive on abnormal cases at the same time. We marked the point where the alarm is set out in red, as shown in

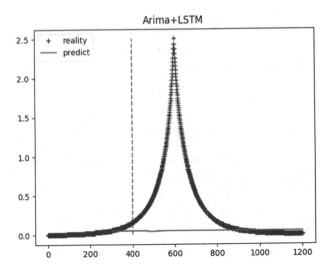

Fig. 8. LSTM-ARIMA test on simulated dataset

Fig. 10. Alarms should be set out before 50% where the abnormal series reaches its peak. The smaller alerting time is, the better the model performs. False prediction is the offset rate of the prediction value and reality value.

Additionally, we measure time consuming by operating time. Combined with different structure of algorithms in Sect. 4.3, it can also indicate computing complexity. Operating time is the prediction time for the single latest observation file. Results can be drawn from Table 4 for mini-GWAC dataset:

(1) GRU trains faster than LSTM for short-timescale gravitational microlensing events prediction.
(2) Alarm timeliness differs little among tested methods. The accuracy is slightly better for hybrid models. LSTM behaves more robust than GRU.
(3) GRUs are simpler and thus easier to modify, for example adding new gates in case of additional input to the network. This results in less training time and computing complexity.
(4) ARIMA can reach smaller alerting time and operating time, yet costing high false prediction rate. By sacrificing 15% operating time, hybrid models of ARIMA and LSTM or GRU can achieve improved 14.5% and 13.2% accuracy respectively (Table 5).

Table 5. Evaluation results of different algorithms

Algorithms	Accuracy/Alerting time	Execution time
ARIMA	81.60%/41.7%	0.349 s/star
LSTM	93.72%/42.6%	0.478 s/star
GRU	93.28%/43.3%	0.440 s/star
ARIMA-LSTM	96.11%/42.2%	0.406 s/star
ARIMA-GRU	94.83%/42.8%	0.413 s/star

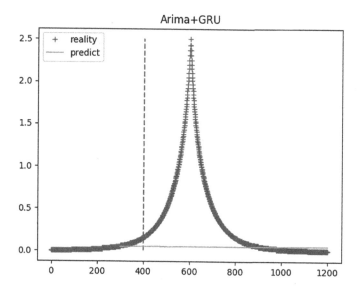

Fig. 9. GRU-ARIMA test on simulated dataset

5.3 Comparison Between LSTM and GRU

The key difference between a GRU and an LSTM is that a GRU has two gates (reset and update gates) whereas an LSTM has three gates (input, output and forget gates). The GRU unit controls the flow of information like the LSTM unit, but without having to use a memory unit. It just exposes the full hidden content without any control. In the algorithm we apply to mini-GWAC dataset, the number of units are set to be 128. See Table 1 for the details of the model sizes.

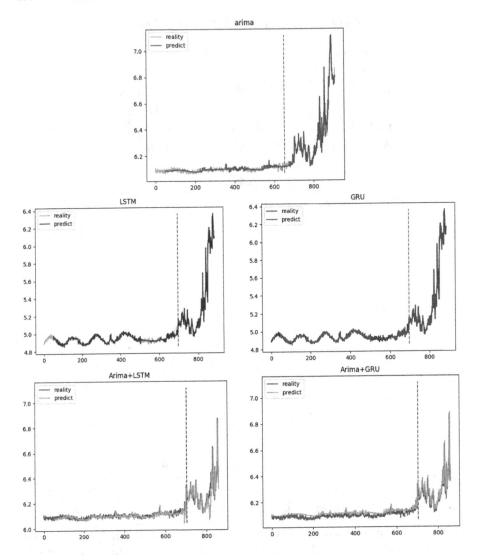

Fig. 10. Test on Star11 of different algorithms

6   Conclusion

Given by the significant role of abnormal activities like micro-lensing for asteroid, computer-aided forecasting is a heavily studied problem in the area of early detection. Like many other studies of this issue, time series are trained and tested by applying neural network which can yield high accuracy of over 90%. However, in the case of mini-GWAC dataset, or for future GWAC dataset, time consuming and computing complexity should be considered as more important criterion for a model. This paper proposed an algorithm of hybrid ARIMA-LSTM and ARIMA-GRU to replace single RNN approach:

ARIMA can reach smaller alerting time and operating time, yet costing high false prediction rate. By sacrificing 15% operating time, hybrid models of ARIMA and LSTM or GRU can achieve improved 14.5% and 13.2% accuracy respectively. Proposed hybrid methods perform better than ARIMA and RNN in the case of mini-GWAC short-timescale gravitational microlensing events prediction.

Additionally, it provides difference between these two methods to indicate strength and weakness of LSTM and GRU which both generate from RNN. This can help with future work of applying different methods to different backgrounds and datasets. Overall, we argue that the proposed work can provide as an essential approach in computer-aided forecasting of time series. Future work contains improvement on strengthening the model and its performance, an experiment test on GWAC dataset, minimizing time consuming especially in the process of neural network training.

Acknowledgement. This research is supported in part by Key Research and Development Program of China (No. 2016YFB1000602), "the Key Laboratory of Space Astronomy and Technology, National Astronomical Observatories, Chinese Academy of Sciences, Beijing, 100012, China", National Natural Science Foundation of China (Nos. 61440057, 61272087, 61363019 and 61073008, 11690023), MOE research center for online education foundation (No 2016ZD302).

References

1. Mayer-Schönberger, V., Cukier, K.: Big Data: A Revolution that Will Transform How We Live, Work, and Think. Houghton Mifflin Harcourt, Boston (2013)
2. Zhang, G.P.: Time series forecasting using a hybrid arima and neural network model. Neurocomputing **50**, 159–175 (2003)
3. Konar, A., Bhattacharya, D.: An introduction to time-series prediction. Time-Series Prediction and Applications. ISRL, vol. 127, pp. 1–37. Springer, Cham (2017). https://doi.org/10.1007/978-3-319-54597-4_1
4. Hochreiter, S., Schmidhuber, J.: Long short-term memory. Neural Comput. **9**(8), 1735–1780 (1997). Massachusetts Institute of Technology
5. Graves, A., Schmidhuber, J.: Framewise phoneme classification with bidirectional LSTM and other neural network architectures. Neural Netw. **18**(5–6), 602–610 (2005)
6. Wyrzykowski, L., et al.: OGLE-III microlensing events and the structure of the Galactic bulge. The American Astronomical Society (2015)
7. Udalski, A.: AcA **53**, 291 (2003)
8. Udalski, A., Szymanski, M.K., Szymanski, G.: AcA **65**, 1 (2015)
9. Feng, T., Du, Z., Sun, Y., et al.: In 6th 2017 IEEE International Congress on Big Data (Honolulu, HI), p. 224 (2017)
10. Lipton, Z.C., Berkowitz, J., Elkan, C.: A critical review of recurrent neural networks for sequence learning. arXiv. 1506.00019
11. Farmer, J.D., Sidorowich, J.J.: Predicting chaotic time series. Phys. Rev. Lett. **59**, 845 (1987)

12. Bi, J., Feng, T., Yuan, H.: Real-time and short-term anomaly detection for GWAC light curves. Comput. Ind. **97**, 76–84 (2018)
13. Cho, K., van Merrienboer, B., Bahdanau, D., Bengio, Y.: On the properties of neural machine translation: encoder-decoder approaches. arXiv preprint arXiv:1409.1259 (2014)
14. http://www.statisticshowto.com/akaikes-information-criterion/
15. Wan, M.: An application research of column store MonetDB database on GWAC large-scale astronomical data management. National Astronomical Observatories, Chinese Academy of Sciences (2016)

Application of Blockchain Technology in Data Management: Advantages and Solutions

Liangming Wen[1,2], Lili Zhang[1], and Jianhui Li[1(✉)]

[1] Computer Network Information Center, Chinese Academy of Sciences, Beijing 100190, People's Republic of China
lijh@cnic.cn
[2] University of Chinese Academy of Sciences, Beijing 100049, People's Republic of China

Abstract. Blockchain is considered to be the key technology to lead the transformation of Information Internet to Value Internet. More and more organizations are exploring the industry applications of blockchain, and how to apply blockchain in data management has become one of the focuses of discussion. This article focuses on blockchain and provides a detailed analysis on its core components, technologies and applications. Then, combs the problems faced by data management in quality, security, sharing and so on, and analyzes the application advantages of blockchain technology in data management, and put forwards a data collaborative management model based on blockchain, which has the characteristics of decentralization, collective maintenance, automatic execution, and non-tamperable. The model covers user authentication, data verification, data logging, data sharing and other processes, and is equipped with a data management incentive system, which can achieve convenient, secure and fast data management. Applying blockchain technology to data management can further improve the effectiveness of data management and improve the quality of data, and create a positive data sharing environment.

Keywords: Blockchain · Data management · Smart contracts · Consensus mechanism · Data sharing

1 Introduction

At present, global technological innovation has entered the era of "Fourth Paradigm" driven by big data. Data resources have become one of the leading forces to reshape the world. With the data, we have mastered the resources and initiative of development [1]. Although the data contains great value and has tremendous potential impact on businesses and societies, it also encounters many problems in the process of utilization, such as scattered data distribution, low data utilization, increased data waste, and frequent data leakage. Under the big data environment, data application innovation and risk avoidance, data sharing and privacy protection have become the important topics in the current data management field. Building an effective data management model and optimizing data management processes has become a hot spot for researchers [2]. The concept of data management also came into being. Data management is a general term for the process of creating, collecting, organizing, storing, and applying data

© Springer Nature Switzerland AG 2019
J. Li et al. (Eds.): BigSDM 2018, LNCS 11473, pp. 239–254, 2019.
https://doi.org/10.1007/978-3-030-28061-1_24

resources, covering almost the entire life cycle of data. Through data management, the conflicts of interest between different groups are balanced, the application needs of all groups are met, and efficient and reliable data management strategies are proposed, which helps to maximize the value of data resources.

The blockchain technology has received interests in recent years, it is originating from Bitcoin and has made it have the characteristics of collective consensus maintenance, decentralization, de-trust, non-tamperable, security and reliability by adopting distributed consensus mechanism, chain block structure, asymmetric encryption algorithm and so on [3], and it has been applied in digital bill verification, supply chain management, food data traceability, intellectual property protection, cross-border trade, etc. Blockchain technology provides a major paradigm shift in business process optimization, data exchange and interoperability in related industries [4], and also provides a new path for data management [5]. The purpose of this paper is to analyze the main problems existing in data management and the key technologies that make up the blockchain, trying to answer the question of how to use of the advantages of blockchain technology to make up for the shortcomings of data management.

This article focuses on the technical aspects of blockchain and their potential benefits to data management. The remainder of this article is organized as follows. Section 2 introduces the concept and development process of blockchain technology. Section 3 Draws various difficulties in data management in terms of quality, security and sharing. Section 4 discusses the application advantages of blockchain technology in data management. Section 5 provides a blockchain-based data collaborative management scheme. Finally, Sect. 6 concludes the article.

2 Blockchain Technology Overview

The earliest known description of the blockchain is the article "*Bitcoin: a peer-to-peer electronic cash system*" written by Satoshi Nakamoto, in which an electronic payment system that does not require a credit intermediary is proposed. This system solves the double payment problem [6]. At present, the definition of blockchain can be divided into two types: Broadly speaking, blockchain is a decentralized infrastructure and distributed computing paradigm [7], which uses distributed node consensus algorithm to generate and update data, and uses encrypted chain block structure to verify and store data, and use smart contracts to edit and manipulate data. Narrowly speaking, blockchain is a decentralized shared ledger [8], which combines the blocks into a specific structure in a chronological order, and cryptographically ensures that the data cannot be falsified and forged. In summary, the blockchain contains three basic concepts: **transactions**, **blocks**, and **chains**. The transaction is the operation of the book, the block is all the transactions and status records that occur within a certain period of time, and the chain is the log record of the blocks in series according to the order of the transactions. The essence is that the asymmetric encryption algorithm, the consensus mechanism, distributed technologies such as distributed storage are integrated into a distributed database system, and all nodes work together to ensure the normal operation of the system [9].

From the perspective of blockchain technology development, the application of blockchain has gone through three stages [10]. (1) Blockchain 1.0 era, the programmable digital currency represented by bitcoin is the initial application field of the blockchain, and is mainly used for transaction record certification. (2) Blockchain 2.0 era, the blockchain is gradually applied to programmable digital assets and smart contracts, and its decentralized and non-tamperable value is reflected. (3) Blockchain 3.0 era, blockchain autonomous organizations and autonomous companies emerged, blockchains +medical, energy, education, etc., humans entered the programmable society. From the perspective of data analysis, Chen and Zheng [11] believe that the structure of the blockchain can be described as three horizontal and one vertical, three horizontal represent the three development stages of the blockchain, and one vertical represents the supporting operation environment of the blockchain - P2P distributed network. At present, it is generally believed that blockchain technology is in the early stage of the 3.0 era, and it can be found from its application in information traceability and network security. What needs to be stated is that the above stages are parallel rather than evolutionary development [12]. On the one hand, application scenarios such as digital currency and digital finance are far from mature, and there is still a long way to go before the vision of global trade integration. On the other hand, the "blockchain+" all kinds of production and life applications are actually integrated with various technologies such as the Internet of Things (IoT) and Artificial Intelligence (AI), and its implementation is still inseparable from the support of digital currency, smart contracts and so on.

3 Difficulties in Data Management

3.1 Low Data Quality

Data quality plays an important role in guiding decision support, process management, and collaborative needs. High-quality data can provide reliable and accurate services [13], but low-quality data will not only cost extra costs but also mean bad business decisions. Therefore, the timeliness, accuracy, completeness, consistency, etc. of the data must be guaranteed [14, 15]. However, in the process of data production and dissemination, the data is not credible due to various factors. Such as the lack of unified data management standards, the deviation in the understanding of the data by various groups [16], and the independence of the process of data collection, processing, storage, remittance, and evaluation, etc., resulting in poor data consistency, difficult to use for analysis and application. In addition, due to the loose regulatory mechanism and the lack of practical verification methods, the data collection process is prone to privately forged, falsified and distorted data, and the data set is heavily doped with useless data, resulting in extremely low data reliability and reduced data accuracy and integrity. In addition, due to the limitations of data acquisition technology, the latest data is not collected, or the collected data is not disclosed in time, for the meteorological, transportation and other industries, the data has lost its timeliness.

3.2 Data Security Is Threatened

The future world will realize the "Internet of Everything". This kind of strong connection also brings people's concerns about data security while facilitating human production and life. Data security issues are mainly reflected in data storage security and personal privacy. Regardless of whether people are willing or not, human data footprints can be recorded almost everywhere, and personal data is always collected and used inadvertently. The so-called "privacy" has become public and transparent, and once this data is leaked, it will cause huge privacy risks. The far-reaching influential events was the US "Prism Project" (PRISM), which was exposed in June 2013, and in March 2018, Facebook was exposed with more than 50 million user information leaked. In addition, big data processing technology emphasizes the relevance of data, so that information fragments that seem chaotic and disorderly can extract valuable information with the help of data mining. In order to solve the data security problem, the United States released *"The Federal Big Data Research and Development Strategic Plan"* in May 2016, China also passed the *"Network Security Law of the People's Republic of China"* in November 2016, and the *"General Data Protection Regulations"* (GDPR), which came into effect in May 2018 by EU, gave the data subject the right to know, access, objection, personal data portability and forgotten [17].

3.3   Data Is not Shared

Open and sharing data can produce greater application value, and there are two reasons to explain this conclusion [18]. Firstly, modern scientific practice requires repeatability of experimental results. Secondly, the distribution of data sources between "Big Science" projects is scattered and they need to be shared with each other. The real world data is mostly stored by various departments, and the degree of data interconnection and open sharing is limited. It is mainly manifested in the copyright infringement caused by the uneven distribution of data and the ambiguity of data rights due to lack of incentive mechanism. Due to the lack of incentives, there is no atmosphere for sharing data between organizations and data owners are not motivated to share data. Low data sharing raises the "Data Gap" problem. Organizations or individuals in the core frontier are more likely to generate, acquire, and use data, while most ordinary users can obtain not only a small amount of data but also relatively lower quality data. In addition, all kinds of data such as personal social media behavior data, smart wearable device data, navigation software data, etc., should be owned by individuals or information service providers? Do telecommunications, finance, medical and other institutions have the right to save and use public data? Due to the ambiguity of ownership, some problems have arisen: On the one hand, the benefits generated by the use of data are difficult to distribute. On the other hand, not only the awareness of data owners to protect their own data copyright is weak, but also the awareness of data users to respect the copyright of others' data is weak.

4   Application Advantages of Blockchain Technology in Data Management

4.1   Decentralization Allows All Nodes to Participation

For a long time, almost all internet applications have adopted a centralized model, that is, the application server is owned by a specific organization or individual, the data in the database can only be viewed and analyzed by internal personnel, which leads to excessive dependence on the central node and information opacity. The fundamental feature of the blockchain is decentralization, there is no top-down hierarchy or centralized organization in the system, instead, the peer-to-peer network is used to realize the coordination and cooperation of the macro system [19]. This kind of organization represents the evolution of the system structure from a fully centralized model to a completely decentralized model, enabling strange nodes to achieve credible value transmission without relying on third-party trusted institutions [20], and reflects the technical characteristics of "equality and freedom" [21]. The public chain represented by Bitcoin, Ethereum, etc., has no restrictions on the joining and exiting of nodes. All nodes joining the blockchain jointly manage the data resources in the maintenance chain, and can easily obtain the data and participate in the data authenticity proof. Bitcoin, Ethereum and other public chains do not have any restrictions on the joining and exiting of nodes. All nodes joining the blockchain jointly manage the data resources in the maintenance chain, and can easily obtain the data and participate in the proof of data authenticity.

4.2   Blocks and Chain Structure Supports Data Traceability

In order to achieve data immutability, the blockchain adopts a chain structure in blocks [22], as shown in Fig. 1. A block contains of the block header and the block body: the block header encapsulated a lot of information, such as the current block version, Merkle Root, time stamp, difficulty value (nBits), random number (Nonce), and parent block signature value (Parent Block Hash) (see Table 1 for details). All the transaction information in the current block is stored in the block body and the transaction is not only a means to change the blockchain data but also a data base of the blockchain.

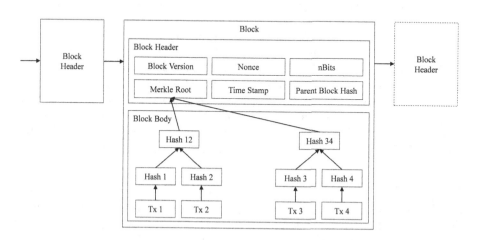

Fig. 1. Data structure of blockchain.

Table 1. Structure of block header.

Information	Information description
Block Version	Current block version number, storage related protocol for the block system
Merkle Root	The hash value of all the transaction in the block
Time Stamp	Block generation time
nBits	The target threshold to be solved for generating the block
Nonce	Current proven algorithm solution
Parent Block Hash	The hash of the previous block, which records all transactions in the block

Transactions within the block are organized by the Merkle tree [24]. Taking the Bitcoin system as an example, when a transaction occurs, the nodes queue the transactions according to the transaction time order, and obtain the hash value of each transaction through Hash operation, and splice different hash values to calculate the new hash value again, which is continuously from bottom to bottom and the Merkle Root is obtained by splicing the Hash operation. Merkle Root represents all transaction information for a period of time and can be regarded as the signature of the entire tree. When any change occurs in the transaction, the Merkle Root obtained by repeating the above operation must be different from the previous value. So as long as Merkle Root is retained, it can be verified whether each transaction has been tampered with [25]. When the node receives the transaction information sent by the system, it first verifies the legality of the transaction. Only when the transaction meets certain verification requirements can the verification be passed. The legitimate transaction is broadcasted to the neighboring node and stored locally, and the illegal transaction is direct rejected [26]. The successor block wants to establish contact with the leading block, must obtain the legal Nonce and pass the verification, and the subsequent block records the Parent Block Hash and joins the chain. As the block continues to increase, a data structure chain is formed from the ordered links. The information contained in the chain is continuously confirmed and locked. To tamper with a transaction, all blocks after the transaction must be calculated, and the chain after recalculation is accepted by other nodes, and to do this requires the tamper to have more than 51% of the system's computing power. Thus, a complete blockchain stores the history information of all nodes. As long as the version number of the last block is saved, it can be verified whether each transaction has been tampered with.

4.3 Encryption Algorithm Guarantees Data Security

In order to prevent the data transmission process from being attacked, the blockchain uses an encryption system to ensure data security. A typical encryption system generally includes encryption and decryption keys, encryption and decryption algorithms, ciphertext, plaintext and other elements. A basic encryption and decryption process is shown in Fig. 2.

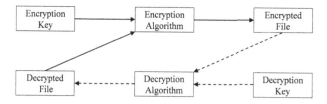

Fig. 2. Encryption and decryption processes.

In the encryption process, the plaintext is encrypted by the encryption key and the encryption algorithm to obtain the ciphertext; during the decryption process, the ciphertext is decrypted by the decryption key and the decryption algorithm to obtain the plaintext. According to whether the keys used are the same, the encryption process can be divided into Symmetric Cryptography and Asymmetric Cryptography. Symmetric encryption's keys are the same, Asymmetric encryption's keys are different, and blockchain systems mostly use Asymmetric encryption algorithm. The asymmetric encryption algorithm uses private key and public key two keys [27]. The private key is generally generated by a random algorithm, and the public key is generated by a private key. Each node in the system has a unique key, the public key is broadcast to the entire network, and the private key is saved by the node. In the transaction, if the node uses the private key to sign and encrypt the data, other nodes need to use the public key to decrypt the data to confirm the authenticity of the data source. Similarly, if the public key is used for encryption, the private key needs to be used for decryption. The security of asymmetric cryptographic algorithms relies on complex mathematical problems such as large-scale factorization, discrete logarithms, and elliptic curves to ensure, the representative algorithms include RSA, Diffie-Hellman, ElGamal, Elliptic Curve Cryptography (ECC), ShangMi 2, etc. On the basis of asymmetric encryption, digital signature technology is also derived to verify the integrity of data content and confirm the data source [28]. The digital signature algorithms used in common scenes include Digital Signature Algorithm (DSA), Elliptic Curve Digital Signature Algorithm (ECDSA), etc. The digital signature algorithms for special scenes include Blind Signature (BS), Multiple Signature (MS), Group Signature (GS), Ring Signature (RS) and so on.

4.4   Smart Contracts for Automatic Verification

Smart sontracts were originally defined as a set of commitments defined in digital form and agreements by contract participants to implement these commitments [29]. The emergence of blockchain technology redefines and implements smart contracts. As a core element of the contract layer in the blockchain infrastructure, smart contracts can be embedded in any tangible or intangible asset transaction to form software-defined systems and assets that can automatically process data and store delivery values. In essence, a intelligent contract is a set of event-driven programmatic rules and logic. It is a program code that can be automatically executed by technical means [30]. Its operating mechanism is shown in Fig. 3 [12].

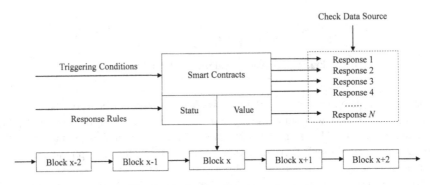

Fig. 3. Rationale of smart contracts.

Smart contracts define trading logic and rules for accessing data. Usually, specialized programmatic rules and logics are set for specific scenarios that are attached to the transaction data in the form of program code after being signed by the parties involved. After the program code is propagated throughout the network and verified by the system, it is compiled into an opcode stored in the blockchain and given a storage address. The external application must call the intelligent contract and execute the transaction and access the data according to the contract rules. Similar to the trigger in the data management system, when the preset condition is reached, the entire network node automatically executes the opcode and writes the execution result to the blockchain. But unlike triggers, smart contracts and all their processing results are stored in the blockchain and synchronized across all nodes in the system to ensure that all nodes see the same smart contracts. In order to prevent the entire blockchain system from being attacked by vulnerabilities and malicious code, smart contracts run in isolated sandbox environments rather than directly on blockchain nodes, between contracts and contracts, contracts and main chain systems, the sandbox environment is effectively isolated. Since the intelligent contract is a piece of program code, once it is started, it will be automatically executed and cannot be intervened. It achieves the goal of "Code is Law", which can effectively guarantees the fairness and security of the data.

4.5 Consensus Mechanism Ensures Data Consistency

In a distributed system, if there is a delay between different nodes, the communication network will be interrupted, and even malicious nodes falsify information, which seriously affects the consistency of the replica data in the system. In order to solve this Byzantine Fault-Tolerant (BFT) problem caused by the participation of multiple nodes, the blockchain adopts a consensus mechanism to ensure that all nodes in the system can participate in the production and verification process of the data block. The consensus algorithm is used to filter the specific node to exercise the accounting power, and the newly generated block must be verified by most nodes before it can be written into the ledger. A complete consensus process can be divided into four stages [31]: Leader Election - Block Generation - Data Validation - Chain Updation. The basic process is shown in Fig. 4.

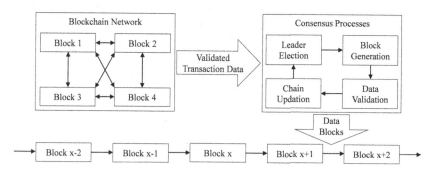

Fig. 4. Consensus processes of blockchain.

After receiving the verified transaction data generated by the nodes in the blockchain network. Firstly, the accounting node needs to be selected from all the transaction nodes (*i.e.*, Leader Election). Secondly, the accounting node packages the transaction data into the block and broadcasts to all nodes, and the Block Generation operation is completed. Then, other transaction nodes except the accounting node verify the transaction data received by the broadcast (*i.e.*, Data Validation). After the data is verified, the accounting node adds the encapsulated data block to the chain, and the Chain Updation operation is completed.

As the core algorithm supporting the consensus process, the consensus mechanism is also constantly improving along with the development of various blockchain platforms. The current three mainstream consensus mechanisms are Proof of Work (PoW), Proof of Stake (PoS) and the Delegated Proof of Stake (DPoS). Bitcoin is the first to apply PoW, and each node competes to solve a complex mathematical problem based on its own computing resources, the fastest problem solving node will get block accounting rights or rewards [32]. One of the disadvantages of PoW is that it wastes a lot of computing power. The PoS used by Peerchain (PPC) solves this pain point [33]. PoS realizes the rights allocation based on the equity stock. The node with the highest interest in the system will get the block accounting right. The interest is determined by the coin age, which is the product of the specific amount of coins and its holding time. The less the age of the currency, the more difficult it is to calculate. Different from the above two, Bitshares adopts DPoS similar to "board decision", that is, each node in the system can authorize the shares held by it to a representative, and the first few nodes that obtain the most authorized votes and are willing to be represented will enter the "Board of Directors" and in turn to carry out the operations of package settlement, signing and generating new blocks, etc., and the authorized node can get rewards from the transaction [34]. These mechanisms describe the process by which multiple nodes in a distributed system agree on a proposal.

5   Blockchain-Based Data Management Scheme

5.1   Data Collaborative Management Model

The Data Management Association International (DAMA) defines data management as the ability to plan, control, and deliver data assets [35]. Combined with the previous summary and summary of data management problems, this article constructs a data collaboration management framework based on blockchain from the perspective of data record storage, data open sharing and data security, as shown in Fig. 5.

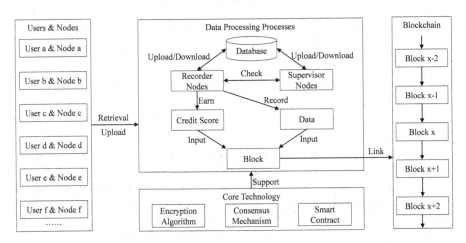

Fig. 5. Data collaborative management model based on blockchain.

The general structure of the model consists of core technology components, user management components, data processing components and blockchain components, and mainly involving user authentication, data validation, data logging, data access and other processes. Based on the management model, data lifecycle management can be realized, ensuring the safe, reliable and effective sharing of data, and ultimately achieve the management goals of strategic consistency, risk controllable, operational compliance and value realization [36].

5.2   User Authentication Process

The blockchain adopts a decentralized concept that allows full participation. In order to better manage users, this article designs a user authentication process, which is carried out outside the system chain, mainly including identity authentication module, rights management module and supervision management module, which is responsible for verifying user identity, supervising user behavior, managing credit scores, and so on. The node becomes a registered user through identity verification, and the registered user becomes a candidate node through identity authentication. The registered user and the candidate node can participate in the subsequent data processing process. The specific authentication process is as follows:

Step 1. The user fills in the registration information and submits the registration application, and the system calls the identity verification module to verify the registration information.

Step 2. The rights management module gives feedback the permission status to the user node.

Step 3. The supervisory management module feeds back the supervisory information to the user node.

Step 4. Users participate in data management activities by calling smart contracts.

In the *Step 1*, if it is already a registered user, the system returns to the "registered" state. If it is not already registered, the registration operation is performed and the operation is completed and returns to the "registration success" state. If the submission information is incomplete, the "registration failure" is returned. The registered user can continue to apply for identity authentication to become a candidate node and a supervisory node. In the *Step 2*, different types of users have different permissions, the registered user has the lowest authority, the supervisory node has the highest authority, and the authority of the candidate node is in an intermediate state. In the *Step 3*, the supervisory information refers to the current violations and security of the node. Only after completing the above authentication steps can each node join the system and participate in data management. Adhering to the concept of openness, users can log out of the system at any time, and don't need to apply for registration when logging in again, directly enter with the registered account and password.

5.3 Data Validation Process

Since the system consists of many nodes. On the one hand, the quality of the data uploaded by each node is uneven. On the other hand, some nodes may deliberately upload a large amount of low-quality data in order to obtain the points. Therefore, it is necessary to verify the data when the user submits the data, which is both the need to ensure the quality of the data and the legal rights of other nodes.

The original block data in the system is classified according to certain standardization rules, and different types of data form a data standard library. When new data is uploaded, the intelligent contract calls the standard library to verify the data, and verification results can be divided into several cases as shown in Table 2.

Table 2. Data verification result type.

Data verification result	Applicable conditions
Repeated	Fully consistent data already exists in the data standard library
Analogous	Similar data already exists in the data standard library
Defective	Incomplete data semantics
Invalid	Data is meaningless
Compliant	Complete data semantics, no repetition, non-similar and meaningful

The repeated, analogous, defective, invalid data are regarded as the illegal data. This kind of data is not allowed to join the block system. The system will informs the user of the verification results and give suggestions for modification. After the modification is completed, it can be re-uploaded and verified. Compliant data can be directly added to the system and automatically added to the type standard library by smart contracts. Such data can also be divided into different levels of excellent, good, qualified, etc., and each level corresponds to different reward rules.

The data verification process is of great significance to data management On the one hand, it can control the data quality from the source, and prevent low-quality data from entering the system. On the other hand, it can build a constantly updated data standard library, and provide reference for rights management and access sharing.

5.4   Data Recording Process

Hierarchical architecture or excessive reliance on central nodes can result in poor information transfer and information distortion. The decentralized thinking and consensus mechanism of the blockchain enables all participating nodes to be supervised while ensuring more users' participation, thus ensuring data consistency. The verified data recording process is as follows:

Step 1. The candidate node encrypts the collected data with a public key and sends an upload request to the data record node.

Step 2. The record node performs public key verification, confirms the permissions of the candidate node, and responds to the upload request.

Step 3. The candidate node signs the data digest with the private key and encrypts the data again with the public key.

Step 4. The record node decrypts the data with the private key, unlocks the signature with the public key, and compares the data digest with the data hash value. After the verification is passed, the data digest and the signature of the candidate node are recorded in the block and the encrypted data is stored in the database.

Step 5. After a period of time (10 min or less), the recording node calculates the Merkle root value in the block and randomly broadcasts the block to the entire network.

Step 6. The supervised node and the candidate node check the block information, and after the verification is passed, the value record node sends the confirmation information.

Step 7. The recording node records the newly generated block into the blockchain, and the data recording operation is completed.

Each data record contains metadata, public keys, and data digests, where the metadata stores descriptive information about the data, the public key is used to confirm the identity and permissions of the uploading node, and the data digest is used to verify data integrity and index data.

5.5   Data Sharing Process

The traditional data sharing process is limited by trust and authority. It requires a lot of resources for data verification and permission review, which seriously affects the transmission of data value. The asymmetric encryption algorithm not only solves the trust problem between nodes, but also ensures the security of the data transmission process. The blockchain-based data sharing process is directly performed between the two parties, and does not require the participation of third-party nodes. The specific process is as follows:

Step 1. The data sender encrypts the data with a public key and generates a data ciphertext.

Step 2. The data sender hashes the data to obtain a data digest, and the data digest is signed with the private key.

Step 3. The data sender sends the data ciphertext and the digital signature to the data receiver.

Step 4. The data receiver decrypts the digital signature with the public key to obtain a data digest, and verifies the identity of the data sender.

Step 5. The data receiver decrypts the data ciphertext with the private key to obtain the original data.

Step 6. The data receiver obtains the data digest again through hash calculation, and compares the two data digests to verify the data integrity.

Step 7. The data verification is passed, and the transaction information is recognized and recorded in the block; if the data verification fails, return to the first step.

In the sharing process, the public key of each data record bears the task of data determination, such as data authenticity verification, upload node selection, data rights division, etc., all relying on public key information verification, which is not only beneficial for finding data objects but also good for the pursuit of false data.

5.6 Data Management Incentive System

The higher the enthusiasm of nodes participating in data management activities, the higher the level of data management. Setting up an incentive system can creates a good atmosphere for data management. Therefore, digital currency can be issued in the system - data coin, the amount of datacoin can be used as a reference for the allocation of rights. The specific issuance rules for datacoin are shown in Table 3.

Table 3. Datacoin issuance rules.

User behavior	Datacoin increase and decrease rules
Registration & certification	The system rewards the user with datacoin
Browse data	The data viewer's datacoin is not increased or decreased
Use data	The data user's datacoin is deducted, and the data provider's datacoin is decreased
Upload illegal data	The data uploader's datacoin is decreased
Upload compliance data	The data uploader's datacoin is deducted
The block of the data record node signature is correct	The data record node's datacoin is decreased

(*continued*)

Table 3. (*continued*)

User behavior	Datacoin increase and decrease rules
There is an error in the block signed by the data record node	The data record node's datacoin is deducted, and the data supervise node's datacoin is decreased
Result verification between data supervise node and the candidate node is correct	Both the data supervise node and the data record node's datacoin is decreased
There is a difference in the result check between the data supervise node and the candidate node	The majority result node's datacoin is deducted, and minority result node's datacoin is decreased

If a node makes a positive contribution to the system, it will be rewarded. If the node makes a bad behavior, it will be punished. Only the datacoin increase and decrease rules are given in Table 3, but the specific increase or decrease needs to be determined by the application scenario. As the basis of credit and contribution, the datacoin will affect the scope of authority of each participating node in the system. In order to obtain more permissions, each node will actively participate in the system-wide data management behavior. Therefore, on the one hand, the incentive system improves the enthusiasm of each node in the system to upload and verify data, and on the other hand ensures the authenticity of the system data.

6   Discussion and Conclusion

In the era of big data, data management is faced with many problems such as low data quality, threatened data security, and difficulty in data sharing. The reasonable and effective management of data resources will promote the maximization of data value. Blockchain technology is consistent with the real-world needs of data management due to its decentralization, traceability, anti-attack, automatic verification, and data consistency. It has great application potential in the new data management environment. Based on blockchain technology and data management theory, this article designs a blockchain-based data management scheme, which involves the core technology of blockchain, such as chain structure and encryption algorithm, smart contracts, consensus mechanisms, etc. The management solution proposed in this paper includes a data collaborative management model and a corresponding set of management processes and systems, aiming to improve the effectiveness of data management and improve data quality, and create a positive data sharing environment.

Acknowledgment. The work described in this paper is supported by the National Natural Science Foundation of China (No. 91546125), the Chinese Academy of Sciences (No. XXH13505-08-04) and National Science and Technology Infrastructure Center (No. 2018DDJ1ZZ14).

References

1. Guo, H.: Scientific big data: a footstone of national strategy for big data. Bull. Chin. Acade. Sci. **33**(8), 768–773 (2018)
2. Yang, L., Gao, H., Song, J., et al.: Research and application of data governance framework in big data environment. Comput. Appl. Softw. **34**(4), 65–69 (2017)
3. van Rossum, J.: Blockchain for Research [EB/OL]. https://digitalscience.figshare.com/articles/Blockchain_for_Research/5607778/1. Accessed 23 Dec 2018
4. Zhao, Z., Song, J., Pang, Y., et al.: Blockchain Rebuild New Finance, pp. 25–28. Tsinghua University Press, Beijing (2017)
5. Zhang, J., Wang, F.: Digital asset management system architecture based on blockchainfor power grid big data. Electr. Power Inf. Commun. Technol. **16**(8), 1–7 (2018)
6. Nakamoto, S.: Bitcoin: a peer-to-peer electronic cash system [EB/OL]. https://bitcoin.org/bitcoin.pdf. Accessed 07 Dec 2018
7. Zyskind, G., Nathan, O., Pentland, A., et al.: Decentralizing privacy: using blockchain to protect personal data. IEEE Symposium on Security and Privacy, pp. 180–184 (2015)
8. Li, X., Liu, Z.: Study on supply chain intelligent governance mechanism based on blockchain technology. China Bus. Mark. **31**(11), 34–44 (2017)
9. Xia, Q., Zhang, F., Zuo, C.: Review for consensus mechanism of cryptocurrency system. Comput. Syst. Appl. **26**(4), 1–8 (2017)
10. Zhu, J., Fu, Y.: Progress in blockchain application research. Sci. Technol. Rev. **35**(13), 70–76 (2017)
11. Chen, W., Zheng, Z.: Blockchain data analysis: a review of status, trends and challenges. J. Comput. Res. Dev. **55**(9), 1853–1870 (2018)
12. Yuan, Y., Wang, F.: Blockchain: the state of the art and future trends. Acta Automatica Sinica **42**(4), 481–494 (2016)
13. Ding, X., Wang, H., Zhang, X., et al.: Association relationships study of multi-dimensional data quality. J. Softw. **27**(7), 1626–1644 (2016)
14. Zhimao, G.: Research on data quality and data cleaning: a survey. J. Softw. **13**(11), 2076–2082 (2002)
15. Kleindienst, D.: The data quality improvement plan: deciding on choice and sequence of data quality improvements. Electr. Mark. **27**(4), 387–398 (2017)
16. Stanford, N.J., Wolstencroft, K., Golebiewski, M., et al.: The evolution of standards and data management practices in systems biology. Mol. Syst. Biol. **11**(12), 851 (2015)
17. Zhang, L., Wen, L., Shi, L., et al.: Progress in scientific data management and sharing. Bull. Chin. Acad. Sci. **33**(8), 774–782 (2018)
18. Birnholtz, J.P., Bietz, M.J.: Data at work: supporting sharing in science and engineering. In: International Conference on Supporting Group Work, pp. 339–348 (2003)
19. Yuan, Y., Wang, F.: Parallel blockchain: concept, methods and issues. Acta Automatica Sinica **43**(10), 1703–1712 (2017)
20. Zhu, L., Gao, F., Shen, M., et al.: Survey on privacy preserving techniques for blockchain technology. J. Comput. Res. Dev. **54**(10), 2170–2186 (2017)
21. Yuan, Y., Zhou, T., Zhou, A., et al.: Blockchain technology: from data intelligence to knowledge automation. Acta Automatica Sinica **54**(10), 2170–2186 (2017)
22. Shao, Q., Jin, C., Zhang, Z., et al.: Blockchain: architecture and research progress. Chin. J. Comput. **41**(5), 969–988 (2018)
23. Zheng, Z., Xie, S., Dai, H., et al.: An overview of blockchain technology: architecture, consensus, and future trends. In: International Congress on Big Data, pp. 557–564 (2017)

24. Cohen, D.: Merkle Trees and Blockchains [EB/OL]. http://www.cs.tau.ac.il/~msagiv/courses/blockchain/Merkle.pdf. 02 Jan 2019

25. Narayanan, A., Bonneau, J., Felten, E.: Bitcoin and Cryptocurrency Technologies, p. 27. Princeton University Press, Princeton (2016)

26. Ben, E., Brousmiche, K.L., Levard, H., et al.: Blockchain for enterprise: overview, opportunities and challenges [EB/OL]. https://www.researchgate.net/publication/322078519_Blockchain_for_Enterprise_Overview_Opportunities_and_Challenges. Accessed 30 Dec 2018

27. Li, J., Jia, C., Liu, Z., et al.: Survey on the searchable encryption. J. Softw. **26**(1), 109–128 (2015)

28. Yang, B., Chen, C.: Blockchain: Principle, Design and Application, pp. 58–59. China Machine Press, Beijing (2018)

29. Szabo, N.: The Idea of intelligent contracts [EB/OL]. http://www.fon.hum.uva.nl/rob/Courses/InformationInSpeech/CDROM/Literature/LOTwinterschool2006/szabo.best.vwh.net/smart_contracts_idea.html. Accessed 27 Nov 2018

30. Gatteschi, V., Lamberti, F., Demartini, C.G., et al.: Blockchain and intelligent contracts for insurance: is the technology mature enough? Future Internet **10**(2), 20 (2018)

31. Yuan, Y., Ni, X., Zeng, S., et al.: Blockchain consensus algorithms: the state of the art and future trends. Acta Automatica Sinica **44**(11), 2011–2022 (2018)

32. Laurie, B., Clayton, R.: "Proof-of-Work" Proves Not to Work [EB/OL]. https://www.cl.cam.ac.uk/~rnc1/proofwork.pdf. Accessed 15 Dec 2018

33. King, S., Nadal, S.: PPCoin: Peer-to-Peer Crypto-Currency with Proof-of-Stake [EB/OL]. https://peercoin.net/assets/paper/peercoin-paper.pdf. Accessed 13 Jan 2019

34. Asolo, B.: Delegated Proof-of-Stake (DPoS) Explained [EB/OL]. https://www.mycryptopedia.com/delegated-proof-stake-dpos-explained/. Accessed 23 Dec 2018

35. Mosley, M., Brackett, M.H., Earley, S., et al.: The DAMA Guide to the Data Management Body of Knowledge (2009)

36. Song, J., Dai, B., Jiang, L., et al.: Data governance collaborative method based on blockchain. J. Comput. Appl. **38**(9), 2500–2506 (2018)

A Framework of Data Sharing System with Decentralized Network

Pengfei Wang[1,2], Wenjuan Cui[1], and Jianhui Li[1(✉)]

[1] Computer Network Information Center, Chinese Academy of Sciences,
Beijing 100190, China
lijh@cnic.cn
[2] University of Chinese Academy of Sciences, Beijing 100049, China

Abstract. With the development of technology, more and more data have been accumulated. Utilization of data can benefit many applications, such as facilitate human daily life, promote scientific research, and expedite event detection etc. How to share data within a certain group or a specific person securely and effectively has become a hot research topic. Meanwhile, many distributed technologies have appeared with the triggering of many centralized cloud web services. Thus, we propose a block-chain based data sharing framework with IPFS (https://ipfs.io/), Ethereum (https://www.ethereum.org/) and uPort (https://www.uport. me/). As a result of this design, the system could store data in IPFS system and control sensitive data by block-chain, and the authentication can be managed by the uPort ID system. Finally, we evaluate the effectiveness by a case study.

Keywords: Decentralized system · Blockchain · Data sharing

1 Introduction

Enormous and various data are accumulated by different individuals and organizations [17] from different devices in different application scenarios every day. Meanwhile, there is no one application could exist without any other applications, that is, one application is often related to one or more applications. Similarly, application optimization with utilizing its own dataset often can not get the best effect. Thus, data sharing is critical in data utilizations [3], such as data statistics, data mining and machine learning.

Many centralized systems have been built to provide data sharing service. Specifically, the users could upload their datasets into the centralized server and download the others' datasets from the centralized server. All options are set to rely on the centralized server, including registering, searching, uploading, downloading.

On the decentralized scheme, some people have done some interesting and useful work [2,7,20,21]. Especially, the technologies about blockchain have been developed continuously and well known since the appearance of Bit Coin. There

© Springer Nature Switzerland AG 2019
J. Li et al. (Eds.): BigSDM 2018, LNCS 11473, pp. 255–262, 2019.
https://doi.org/10.1007/978-3-030-28061-1_25

are more and more concerns on these technologies, focusing on governance, security, and privacy, etc.

Traditional data sharing systems, as previously mentioned, could satisfy the requirements when the volume of datasets and number of users are limited. With the explosive growth of data, it is necessary to find a secure, efficient and effective way to build a data sharing system.

Decentralized storage system have been proposed to protect personal data [25]. And some P2P distributed systems have been constructed to share data [18].

However, most of the previous work, including the centralized data sharing systems and the traditional P2P distributed systems, have some disadvantages. For example, the centralized data sharing systems often get the less efficiency compared with the decentralized ones because of all the options are depended on the central server. The traditional P2P data sharing systems often cause disorganized managements and insecure environments. Thus, to address the weaknesses of **Risk Resistance Capacity, User Privacy Policy** and **Data Transmission Efficiency** in traditional data sharing systems, we propose a block-chain based data sharing framework with the new decentralized technologies, IPFS and uPort, to provide securely, efficiently and effectively distributed data sharing services.

2   Related Works

This paper is related to the following categories of prior work: data sharing and the application of blockchain in various fields.

Data sharing. Data sharing has been put forward at least from 1980s [10]. There was a discussion about whether the researchers should share their preliminary results in [16]. Data sharing is a complex issue with multiple technical, social, financial and legal facets [11]. And in [11], the authors analyzed the NIH (National Institutes of Health) policy statement and data sharing models to ensure the data sharing is effective and rewarding. Fecher et al. focused on the academic data sharing and wanted to provide evidence for science policies and research data infrastructure [9].

There are two models to share data, central database resources sharing system and peer-to-peer exchange. On central database resources sharing system, every user should access the central server to get the data. More specifically, a person who wants to share data through a centralized system should register on the system to have a user ID, then he/she could upload, download dataset through the central database. [14] focused on discussion on the Archaeological Markup Language (ArchaeoML) and explored approaches to data sharing and data integration which could be used to organize archaeological information. On peer-to-peer exchange situation, the individuals could exchange the local data offline. What's more, we can construct a peer-to-peer network and then share the datasets through the system [15]. For example, in the work [8], the authors proposed future directions for research on P2P systems, and they illustrated some of the trade-offs at the heart of search and security problems in P2P data sharing systems. In [18], the authors designed a peer-to-peer data sharing system. They also implemented and evaluated the PeerDB on a computer cluster. The results showed that PeerDB is efficient. In the work [13], the authors proposed a novel P2P-based MCS(Mobile crowdsensing) architecture, where the sensing data was saved and processed in user devices locally and shared among users in a P2P manner. Then they analyzed the user behavior dynamically and proposed iterative algorithms that were guaranteed to converge to the game equilibrium.

Application with Blockchain. There are many applications in various domains based on blockchain, such as Bitcoin in economic field, voting system [22] in political field and energy supply in resource field, etc.

Decentralized personal data management system was designed in [25]. By using this system, the users could own and control their data. Furthermore, the authors implemented a protocol which made a blockchain work as an automated access-control manager that did not need trust in a third party. In work [19], the authors constructed a framework for cross-domain image sharing based on blockchain. In their framework, the blockchain worked as a distributed data store to establish a ledger of radiological studies and patient-defined access permissions. In the work [1], the authors designed a data sharing framework for electronic health record using ethereum-based blockchain technology.

As mentioned above, the traditional data sharing systems are centralized. The efficiency is limited. The peer-to-peer data sharing system is lack of a kind of effective authorization to manage the datasets. Thus, we propose a blockchain based framework with IPFS and uPort to provide safe and efficient data sharing services.

3   Decentralized Technologies

Distributed technologies have influenced human life in various aspects. For example, from the users' perspective, Twitter[1] is a peer-to-peer alternative social network operating on a decentralized framework, on which users could deliver information and get others' messages; Massive Online Open Courses (MOOCS) is a decentralizing education service system, from which everyone could become accessible to college; And peer-to-peer payment networks have been constructed with online payment framework, such as VenMo and Alipay. Furthermore, the decentralized technologies are more likely to be the Blockchain, Ethereum IPFS and uPort, etc.

Blockchain. A blockchain is a set of records which are called blocks. Given a specific blockchain, each block contains three parts, a cryptographic hash of previous block, a timestamp the block generates and the transaction data. In general, a public blockchain does not have any access restriction. Thus, many algorithms have been used to secure the blockchains, such as Proof of Work (PoW), Proof of Stake (PoS) and the Delegated Proof of Stake (DPoS). For

---

[1] https://twitter.com/.

example, bitcoin uses a proof-of-work system to create the next block, and EOS[2] uses a proof-of-state to create the next block.

Ethereum. Ethereum [6] is a decentralized platform which can run smart contracts, and smart contracts are the kernel technologies of blockchain. Smart contracts could be authored so that algorithmically specify and autonomously enforce rules of interaction through Ethereum [24]. Many tools could be used to develop smart contracts, such as Truffle[3] and Remix[4]. Many decentralized applications could be programmed without third-party interference.

IPFS. IPFS [4] is a peer-to-peer distributed file system that seeks to connect all computing devices with the same system of files. That is, a terminal could be a client and a server at the same time. IPFS takes advantage of the same peer-to-peer file-sharing capabilities of BitTorrent, and has more expanded functionality [23]. In the framework, we can create an instance for each user, encode the data into JSON format, and put the JSON code into IPFS and generate IPFS hash code.

uPort. uPort is an interoperable identity network for a secure, private, decentralized web. Furthermore, uPort is a decentralized infrastructure for claiming identities and receiving verification from other parties in the network [12]. uPort could help us solve the unique identity for each user [5]. Figure 1 shows the interface of uPort application. First, each user should install the uPort app on mobile devices and register through a Dapp. Then, each user can interact with the ethereum network (such as rinkeby net) by the unique ID generated by uPort.

4   Framework Overview

Figure 2 shows the framework overview of the system. Given a specific user, the general operation steps are as follows.

Register uPort. Firstly, each user log in the uPort system to get an unique user ID in the decentralized network. Every user can manage his/her Dapp and ethers, and the user can broadcast his/her public key into the network.

Scan QR Code. Then, a centralized web sever should be constructed in order to manage the users' requests. When the sever receives a request to upload/download a specific dataset, the sever can show a QR code on the screen of the user's device who sends the request.

Authorize Transaction. Once the user scan the QR code through uPort system, the user needs to conduct a confirm operation which is an authorization of a transaction to make the request execute successfully.

---

[2] https://eos.io/.

[3] https://truffleframework.com/.

[4] http://remix.ethereum.org/.

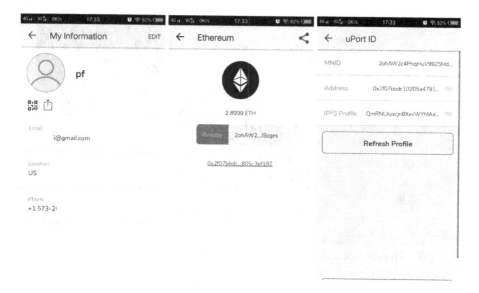

Fig. 1. Example of uPort interface.

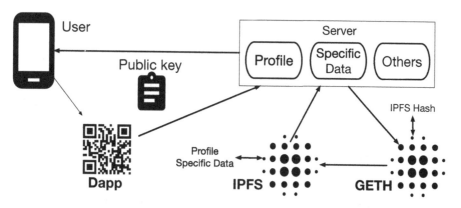

Fig. 2. Framework overview of secure data sharing system.

Execute Operation. If the operation is uploading, the system will charge some ether from the user's account, store the dataset into IPFS and link the generated hash code with the filename, and finally the IPFS's hash code will be stored in Ethereum. If the operation is downloading, the system will charge some ether from the user's account and return the hash-code of store in the IPFS system with a secure way. Then the user can download the data from the decentralized network.

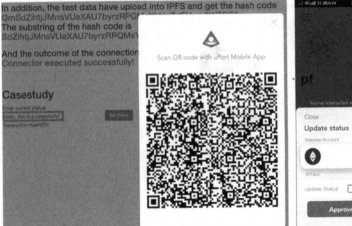

(a) Web interface and the QR code for Uport.   (b) A mobile phone interface for Uport.

Fig. 3. Interface of the simple case.

(a) Example of internal transactions of a transaction   (b) Example of event logs of a transaction

Fig. 4. Schematic diagram of a transaction on Rinkeby net.

5   Case Study

We further use a case to illustrate the effectiveness of this framework. In the framework, users can use the IPFS to store data and use the uPort system to authorize a transaction to be performed. Figure 3 shows the interface of a simple case. When we start the web service, we can upload the test data into the IPFS system, and get a hash code which is the first red string in Fig. 3. When we want to execute a transaction, we can input some information in the "input field" which is circled by a blue rectangle, then we should click the green button and we can see a QR code. Once we scan the QR code in the web page by the uPort system, we can get the information on the mobile as shown in Fig. 3(b).

If the user approve the transaction from the interface of uPort system, the transaction will be processed on Ethernet. As Fig. 4 shows, the transaction information can be seen on the website, such as internal transactions and event logs. And with the control of smart contract, the server can find the hash code of a specific dataset in the IPFS system, then the users could get the dataset in a secure way.

6 Conclusion

In this paper, we try to find a solution to share data securely and effectively in a peer-to-peer network. Specifically, we use the IPFS system to store data, manage the sensitive data by Ethereum, and verify authority through uPort system. With our design, the system can ensure the data transfer safely and each user can upload/download data through a decentralized system. Finally, we evaluate the effectiveness with a case study.

However, there are much work which could be improved. For example, in this framework, we still construct a centralized web service to manage the users' requests, that is, the framework is not an absolutely decentralized structure. Moreover, we only use a simple case study to test the effectiveness of our proposed framework, and more real data sharing experiments cloud be added in the future. Thus, we would follow these available improve-points to continue our work in the future.

Acknowledgment. The work described in this paper was supported by the National Natural Science Foundation of China (No. 91546125), the Chinese Academy of Sciences (No. XXH13505-08-04) and National Science and Technology Infrastructure Center (No. 2018DDJ1ZZ14).

References

1. Adhikari, C.L., et al.: Decentralized secure framework for sharing and managing electronic health record using ethereum-based blockchain technology (2017)
2. Bähnemann, R., Schindler, D., Kamel, M., Siegwart, R., Nieto, J.: A decentralized multi-agent unmanned aerial system to search, pick up, and relocate objects. In: Proceedings of 2017 IEEE International Symposium on Safety, Security and Rescue Robotics (SSRR), p. 8088150. IEEE (2017)
3. Bechky, B.A.: Sharing meaning across occupational communities: the transformation of understanding on a production floor. Organ. Sci. **14**(3), 312–330 (2003)
4. Benet, J.: Ipfs-content addressed, versioned, p2p file system. arXiv preprint arXiv:1407.3561 (2014)
5. Brundo, R., De Nicola, R.: Blockchain-based decentralized cloud/fog solutions: challenges, opportunities, and standards. IEEE Commun. Stand. Mag. **2**(3), 22–28 (2018)
6. Buterin, V., et al.: A next-generation smart contract and decentralized application platform. white paper (2014)
7. Chrysanthakopoulos, G., Nielsen, H.F., Moore, G.M.: Decentralized system services, uS Patent 8,122,427, 21 February 2012

8. Daswani, N., Garcia-Molina, H., Yang, B.: Open problems in data-sharing peer-to-peer systems. In: Calvanese, D., Lenzerini, M., Motwani, R. (eds.) ICDT 2003. LNCS, vol. 2572, pp. 1–15. Springer, Heidelberg (2003). https://doi.org/10.1007/3-540-36285-1_1

9. Fecher, B., Friesike, S., Hebing, M., Linek, S., Sauermann, A.: A reputation economy: results from an empirical survey on academic data sharing. arXiv preprint arXiv:1503.00481 (2015)

10. Fienberg, S.E., Martin, M.E., Straf, M.L.: Sharing Research Data. National Academy Press, Washington, D.C. (1985)

11. Gardner, D., et al.: Towards effective and rewarding data sharing. Neuroinformatics 1(3), 289–295 (2003)

12. Hajialikhani, M., Jahanara, M.: Uniqueid: decentralized proof-of-unique-human. arXiv preprint arXiv:1806.07583 (2018)

13. Jiang, C., Gao, L., Duan, L., Huang, J.: Scalable mobile crowdsensing via peer-to-peer data sharing. IEEE Trans. Mobile Comput. 17(4), 898–912 (2018)

14. Kansa, E.C., Kansa, S.W., Burton, M.M., Stankowski, C.: Googling the grey: open data, web services, and semantics. Archaeologies 6(2), 301–326 (2010)

15. Lee, W., Kim, T.Y., Kang, S., Kim, H.C.: Revised P2P data sharing scheme over distributed cloud networks. In: Kim, K.J. (ed.) Information Science and Applications. LNEE, vol. 339, pp. 165–171. Springer, Heidelberg (2015). https://doi.org/10.1007/978-3-662-46578-3_20

16. Marshall, E.: DNA sequencer protests being scooped with his own data. Science 295(5558), 1206–1207 (2002)

17. McAfee, A., Brynjolfsson, E., Davenport, T.H., Patil, D., Barton, D.: Big data: the management revolution. Harvard Bus. Rev. 90(10), 60–68 (2012)

18. Ng, W.S., Ooi, B.C., Tan, K.L., Zhou, A.: Peerdb: a p2p-based system for distributed data sharing. In: 19th International Conference on Data Engineering, Proceedings, pp. 633–644. IEEE (2003)

19. Patel, V.: A framework for secure and decentralized sharing of medical imaging data via blockchain consensus. Health Inf. J. (2018). https://doi.org/10.1177/1460458218769699

20. Pouwelse, J., Garbacki, P., Epema, D., Sips, H.: The bittorrent P2P file-sharing system: measurements and analysis. In: Castro, M., van Renesse, R. (eds.) IPTPS 2005. LNCS, vol. 3640, pp. 205–216. Springer, Heidelberg (2005). https://doi.org/10.1007/11558989_19

21. Shehabi, A., Stokes, J.R., Horvath, A.: Energy and air emission implications of a decentralized wastewater system. Environ. Res. Lett. 7(2), 024007 (2012)

22. Shukla, S., Thasmiya, A., Shashank, D., Mamatha, H.: Online voting application using ethereum blockchain. In: 2018 International Conference on Advances in Computing, Communications and Informatics (ICACCI), pp. 873–880. IEEE (2018)

23. Swan, M.: Blockchain thinking: the brain as a DAC (Decentralized Autonomous Organization). In: Texas Bitcoin Conference, Chicago, pp. 27–29 (2015)

24. Wood, G.: Ethereum: a secure decentralised generalised transaction ledger. Ethereum Proj. Yellow Paper 151, 1–32 (2014)

25. Zyskind, G., Nathan, O., et al.: Decentralizing privacy: using blockchain to protect personal data. In: 2015 IEEE Security and Privacy Workshops (SPW), pp. 180–184. IEEE (2015)

Agricultural Disease Image Dataset for Disease Identification Based on Machine Learning

Lei Chen and Yuan Yuan[(⊠)]

Institute of Intelligent Machines, Chinese Academy of Sciences, Hefei, China
{chenlei,yuanyuan}@iim.ac.cn

Abstract. Identification and control of agricultural diseases and pests is significant for improving agricultural yield. Food and Agriculture Organization of the United Nations reported that more than one-third of the annual natural loss is caused by agricultural diseases and pests. Traditional artificial identification is not accurate enough since it relies on subjective experience. In recent years, computer vision and machine learning, which require large-scale training samples, have been widely used for crop disease image identification. Therefore, building large training dataset and studying new classifier modeling methods are very important. Accordingly, on the one hand, we have constructed an agricultural disease image dataset which covers many research fields such as image acquisition, segmentation, classification, marking, storage and modeling. The dataset currently has about 15,000 high-quality agricultural disease images, including field crops such as rice and wheat, fruits and vegetables such as cucumber and grape, etc. And it will continue to grow. On the other hand, with the support of this dataset, we investigated a disease image identification method based on different kinds of transfer learning with deep convolutional neural network and achieved good results. The paper has two contributions. First, the constructed agricultural disease image dataset provides valuable data resources for the research of agricultural disease image identification. Secondly, the proposed disease identification method based on transfer learning can provide reference for disease diagnosis where the available labeled samples are still limited.

Keywords: Agricultural disease image dataset · Image identification · Transfer learning · Deep learning · Big data

1 Introduction

Agricultural disease is one of the important factors that affect agricultural yield and food security [11,16]. Food and Agriculture Organization of the United Nations reported that more than one-third of the annual natural loss is caused by agricultural diseases and pests [3]. Due to the wide variety of agricultural diseases, it is easy to misdiagnose by artificial observation and experience judgment. So traditional artificial identification method is not accurate enough.

© Springer Nature Switzerland AG 2019
J. Li et al. (Eds.): BigSDM 2018, LNCS 11473, pp. 263–274, 2019.
https://doi.org/10.1007/978-3-030-28061-1_26

In recent years, computer vision and machine learning have been widely used for agricultural disease image identification. Especially with the introduction and application of deep learning method, some achievements have been made in the filed of agricultural disease image identification. These machine learning based methods usually require large-scale training samples to construct the classification model. Unfortunately, due to the wide variety of crops and diseases, the scale of available agricultural image resources is still too small to construct the ideal disease identification models for most crops. Therefore, building large training dataset and studying new modeling methods for agricultural disease image identification are very important.

Accordingly, in this paper, we introduce an agricultural disease image dataset and propose some disease image identification methods based on different category of kinds of transfer learning with deep convolutional neural networks. On the one hand, we have built an agricultural disease image dataset which covers many research issues of agricultural disease image identification, such as image acquisition, segmentation, classification, marking, storage and modeling. The image dataset currently has about 15,000 high-quality agricultural disease images, including field crops such as rice and wheat, fruits and vegetables such as cucumber and grape, etc. And it will continue to grow. Especially different from the existing agricultural disease graph resources which mostly contain only 3 to 5 typical symptom images [4,7], our dataset consists of the original image data of the same kind of crop diseases with high resolution and high similarity, containing hundreds or even thousands images of each disease, which can be used as training samples for constructing the model of disease image identification. On the other hand, using the image resources of this dataset, we investigated some agricultural disease image identification methods, which combined different transfer learning approaches with deep convolutional neural networks, and achieved good results.

2   Agricultural Disease Image Dataset

2.1   Image Acquisition

All the images in this agricultural disease image dataset are captured by the authors in the field or in the greenhouse of Anhui province. The collected disease images are mainly under natural light conditions and required to meet certain shooting conditions to ensure uniform light. The shooting angle is needed to make the light path perpendicular to the plane of the crop organs as far as possible. And the organ of crop captured occupies the central position of the image. The image acquisition devices used in this paper are as follows:

- Canon digital SLR camera EOS 6D, with EF 17–40 mm f/4L USM lens and EF 100 mm f/2.8L IS USM Macro lens.
- Canon digital camera EOS M6, with EF-M 28 mm f/3.5 IS STM Macro lens.
- SONY digital camera DSC-RX100M3.

We use the best image quality and maximum resolution of each camera (6000 * 4000 and 5472 * 3648 respectively) to capture images. The original format of Canon digital cameras is RAW, which can be converted into JPG format by using the Canon software Digital Photo Professional on computer. The SONY digital camera is directly captured in JPG format. All cameras use the aperture priority mode, where the aperture can be adjusted to make the depth of field enough, in order to ensure that the captured crop organ is clearer in the image.

2.2 Sample Description

The dataset is managed in the form of image database. The first layer of image data dictionary is stored as a data table in Microsoft SQL Server. Some examples of the fields in this data table are shown in Table 1, where image path is an index to the original image file stored on computer's hard drive. Besides, some key information such as shooting data, time, location, image size, etc. are already contained in the EXIF (Exchangeable image file format) information of the image file. Therefore, they are not repeated in the data table.

Table 1. Examples in the data table of database

Crop	Organ	Disease	Image path	Remark
Cucumber	Leaf	Powdery mildew	huanggua\baifen\IMG1882.jpg	Null
Cucumber	Leaf	Powdery mildew	huanggua\baifen\IMG1883.jpg	Null
Rice	Leaf	Rice blast	shuidao\daowen\DSC18_2083.jpg	Null
Rice	Ear	Rice false smut	shuidao\daoqu\IMG17_5657.jpg	Null

The second layer is the raw image data stored on the computer hard disk, where the first-level folder is established according to the crop category and the secondary folder is established according to the type of disease. Each file is a single image, representing a disease sample. Figure 1 shows some samples of rice blast stored on the hard disk.

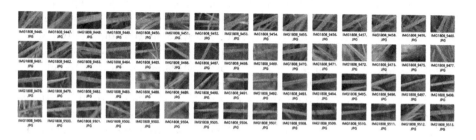

Fig. 1. Image samples of rice blast

There are two main ways to collect agricultural disease images in this study. The first way is artificial inoculation by plant protection experts, that is, inoculating pathogenic microorganisms on target crops to cause disease. In this way, the captured images of target crops can completely guarantee the classification accuracy of disease samples. The other way is to capture crop disease images at various agricultural production bases. In this way, diseases are naturally occurring. First, plant protection experts are needed to identify diseases in these agricultural production bases. And then the disease images are further analyzed and identified by other different plant protection experts to ensure the classification accuracy of the captured images. In addition, the image acquisition work is carried out by professional technicians engaged in image identification of agricultural diseases in according to standard operating procedures and specifications. And then the collected images are manually screened to eliminate the images that do not meet the requirements. The above practices can ensure the quality of our agricultural disease image dataset.

3   Disease Identification Based on Transfer Learning

3.1   From Traditional Methods to Transfer Learning

As mention above, in the past few years, with the adoption of computer vision and machine learning, there has been a great progress in the study of agricultural disease image identification. The most widely studied image classification and identification methods include SVM (support vector machine) [12,17], discriminant analysis [18] and KNN (K-nearest neighbors) [10,21], etc. Although these machine learning traditional methods have made some achievements, they are still subject to certain restrictions: First, the dataset used and handled in these methods are small in scale, less than 300 images. Second, before constructing a classification model, lesion image segmentation and feature extraction are required. However, lesion image segmentation is always not easy for some kinds of agricultural diseases and artificially defined low-level image features may not well represent the characteristics of diseases. Therefore, existing traditional machine learning methods are difficulty to get satisfactory results in agricultural disease image identification.

With the development of machine learning technology, the deep learning method has able to solve the modeling problem of big data. This method can be used in the field of disease image identification without the complicated operations such as lesion image segmentation and feature extraction. However, the current amount of data in some crop disease images is not enough to support deep learning to build the ideal model. Therefore on the one hand, while building agricultural disease image data resources, on the other hand, we also actively study new modeling methods for small-scale data sets. The introduction of transfer learning provides us with new ideas, which is used to improve a learner from one domain by transferring information from a related domain [9,19]. According to the form of transferred information, transfer learning can be categorized as instance-based, parameter-based, feature-based and so on. Considering the particularity of the study of agricultural disease image identification and the scale of data resources, this paper investigated instance-based and parameter-based transfer learning methods respectively.

3.2 Instance-Based Transfer Learning

Instance-based transfer learning is a kind of homogeneous transfer learning. A common method used in this case is to re-weight the instances from the source domain in order to correct the marginal distribution differences. Then these re-weighted instances are directly used in the target domain for training model. These re-weighting algorithms work best when the conditional distribution is the same in both domains [2]. Inspired by this method, we attempt to train the target disease image classification model by using other disease image dataset with large amount of data, that is, using the agricultural disease image dataset we have built. Our method mainly includes the following steps:

1. Image pre-processing, such as image compression, lesion segmentation, etc.
2. Instance-based transfer learning, consisting of weight adjustment of auxiliary data and target data.
3. Training the final classification model.

In key step 2, based on the TrAdaBoost algorithm [1], which constantly adjusts the weights of the target training samples and the auxiliary training samples to get the final classifier, we propose a training set optimization strategy. In this strategy, detailed as the following Algorithm 1, the auxiliary data with smaller similarity to target training data will be filtered by using the KNN algorithm.

Algorithm 1. Training set optimization

Input: An auxiliary dataset D_a; A target training dataset D_t; The number of nearest
  neighbor sample k;
Output: A new auxiliary dataset D'_a;
  1: Calculate the Euclidean distance between each target training data and auxiliary
    data;
  2: Sort all distances to determine the k sample points with the smallest distance;
  3: Calculate the frequency of the categories of the k sample points;
  4: Use the highest frequency category C_p as the predictive category of D_a;
  5: Compare the true category of the auxiliary dataset D_a with C_p to remove the
    predicted wrong data from D_a;
  6: **return** The new auxiliary dataset D'_a.

In the process of agricultural disease image identification, first we choose some other disease image dataset as the auxiliary data, which will be filtered by the proposed training set optimization KNN algorithm. And then the TrAdaBoost algorithm is adopted to adjust the weight of training data in order to construct the final disease image classification model.

The experimental data contains disease image datasets of six crops, that is, cucumber target spot (CTS), cucumber downy mildew (CDM), cucumber bacterial angular spot (CBAS), rice blast (RB), rice brown spot (RBS) and rice sheath blight (RSB). The specific data of four groups of experiments are shown in the following Table 2.

Table 2. Experimental data of instance-based transfer learning

Group	Auxiliary dataset	Feature vector	Target dataset	Feature vector
1	CTS & CDM	1472	RBS & RB	540
2	CDM & RB	1472	CTS & CBAS	540
3	RSB & RB	1472	CBAS & CDM	540
4	CTS & CDM	1472	RSB & RB	540

As the evaluation criteria of experiments, the classification accuracy A_c is defined as follows:

$$A_c = \frac{N_r}{N_t} \times 100\% \tag{1}$$

where N_r is the number of test lesions that are correctly classified and N_t is the total number of test lesions. Table 3 shows the experimental results of the comparison between our method with the traditional SVM method and original TrAdaBoost algorithm when the ratio of target data to auxiliary data is 4%.

Table 3. Experimental results (A_c) of instance-based transfer learning

Group	SVM	TrAdaBoost	Ours
1	91.23	93.20	95.27
2	89.40	94.12	96.48
3	80.79	90.37	92.42
4	67.46	82.15	83.18

From Table 3 we can see that the results of original TrAdaBoost algorithm are better than traditional SVM method, indicating that transfer learning can find information that is helpful to target data modeling in auxiliary datasets, so as to improve classification effect. And the results of our method are better than original TrAdaBoost algorithm, showing that the optimization strategy of training set is effective. For more details, please refer to the previous work of Fang et al. [2].

Obviously, when the size of target disease image data is insufficient to train the ideal classification model, the proposed instance-based transfer learning can achieve some improvements. This method will be more effective when the similarity between auxiliary data and target data is high. However, it is also difficult to find such auxiliary data with high similarity in practice.

3.3 Parameter-Based Transfer Learning

Parameter-based transfer learning is also a kind of homogeneous transfer learning. This method transfers knowledge through shared parameters of source and target domain learner models or by creating multiple source learner models and optimally combining the re-weighted learners to construct an improved target learner. In this case the similarity between auxiliary data and target data is not required to be very high. And the open-source ImageNet dataset is often used as auxiliary data to train the required parameters, which will be fine-tuned and adopted in the target data modeling [6, 13, 15]. However, this dataset is not suitable as an auxiliary data for the issue studied in this paper, since the difference between our target data and ImageNet is still large and the scale of target data is still small. So another open-source dataset called PlantVillage [8, 14], which has relatively better similarity to our agricultural disease image dataset, is selected as the auxiliary data in this section. Generally speaking, we combine parameter-based transfer learning with two popular deep learning architectures AlexNet [6] and VGGNet [13] to classify different kinds of agricultural diseases images. Figure 2 shows the whole network architecture based on parameter-based transfer learning.

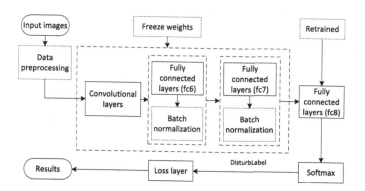

Fig. 2. The whole network architecture based on parameter-based transfer learning

Concretely, two techniques called batch normalization and DisturbLabel are introduced to reduce the number of training iterations and over-fitting in the training procedure. Batch normalization ensures that the inputs of layers always fall in the same range even though the earlier layers are updated. And it can obviously reduce the number of training iterations and regularize the model [5].

DisturbLabel is actually a regularization method on the loss layer. It randomly chooses a small subset of training data and intentionally sets their ground-truth labels to be incorrect in order to prevent network from over-fitting and improve the training process [20]. After constructing the pre-trained model by using the PlantVillage dataset, we fine-tune this model with our relatively small dataset preprocessed by adopting DisturbLabel on the loss layer.

The experiments are conducted on the open-source machine learning framework called TensorFlow. Table 4 gives the experimental target data that all come from the agricultural disease image dataset we built.

Table 4. The size of experimental target data

Disease	Raw dataset	Processed dataset
Cucumber powdery mildew	201	763
Cucumber downy mildew	198	780
Cucumber bacterial angular spot	144	576
Rice sheath blight	893	3559
Rice blast	693	2737
Rice brown spot	201	795
Rice bacterial blight	50	196
Rice false smut	49	186
Total	2429	9592

In Table 4, the raw dataset is obtained by simply cropping the central lesion area in the original image and the processed dataset is obtained by cropping from the original images according to some rules, as shown in Fig. 3. Note that the images without disease spots will be filtered out.

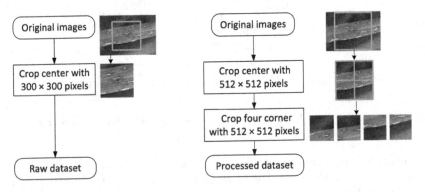

Fig. 3. Two kinds of target data from our agricultural disease image dataset

As the auxiliary data, PlantVillage dataset, which includes more than 50,000 images, is used to construct the pre-trained model, where the dataset is split into two sets, namely 80% is training set and the left 20% is validation set. Each of these experiments runs for 100 epochs. Transfer learning is carried out based on the pre-trained model. Each target dataset is also split into two sets, 80% is training set and the left 20% is validation set. And we use DisturbLabel algorithm on the loss layer to reduce the over-fitting problem. Then we compare the results of AlexNet and VGGNet by training models on the raw dataset and processed dataset respectively. According to the empirical observation, the range of noise rate γ is set from 0.08 to 0.2, the number of iterations is set to 300 epochs and dropout rate is set to 0.5. We use the average accuracy A_{ave}, calculated as follows, to evaluate the experimental results:

$$A_{ave} = \frac{1}{n_c} \sum_{i=1}^{n_c} \frac{n_{ai}}{n_i} \times 100\% \tag{2}$$

where n_c is the training number of each epoch, n_{ai} is the number of the sample predicted accuracy of each training and n_i is the number of the sample of each training. The experimental results are given in Table 5. For raw dataset, the highest validation accuracies of two networks are 94.97% and 95.14% respectively when γ is 0.15. For processed dataset, the highest validation accuracies are 95.93% and 95.42% respectively when γ is 0.08. The overall experimental results of processed dataset are better than raw dataset.

Table 5. Accuracies of different γ on AlexNet and VGGNet

Model	Dataset	$\gamma = 0.08$	$\gamma = 0.1$	$\gamma = 0.15$	$\gamma = 0.2$
AlexNet	Raw dataset	94.14	91.81	**94.97**	92.39
AlexNet	Processed dataset	**95.93**	95.66	95.42	94.84
VGGNet	Raw dataset	91.79	95.05	**95.14**	92.15
VGGNet	Processed dataset	**95.42**	94.97	94.89	94.57

In addition to the original network's own optimization approaches such as dropout layer and L2 regularization, we also combine DisturbLabel with batch normalization to improve the final results. The experimental results are shown in Table 6, indicating that comparing with two original networks for fine-tuning the pre-trained model, our method can achieve better results on both raw dataset and processed dataset.

Furthermore, as shown in Fig. 4, we can see that our method can converge well within 300 epochs. And the performance of the training on the processed dataset is more stable than raw dataset. The reason may be that the images in processed dataset is more than the images in raw dataset.

Besides, the effect of batch normalization and DisturbLabel on results were explored. Figure 5 shows the method only with the batch normalization (BN)

Table 6. Accuracies of different transfer learning approaches

Model	Raw dataset	Processed dataset
Our method	**94.97**	**95.93**
AlexNet fine-tuning	91.00	94.92
AlexNet from scratch	89.24	94.73
Our method	**95.14**	**95.42**
VGGNet fine-tuning	82.60	86.13
VGGNet from scratch	84.48	85.20

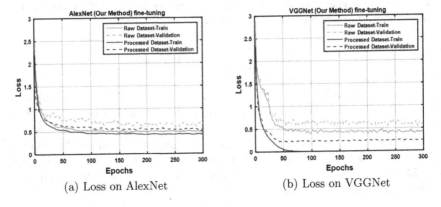

(a) Loss on AlexNet   (b) Loss on VGGNet

Fig. 4. Comparison of loss on two datasets of AlexNet and VGGNet

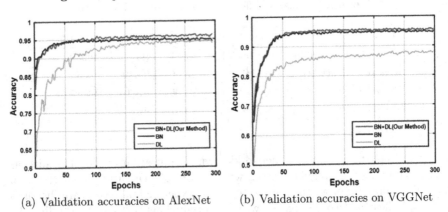

(a) Validation accuracies on AlexNet   (b) Validation accuracies on VGGNet

Fig. 5. The effect of batch normalization and DisturbLabel

and the method only with DisturbLabel (DL) for two networks on the processed dataset. As expected, adding the batch normalization to the fully connected layers can lead to a faster convergence than only using DisturbLabel. Because batch normalization can lead to in a significant reduction in the required number of training iterations.

4   Conclusion

The paper introduced an agricultural disease image dataset, which can be used as the training set of machine learning method to construct the disease image identification model. So the constructed agricultural disease image dataset can provide valuable data resources for the research of agricultural disease image identification. The dataset currently has about 15,000 high-quality agricultural disease images, including field crops such as rice and wheat, fruits and vegetables such as cucumber and grape, etc. And we are still continuing to build this dataset and more kinds of crops will be included soon.

Meanwhile, with the support of this dataset, the paper proposed the disease image identification method based on different kinds of transfer learning with deep convolutional neural network. The experiments revealed that our method can achieve better results than traditional machine learning method such as SVM and original deep learning architecture such as AlexNet and VGGNet. Especially in parameter-based transfer learning, we didn't carry out some complicated image preprocessing such as lesion image segmentation and feature extraction. Compared with traditional machine leaning methods, this is very simple and attractive.

Obviously, the work in this paper is still preliminary. In next work, while continuing to build the agricultural disease image dataset, how to select the more suitable auxiliary dataset and choose more appropriate transfer learning approaches will be studied. And comparisons with more deep learning architectures will also be considered.

Acknowledgments. The authors would like to thank the anonymous reviewers for their helpful reviews. The work is supported by the 13th Five-year Informatization Plan of Chinese Academy of Sciences (Grant No. XXH13505-03-104) and National Natural Science Foundation of China (Grant No. 31871521).

References

1. Dai, W., Yang, Q., Xue, G., Yu, Y.: Boosting for transfer learning. In: Proceedings of the Twenty-Fourth International Conference on Machine Learning (ICML 2007), Corvallis, Oregon, USA, 20–24 June 2007, pp. 193–200 (2007)
2. Fang, S., Yuan, Y., Chen, L., Zhang, J., Li, M., Song, S.: Crop disease image recognition based on transfer learning. In: Zhao, Y., Kong, X., Taubman, D. (eds.) ICIG 2017. LNCS, vol. 10666, pp. 545–554. Springer, Cham (2017). https://doi.org/10.1007/978-3-319-71607-7_48
3. Harvey, C.A., et al.: Extreme vulnerability of smallholder farmers to agricultural risks and climate change in Madagascar. Philos. Trans. R. Soc. Lond. **369**(1639), 20130089 (2014)

4. International, C., (RU), W.: Crop Protection Compendium. Blackwell Verlag GmbH (2005)
5. Ioffe, S., Szegedy, C.: Batch normalization: accelerating deep network training by reducing internal covariate shift, pp. 448–456 (2015)
6. Krizhevsky, A., Sutskever, I., Hinton, G.E.: ImageNet classification with deep convolutional neural networks. In: International Conference on Neural Information Processing Systems, pp. 1097–1105 (2012)
7. Lv, P.: Diseases and Pests Original Color Drawing of Chinese Grain Crops, Economic Crops and Medicinal Plants. Yuanfang Press (2007)
8. Mohanty, S.P., Hughes, D.P., Salathé, M.: Using deep learning for image-based plant disease detection. Front. Plant Sci. **7**, 1419 (2016)
9. Pan, S.J., Yang, Q.: A survey on transfer learning. IEEE Trans. Knowl. Data Eng. **22**(10), 1345–1359 (2010)
10. Prasad, S., Peddoju, S.K., Ghosh, D.: Multi-resolution mobile vision system for plant leaf disease diagnosis. Sig. Image Video Process. **10**(2), 379–388 (2016)
11. Sanchez, P.A., Swaminathan, M.S.: Cutting world hunger in half. Science **307**(5708), 357–359 (2005)
12. Semary, N.A., Tharwat, A., Elhariri, E., Hassanien, A.E.: Fruit-based tomato grading system using features fusion and support vector machine. In: Filev, D., et al. (eds.) Intelligent Systems'2014. AISC, vol. 323, pp. 401–410. Springer, Cham (2015). https://doi.org/10.1007/978-3-319-11310-4_35
13. Simonyan, K., Zisserman, A.: Very deep convolutional networks for large-scale image recognition. Comput. Sci. (2014)
14. Srdjan, S., Marko, A., Andras, A., Dubravko, C., Darko, S.: Deep neural networks based recognition of plant diseases by leaf image classification. Comput. Intell. Neurosci. **2016**(6), 1–11 (2016)
15. Szegedy, C., et al.: Going deeper with convolutions, pp. 1–9 (2014)
16. Tai, A.P.K., Martin, M.V., Heald, C.L.: Threat to future global food security from climate change and ozone air pollution. Nat. Clim. Change **4**(9), 817–821 (2014)
17. Tian, Y.W., Li, T.L., Zhang, L., Wang, X.J.: Diagnosis method of cucumber disease with hyperspectral imaging in greenhouse. Trans. Chin. Soc. Agric. Eng. **26**(5), 202–206 (2010)
18. Wang, X., Zhang, S., Wang, Z., Zhang, Q.: Recognition of cucumber diseases based on leaf image and environmental information. Trans. Chin. Soc. Agric. Eng. **30**(14), 148–153 (2014)
19. Weiss, K.R., Khoshgoftaar, T.M., Wang, D.: A survey of transfer learning. J. Big Data **3**(1), 9–48 (2016)
20. Xie, L., Wang, J., Wei, Z., Wang, M., Tian, Q.: DisturbLabel: regularizing CNN on the loss layer. In: Computer Vision and Pattern Recognition, pp. 4753–4762 (2016)
21. Zhang, S.W., Shang, Y.J., Wang, L.: Plant disease recognition based on plant leaf image. J. Anim. Plant Sci. **25**(3), 42–45 (2015)

Simbaql: A Query Language for Multi-source Heterogeneous Data

Yuepeng Li[1,2], Zhihong Shen[1], and Jianhui Li[1(✉)]

[1] Computer Network Information Center, Chinese Academy of Sciences,
Beijing 100190, China
{liyuepeng,lijh}@cnic.cn
[2] University of Chinese Academy of Sciences, Beijing 100049, China

Abstract. In a data-driven era, scientific discovery by querying integrated heterogeneous data is becoming a popular approach. However, most current search engines retrieve data using SQL, which only addresses part of the common data processing use cases in data discovery for scientific research. Besides, the expressive power of SQL and other popular languages can't describe some simple but useful second order logic queries. Therefore, we first proposed an abstract data processing model which contains four components: data set, data source, data model, and analysis tool. Then, we introduced a unified data model and SimbaQL query language to describe the steps of data processing. At last, we studied two cases by describing the data processing using SimbaQL, and it turns out that SimbaQL can describe these tasks properly.

Keywords: Query language · Heterogeneous data · Scientific data

1 Introduction

Since the concept of "Big Data" was proposed in the 2011 Mckinsey annual report, the definition of "Big Data" has been in dispute on academia and industry [1]. Most of the dispute came from the characteristics of big data in different domains. And in scientific research, data sharing is an obvious different characteristic with other domains [2]. For example, most of nowadays significant scientific programs, such as Human Genome Project (HGP), Large Hardon Collider (LHC), etc. share data around the world. And scientists make discoveries from the open data, including the discoveries of gravitational wave and god particle which win the Nobel prize. Generally, several open source data will be involved in a scientific research. For example, the World Data Center for Microorgannisms (WDCM) [3] published a knowledge graph that is built from 36 open source data including Taxonomy, Genbank, Gen, etc. These data has various formats, such as file, relational database etc. BigDAWG [4] is a polystore database, which manages the medical record, sequence data, image, radio etc. for medical research. The applications mentioned above have a similar characteristic that all of them

J. Li et al. (Eds.): BigSDM 2018, LNCS 11473, pp. 275–284, 2019.
https://doi.org/10.1007/978-3-030-28061-1_27

need search heterogeneous data in data processing. However, the query language of current tools for heterogeneous data cannot describe the whole life cycle of these data processing. Therefore, we proposed a query language for multi-source heterogeneous data, which can import/export data to database, transform data from different data models and integrate multi-source data.

2   Related Work

Multiple types of databases have been used to manage data in nowadays system, such as relational database, graph database, document database, key-value database and so on. These databases build data models to describe the data formally, and retrieval data using the operations or query language of data model. For example, relational databases query the relational data model by SQL; graph databases describe data by property graph model and RDF, and query the graph by Cypher or SparQL; Xquery language search object model for XML file [5].

When it comes to heterogeneous data, there are three methods to access data: data integration, multi-model database, polystore [4]. Data integration [6] is defined as a three tuples (G, L, M), where G is global data model, L is local data model, and M is the mapping from local data models to global data model. Data integration queries local data models using global data model query language. For example, popular-used SQL query engines, such as Impala, Spark-SQL, Presto, query multiple data sources using SQL language. Besides, data warehouse, federated database [7] are also methods of data integration. Multi-model database [8] stores different types of data using one unified data model, and query data using unified language, such as OrientDB [8], Ag- groDB, etc. Polystore database stores data in different databases. When query by language A, polyestore database execute the query in A database. If the target data is not stored in A database, then polyestore will CAST the target data to A database [9].

However, above methods for heterogeneous data have several defects. First, data integration mainly use SQL as global query language, but NoSQL databases are de- signed to against SQL, therefore SQL is obviously not suitable for querying different data model. Second, multi-model databases store data in one back-end, which is not suitable for multi-source data. Besides, although polystore databases can query data in different ways, but it require users to learn many database languages, which means polystore just transform data between different data models and cannot achieve the goal of unified search. And current polystore query language only supports one time transform, and cannot query the intermediate results many times.

As regard as the query language for heterogeneous data, current query language have another two defects. On one side, current query language is not designed for data integration specially. For example, in the processing of build knowledge graph, WDCM wants to create the relation between species's information stored in relational database and the gene sequence information stored in Genbank file. However, SQL failed to treat species and genbank file as a whole to extract and integrate data. On the other side, current languages are good at dealing with different type of data, and most of them can only solve first logic problems [10–12]. For example, relational data model describe relationships between objects using foreign key, which means we can only get the relationship between objects by computing. As for Cypher [13], it can only query related data stored in database. The following query is illegal

```
match (p:Person), (p2:Person)
where p.name=p2.name
```

To solve the problems mentioned above, we first proposed an intermediate data model which can transform between common data models. Then, we defined the operations and query language on the data model. At last, we studied the query language by two scientific cases. The main contributions of this paper are as followed. First, we abstract the scientific data processing by data set, data source, data model, analyzing tools. Second, we designed data model, Linked-Document-Model, and query language, SimbaQL, for data integration specialty. Third, we enhance the expressive power by introducing the operation of describing the property of LDM.

3 Methodology

In big data era, scientific researches are mainly driven by data. If one scientist wants to find which factor is the most important to a country's happiness, he need to get the published GDP data, agriculture data, political data and so on. These data may be located in files, web service, database and so on. Then, he needs to import theses data into databases, and extract interested part from database. Finally he will integrate all the data to the format that analyze tools can recognize.

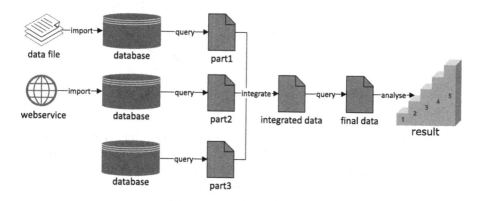

Fig. 1. Data processing in scientific research.

Above data processing is not alone. As Fig. 1 shows, the last step in scientific discoveries is analyzing data. However, most of scientist's work are preparing data from heterogeneous sources. In the process of preparing data, three types of components are involve: *data set, data source and analyzing tools*. Data set, such as file, can only store data. Data source, such as database, retrieval engine, can not only store data, but also manage data. Analyzing tools, such as Spark, can only process data. We can divide the data process into the following steps:

1. *Find open source data set or build his own data set.*
2. *Import data set into proper data source.*
3. *Search interested data and integrate them.*
4. *Reprocess the integrated data to the format that analyzing tools can recognize.*
5. *Analyze the data by tools.*

In this paper, we want to design a query language that can accomplish the tasks in 1- 4 steps. For the first step, we define different data set types, such as file, http, web- service, to get the target data. For the second step, we define the import operation of data set. For 3–4 steps, we defined intermediate data model LDM and its mapping rules and operations. Besides, we define the description operation to enhance the expressive power of SimbaQL, and export operation to push the LDM data to analyzing tools. Our abstract model of data processing in scientific research is shown as Fig. 2.

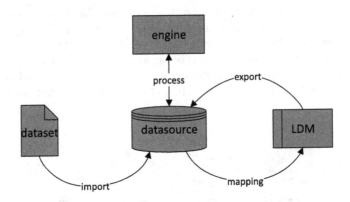

Fig. 2. Data processing in scientific research.

As Fig. 3 shows, heterogeneous data query system that implement SimbaQL contains three components: SimbaQL parser, execute engine and data source. When user input a query, the parser will translate it into data set, data source, and operations of LDM. For example, if we want to query a csv file, the system may import the file data set to relational database, and then perform the query. The system will invoke the query en- gine of this data source when a query only involve a single data source. And If one or many data sources cannot solve the query problem, then execute engine will make a compensate computing for this query.

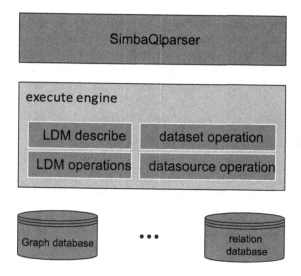

Fig. 3. Data processing in scientific research.

4   Simba Query Language

In this section, we proposed a unified data model LDM, which contains four types of operations. After that we defined the mapping rules between four often-used data mod- els and LDM. At last, we introduced the informal SimbaQL language grammar.

Linked-Document-Model. The goal of data model is to describe the objective things and relationship between things formally. There are two methods to describe a thing: schema and schemaless. Schema is more suitable for computing, but schemaless way is more suitable for altering thing dynamically. Similarly, there are two methods to describe the relationships between things: storage and computing. Relational data model use computing to get relationship, but the relationship has less semantic. Graph model use storage to get relationship, but it costs storage space.

To satisfy unified query for heterogeneous data, we want LDM has the following three characteristics.

1. *Support both schema and schemaless ways to describe things.*
2. *Support both storage and computing ways to describe relationship.*
3. *Support describing the data processing in scientific research.*

Link and Document. Document is a semi-structured document which has multiple attributes. Each document must has unique key to identify this document. We can add attributes to a document dynamically, and we can also require a document must have some attributes. Generally, we are talking about document set which must have a label name, and one document can belong to multiple document sets. Link is a special document which must have two attributes *(from, to)*. Similar with document set, link set must have a label name. We can define a link set by two ways. For example, we can define the "knows" link by storage way,

$$knows = [(from : 1, to : 2), (from : 2, to : 3)]$$

and we can also define the link by computing way

$$knows = \{(from, to) | from \in customer.id \quad and \quad to \in order.customerid$$
$$and \quad customer.id = order.customerid\}$$

Mapping Rule. LDM is a tuple defined as followed. The first item of this tuple is

$$ldm = (\{document\ sets\}, \{link\ sets\})$$

a set of document set, and the second item is a set of link set. A LDM can stand for a data source where the document set and link set in LDM is mapped to the components of data source. For example, table in relational database is mapped to a document set, and foreign key is mapped to a link set. As Table 1 shows, we define the mapping rules between relational data model, graph data model, key-value data model, document data model and LDM.

Table 1. LDM mapping rules.

LDM	Relational model	Key-value mode	Document model	Graph model
attribute	attribute	key	attribute	attribute
document	record	pair	collection	vertex
document set	table	——	——	——
link	——	——	——	edge
link set	foreign key	——	——	——

Operations. There are four types of operations on LDM: set operation, link set operation, document set operation and transform operation. The detail of these operations are shown as Table 2. Besides, we can access the data in LDM similar to url, the url is defined as followed.

$$< datasource > . < document > . < links > . < index > . < attributes >$$

For example, wdcm.species.gen stands for the gene of the species.

Table 2. LDM mapping rules.

Operation	Description
Set Operation	
+(Union)	Union two LDM into one LDM. Combine the documents and Links with same label
∩(Intersect)	Find the same documents and links between two LDMS
−(Minus)	Find the documents and links which are in LDM1 and not in LDM2
Link Set Operation	
×(Create Link)	Create links between documents based on given conditions
δ_{link}(Projection)	Project a LDM into a link means filter the data in document which exist some link
π_{link}(Selection)	Select the links which has the feature such as name, count, etc
Document Set Operation	
δ_{atr}(Projection)	Aimed at attributes which have some features. If the documents does not have an attribute, the document is not in the result LDM
π_{atr}(Selection)	Aimed at attributes which have some features. If the attribute of a document does not satisfy the condition, the document is filtered
Transform Operation	
todocs	Transform the LDM to different type of documents
tograph	ransform a LDM into a graph
totables	Transform a LDM into different relation tables, including the Link

Description. We use second order logic to describe what characteristic a LDM should have. And the way we describe characteristic is by rules. For example, we first define a social network, which has *person* document set and *knows* link set.

$$soc = (\{person\}, \{knows\})$$

We can require the person set satisfy a transitive closure characteristic: suppose that *p1,p2,p3* are documents in person, if *p1* knows *p2*, and *p2* knows *p3*, then *p1* must knows *p3*. This closure can be described as followed,

```
soc.person p1,p2,p3;
p1.knows=p2 and p2.knows=p3 -> p1.knows=p3;
p1,p2,p3 >>
```

The last line stands for which part of the set that *soc.person* should contain. Another case is the person set stratify the condition that it contains all the person that "simba" indirectly knows. The characteristic is defined as followed

```
define x.indirknows=y: x.knows.knows=y or
x.indirknows=z and z.knows=y;
soc.person p1,p2;
p1.name= ""simba;
p2->p1.indirknows=p2;
p2>>
```

The keyword 'define' defines a new link named 'indirknows'.

4.1 SimbaQL Grammar

As described in Sect. 3, SimbaQL need to describe the whole data processing in scientific research. We divide the grammars into four parts: data set grammar, data source grammar, LDM operation grammar, and LDM description grammar. We use keyword *'create'* to define a data set or data source, and a data set can call the *'importto'* function to import a data set to data source. Data source can call *'query'* function to search the local database. At the same time, data source is also a LDM, thus can perform the operations of LDM. For example,

```
create genfile = File(*);
genbank = genfile.importto'('mongodb);
a_class_gen = genbank.query({name: "a_class"});
```

The most used three operations in LDM are selection, projection, union. The grammars for theses operations are as follows,

```
a_class_gen = genbank.gen(name= "a_class");
b_class_gen = genbank.gen(name= "b_class");
ab_class_gen = a_class_gen + b_class_gen;
ab_class_gen = ab_class_gen[name, sequence];
```

For the LDM description, we use keyword *'st'* to represent this LDM should satisfy the following characteristic. For example

```
newsoc = soc st {
  soc.person p1,p2;
  p2->p1.knows=p2;
  p2>>
}
```

5 Case Study

In this section, we apply SimbaQL in two actual scientific research cases. The first case is the data processing when building the WDCM knowledge graph.

The knowledge graph contains the information of microorganism species, protein, gene and the relationship between them. In this process, WDCM creates a relational database to store the species's information, and they want to build the relationship between species and their gene sequence which can get from the file of genbank. This process can be implemented by SimbaQL as follows:

```
create  species=JDBC(*);
create  genbank=File(*);
genbank=genbank.import'('mongo);
genbank=genbank.gen[sorcename,sequence];
wdcm=species+genbank;
wdcm.spece*wdcm.gen{wdcm.spece.name
.contains(wdcm.gen.name)};
wdcm.importto'('gstore);
```

The second case is that National Nature Science Foundation of China has foundation information stored in relational database, and they want to know the mapping between paper and foundation. However, the papers published by a foundation are listed in the summary report. Therefore, they need to extract the papers from the reports, and integrated the data. The paper can be extracted by regular expression, and this process can be described by SimbaQL as follows:

```
create  found=JDBC(*);
create  reports=File(*);
reports=reports.importto'('path_search);
papers=reports.query(paperegx);
found_paper=found+papers;
found_paper.found*found_paper.paper{found_paper.
found.name=found_paper.paper.filename};
found_paper.import'('gstore);
```

6   Conclusion and Future Work

We introduced SimbaQL for multi-source heterogeneous data query which is a common data processing in scientific research. First, we propose an abstract model of data processing, which consists of data set, data source, data model, and analyzing tools. Then, we defined an intimidate data model LDM)to map the data between heterogeneous data, and operations on LDM to query and integrate data. We study two actual cases using SimbaQL, and it turns out that SimbaQL could describe the common data processing in scientific research. In the future, we will give the formal definition of SimbaSQL, prove the expressive power of language, and implement the query engine for heterogeneous data.

References

1. Ward, J.S., Barker, A.: Undefined by data: a survey of big data definitions, arXiv preprint arXiv:1309.5821 (2013)
2. Lu, J., Holubová, I., et al.: Multi-model data management: what's new and what's next? (2017)
3. World data centre for microorgannisms. http://www.wdcm.org/
4. Duggan, J., et al.: The BigDAWG polystore system. ACM Sigmod Rec. **44**(2), 11–16 (2015)
5. Chamberlin, D.: XQuery: an xml query language. IBM Syst. J. **41**(4), 597–615 (2002)
6. Lenzerini, M.: Data integration: a theoretical perspective. In: Proceedings of the Twenty-First ACM SIGMOD-SIGACT-SIGART Symposium on Principles of Database Systems, pp. 233–246. ACM (2002)
7. Sheth, A.P., Larson, J.A.: Federated database systems for managing distributed, heterogeneous, and autonomous databases. ACM Comput. Surv. (CSUR) **22**(3), 183–236 (1990)
8. Baranyi, P.: Multi-model database orientdb. https://orientdb.com/
9. Gadepally, V., et al.: Version 0.1 of the BigDAWG polystore system, arXiv preprint arXiv:1707.00721 (2017)
10. Libkin, L.: Expressive power of SQL. Theor. Comput. Sci. **296**(3), 379–404 (2003)
11. Angles, R., Gutierrez, C.: The expressive power of SPARQL. In: Sheth, A., et al. (eds.) ISWC 2008. LNCS, vol. 5318, pp. 114–129. Springer, Heidelberg (2008). https://doi.org/10.1007/978-3-540-88564-1_8
12. Bagan, G., Bonifati, A., Ciucanu, R., Fletcher, G.H., Lemay, A., Advokaat, N.: Generating flexible workloads for graph databases. Proc. VLDB Endow. **9**(13), 1457–1460 (2016)
13. Cypher-the neo4j query language. http://www.neo4j.org/learn/cypher

Changing Data Policies in China: Implications for Enabling FAIR Data

Lili Zhang[1](✉), Robert R. Downs[2], and Jianhui Li[1]

[1] Computer Network Information Center,
Chinese Academy of Sciences, Beijing, China
zhll@cnic.cn
[2] Center for International Earth Science Information Network,
Columbia University, Palisades, NY, USA

Abstract. As fundamental resources of research activities, data is vitally important for scientific progress and general social society. Thus, open data practices are becoming more prevalent and the adoption of the FAIR (Findable, Accessible, Interoperable, and Reusable) principles is fostering open access data from four general perspectives. This paper firstly analyzes the general benefits and necessities of open data and the FAIR principles. Then, data policies in China are described from four views. Subsequently, challenges and opportunities for data usability across disciplinary boundaries and levels of expertise are described. Finally, categories are presented for how data policies in China enable FAIR data in terms of the four views of Chinese data policies. Above all, FAIR data is a good beginning, and for FAIR open data, we still need more efforts on intrinsic data culture, trustworthiness, sustainability, and multilateral cooperation among various stakeholders, as well as consistent and effective approaches for adopting data policies.

Keywords: Data policy · FAIR · Open data · Research data · China

1 Introduction

Open access to scientific data is necessary for scientific progress. Data is a necessary and fundamental resource for engaging in scientific activity. Reports of scientific results are often based on data that have been collected for the study being reported or on data that have been reused from the study that collected the data. Making data publicly available to others as open data, without restrictions on access or on who can use the data or on the ways in which the data can be used, fosters science. Enabling the use of open data also facilitates the application of science to other fields, such as engineering, planning, policy development, and education.

Open data and open access practices contribute to open science. Recognizing the need for guidance in managing data, principles for Findable, Accessible, Interoperable, and Reusable (FAIR) data have been proposed (Wilkinson et al. 2016). Making open data findable should enable access by a variety of potential users, not only from the discipline in which the data were created, but also for members of other scientific disciplines and for individuals with varying levels of expertise. Access to data should

J. Li et al. (Eds.): BigSDM 2018, LNCS 11473, pp. 285–290, 2019.
https://doi.org/10.1007/978-3-030-28061-1_28

be facilitated for diverse audiences, beyond those who are members of the data collection team or those who work in the same discipline. Researchers from a variety of disciplines should be able to obtain data products and services of interest and students of such disciplines also should be able to access the data. The intellectual property rights for using the data should not be restrictive, and the information about such rights also should be accessible and understandable by potential users. Enabling interoperability of open data fosters capabilities to compare and combine data, providing users with opportunities to conduct studies on new topics and advance innovation (Borgman 2012). Interoperable open data makes it possible to conduct interdisciplinary research and research within different disciplines other than the field or subfield of those who collected the data. Ensuring capabilities for the reusability of open data is critical for the reproduction of previous findings and the replication of studies. In addition to its importance for advancing science within the field in which the data were collected, enabling the reusability of open data also offers opportunities for researchers from other fields to use the data or employ similar methods for studies within their own disciplines. Students also could have the opportunity to use the data or improve their understanding of data collection methods.

2   Data Policies in China

In China, data policies generally encompass rules from national, institutional as well as disciplinary level practices. The newly launched national-level legislation, "Measures for Managing of Scientific Data", guides lifelong data activities across the country (2018). Key concepts in the rules are "lifelong data stewardship, safe guard for key data assets, open public funded data by default as well as intellectual property protection" (Feng 2018). As national guidance, "Measures for Managing of Scientific Data" provides general principles, clear explanations of all stakeholders' responsibilities, and specific guidelines for data stewardship. Difficult issues, such as confidentiality and security, as well as incentive mechanisms and punishment against the rules, are also specified. Adopting such rules can foster data access within and across the country. Institutional rules for data generally focus on proprietary interests, such as cybersecurity and intellectual property protection. Disciplinary rules can focus more on data stewardship, since those working within a field often share use cases, reflecting common data management needs. For example, the geosciences have developed a variety of data stewardship maturity measures (Peng 2018).

Content analysis of data policies at different levels in China revealed the following four views:

(1) Management of Data by Levels

Classification of data and different data management strategies reflect the roles of stakeholders, such as data producers, data owners as well as data re-users. Sharing is not sacrificing. Data sharing facilitates long-term sustainability when trust and incentives coexist.

(2) Lifelong Data Stewardship

Disciplinary data policies are often generated from bottom-up needs and call for unambiguous rules for everyday data activities. Data lifecycle rules governing data capture, data processing, data storage and sharing as well as preservation efforts are the usual practices.

(3) Data Publishing and Reusing

Data publishing offers additional options for providing value-added services. Repository publishing is recommended for releasing data. The Ministry of Science and Technology, The National Natural Sciences Foundation, and the Chinese Academy of Sciences, for example, all support long-term data projects for data production and publishing through professional data platforms. Also, China Scientific Data (www. csdata.org), known as the first multidisciplinary data paper journal in China, has taken the lead to publish datasets together with data papers. GigaScience publishes large-scale biological datasets in their self-constructed repository GigaDB (Edmunds et al. 2016) and Global Change Research Data Publishing & Repository endeavors to publish geoscience datasets.

(4) Long-term Data Preservation

For non-reproducible and valuable datasets, long-term preservation is necessary. Sustainable long-term data preservation requires the installation of labor-saving technology and systematic maintenance services within a cost-effective budget. Thus, barring unplanned events, flexible business models, current data technologies, and trustworthy preservation strategies can contribute to a comprehensive design for implementing data policies effectively.

3   How Data Policies Enable FAIR Data

As mentioned above, data management can be complex as data usage varies across different disciplines, institutions, and stakeholders. Figure 1 depicts the intersection of opportunities for data usability across disciplinary boundaries and levels of expertise. Enabling FAIR use of data can help facilitate the use of scientific data across disciplinary boundaries and levels of expertise.

	Discipline	
Expertise	**a.** Not Trained in Discipline of Data Set Experienced in Research	**b.** Trained in Discipline of Data Set Experienced in Research
	c. Not Trained in Discipline of Data Set Not Experienced in Research	**d.** Trained in Discipline of Data Set Not Experienced in Research

Fig. 1. Data usability opportunities across disciplinary boundaries and levels of expertise

Cell "a" describes less data-intensive practices or experience within the discipline of the data. This could reflect research across disciplines or experience within disciplines where creative ideas and logical theories, other than data, are prioritized. Such practices do not necessarily reduce the importance of data use. On the contrary, more attention to data use, including interdisciplinary data use, can contribute to such areas of research and lead to new discoveries. Data policies with explicit guidelines and sufficient funding for enabling data access could facilitate such use, narrowing the gap between "less data-intensive" fields and those that are more data-intensive.

Situations described in cell "b" may be more common for data-intensive disciplines, such as in astronomy and space sciences, high energy physics, and other fields that produce and share large scale datasets and develop mature data policies. Guoshoujing Telescope (the Large Sky Area Multi-Object Fiber Spectroscopic Telescope, LAMOST), known as the National Major Scientific Project built by the Chinese Academy of Sciences, has released over 739,006 spectra with 438 plates observed during its 6th data release so far (Lamost 2018). They provide the best exemplars for data stewardship in line with current data policies. As China's first X-ray astronomy satellite, the Hard X-ray Modulation Telescope (HXMT) classified data into three categories, according to data processing levels, and made specific open data rules for these categories (HXMT 2018).

Cell "c" focuses primarily on beginners and new students in science. For such audiences, data science training should be prioritized as this group represents the future of science. Improving data literacy and understanding of data stewardship could lead to better data management, over time, and help establish harmonious data ecology within their scholarly communities. The Chinese Academy of Sciences, the Minister of Science and Technology, as well as the Minister of Education, etc., and international organizations, such as CODATA China (www.codata.cn), and WDS China (www.wds-china.org) actively support data science training activities.

Cell "d" reflects disciplinary expertise with less experience in research, generally. Enabling interdisciplinary data use could reduce potential challenges for data discovery and access. Likewise, data policies that facilitate interoperability could inspire the integration of data across disciplinary boundaries where the expertise from one discipline is applied to an integrated data product. Adopting policies that create opportunities for cross-disciplinary use and knowledge transfer, by offering ample data documentation, also could foster data use across disciplines.

While Fig. 1 describes general challenges for data use across disciplines and levels of expertise, the four views on data policies in China can provide insight into how such policies support FAIR data in China. By classifying these views in terms of the FAIR characteristics, their applicability becomes more apparent. By data classification, it's more clear for us to know the lists of available datasets, thus it surely makes data findable and accessible. Likewise, support for lifecycle data stewardship can be observed through all four characteristics of FAIR. Mature metadata and data cataloging can improve the findability of data. Steady data curation helps facilitate accessibility. Development of general standards and implementation of efficient techniques improves interoperability by making data readable by machines and humans (Faniel and Zimmerman 2011). Exploration of user friendly tools aids the reusability of data, especially across disciplines. And for data publishing, fostering automatic and manual quality control can improve the reuse of data. Moreover, long-term data preservation goes even further by improving the usability of data now and in the future. In summary, categorizing the four views of data policies in terms of the characteristics of FAIR data, as depicted in Table 1, shows how the policies can propel capabilities for enabling FAIR data.

Table 1. Data policies enabling FAIR data

Data policies focuses	Findable	Accessible	Interoperable	Reusable
Management of data by levels	√	√		
Lifelong data stewardship	√	√	√	√
Data publishing and reusing				√
Long-term data preservation				√

4 Discussion and Conclusion

Open data is vital for progress in modern scholarly communities and has become the overwhelming trend for the global village. The FAIR principles provide us with implications for data culture (Barbui et al. 2016), data ecology, and science community practices (Kaye et al. 2009).

In China, data policies and practices are equally important since they both contribute to the implementation of open FAIR data. They are both actively improving the comprehensive policy framework for governing and reusing data. The data policies enable data to be more FAIR by guiding and supporting improved data practices through funding for data programs, sound data stewardship practices, and data education and training for broad audience.

FAIR data is a good start, but only FAIR is just the beginning. Support for data users should provide affordances that encourage their full use of data from many currently available sources. Likewise, data producers and owners should be recognized and rewarded appropriately to ensure that data sharing becomes an ongoing practice that is intrinsic to the culture of science. Data sharing is based upon mutual interests and trust. Furthermore, trustworthiness (CoreTrustSeal n.d.) and sustainability (OECD 2017) are also fundamental to long-term data use. Sustainability calls for suitable business models to cover the data stewardship costs of today and those that will be incurred in the future. Such models also should include mechanisms that can survive unexpected conditions and maintain the capabilities and expertise needed to continue enabling access to open FAIR data. Moreover, cooperation between stakeholders necessarily contribute to a win-ecology for long-term open data (Longo and Drazen 2016). And above all, for the data policies themselves, consistency and simplicity as well as effectiveness and efficiency are also vitally important for FAIR data (Doshi et al. 2016).

Acknowledgements. This work is an outcome of the project of "International and National Scientific Data Resources Development Report in China 2018" (No. 2018DDJ1ZZ14) supported by the National Science and Technology Infrastructure Center, and the project "Decision Making Oriented Massive Data Resources Sharing and Governance" (No. 91546125) supported by National Natural Science Foundation of China, and "Big data resources pool and system portal" (No. XDA19020104) funded by the Chinese Academy of Sciences.

References

Borgman, C.L.: The conundrum of sharing research data. J. Am. Soc. Inf. Sci. Technol. **63**(6), 1059–1078 (2012). https://doi.org/10.1002/asi.22634

Barbui, C., Gureje, O., Puschner, B., et al.: Implementing a data sharing culture. Epidemiol. Psychiatr. Sci. **25**, 289–290 (2016). https://doi.org/10.1017/S2045796016000330

CoreTrustSeal: Data Repositories Requirements (n.d.). https://www.coretrustseal.org/why-certification/requirements/. Accessed 8 Jan 2019

Doshi, J.A., Hendrick, F.B., Graff, J.S., et al.: Data, data everywhere, but access remains a big issue for researchers: a review of access policies for publicly funded patient-level health care data in the United States. EDM Forum Community **4**(2), 1–20 (2016). https://doi.org/10.13063/2327-9214.1204

Edmunds, S.C., Li, P., Hunter, C.I., et al.: Experiences in integrated data and research object publishing using GigaDB. Int. J. Digit. Libr. (2016). https://doi.org/10.1007/s00799-016-0174-6

Faniel, I.M., Zimmerman, A.: Beyond the data deluge: a research agenda. Int. J. Digit. Curation **6**(1), 58–69 (2011)

HXMT: Data policy of HXMT (2018). http://www.hxmt.org/index.php/2013-03-22-08-08-48/docs/319-hxmt-data-polocy-of-hxmt. Accessed 29 Dec 2018. (in Chinese)

Kaye, J., Heeney, C., Hawkins, N., et al.: Data sharing in genomics-re-shaping scientific practice. Nat. Rev. Genet. **10**, 331–335 (2009)

Lamost: Lamost DR 6 V1 (2018). http://dr6.lamost.org/. Accessed 29 Dec 2018

Longo, D.L., Drazen, J.M.: Data sharing. New Engl. J. Med. **374**(3), 276–277 (2016)

General Office of the State Council, China: Measures for Managing Scientific Data (2018). http://www.gov.cn/zhengce/content/2018-04/02/content_5279272.htm. Accessed 9 Jan 2019

OECD: Business models for sustainable research data repositories. OECD Science, Technology and Industry Policy Papers, No. 47. OECD Publishing, Paris (2017). https://doi.org/10.1787/302b12bb-en

Peng, G.: The state of assessing data stewardship maturity – an overview. Data Sci. J. **17**, 7 (2018). https://doi.org/10.5334/dsj-2018-00

Feng, H.: Interpretation of the first national level Measures for managing scientific data. People's Daily, 8 April 2018. http://www.xinhuanet.com/politics/2018-04/08/c_1122647406.htm. Accessed 9 Jan 2019. (in Chinese)

Wilkinson, M., Dumontier, M., Aalbersberg, I., et al.: The FAIR guiding principles for scientific data management and stewardship. Nat. Sci. Data **3**(2016). https://doi.org/10.1038/sdata.2016.18

A Data Quality Evaluation Index
for Data Journals

Lihua Kong[1,2,3(✉)], Yan Xi[3], Yangqin Lang[3], Yang Wang[3],
and Qingfei Zhang[3]

[1] National Science Library, Chinese Academy of Sciences,
100190 Beijing, China
klh@cnic.cn
[2] Department of Library, Information and Archives Management,
School of Economics and Management,
University of Chinese Academy of Sciences, 100409 Beijing, China
[3] Computer Network Information Center, Chinese Academy of Sciences,
100190 Beijing, China

Abstract. As a new model of data sharing, data publication requires data to be reviewed and published together with a data descriptor. The peer review process of data publication is very different from that of traditional academic papers. It is still in a preliminary stage and awaits further discussion. This study proposes a data quality evaluation index system for data publishing, which aims to enable high-efficiency data sharing and make data accessible, intelligible, evaluable, and reusable.

Keywords: Data publication · Data quality · Data journal · Data paper · Peer review

1 Introduction

Data lays the basis for scientific hypothesis, scientific analysis, and theoretical information. It constitutes an important evidence with which peer reviewers evaluate and test research results. In 2012, the Royal Society of Science published "Science as an Open Enterprise" [1], which proposed that, to maintain science's self-correction capacity, support the verification of research outcomes and the discovery of new knowledge, and promote the use of research outcomes in innovation and education, we must achieve data sharing, and make scientific data accessible, assessable, intelligible, and usable. Today, with the rapid development of information technology, research activities are facing new challenges. Scientific data is not only an important output of scientific research and scientific literature, but has become an important content, tool and infrastructure of scientific research and even social progress. Its "First Class Citizen" status [2] increasingly urges the science community to pursue the sharing of high-quality data.

In the past decade, scientific data sharing has received widespread attention. Publishers have taken a series of initiatives, including a number of Data Policy developed by major publishing groups and journals. For example, *PLoS One* implemented a data

© Springer Nature Switzerland AG 2019
J. Li et al. (Eds.): BigSDM 2018, LNCS 11473, pp. 291–300, 2019.
https://doi.org/10.1007/978-3-030-28061-1_29

sharing policy in 2008, which was further revised and improved in 2014. The data policy [3] was implemented on March 3, 2014. It was described that "PLOS journals require authors to make all data underlying the findings described in their manuscript fully available without restriction, with rare exception. Refusal to share data and related metadata and methods in accordance with this policy will be grounds for rejection".

Springer Nature formulated 4 types of research data policy [4], which stipulated that data should be available for reuse by the research community under a Creative Commons attribution license unless otherwise stated. The majority of Springer journals have now adopted one of the policies. For example, *Photosynthesis Research* adopted Type 1 and *Plant and Soil* adopted Type 2. All BioMed Central journals, a growing number of Nature journals, and *Palgrave Communications* supported Type 3. Journals with the most rigorous open data policies supported Type 4, including *Scientific Data* and *Genome Biology* (Fig. 1).

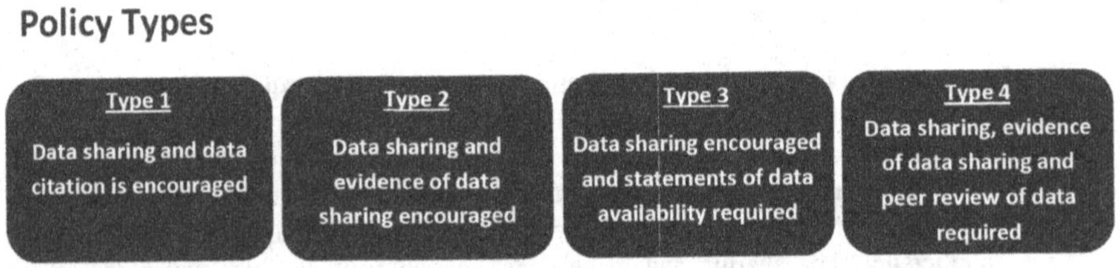

Policy Types

Type 1	Type 2	Type 3	Type 4
Data sharing and data citation is encouraged	Data sharing and evidence of data sharing encouraged	Data sharing encouraged and statements of data availability required	Data sharing, evidence of data sharing and peer review of data required

Fig. 1. The 4 types of research data policy of Springer Nature

As a new model of data sharing, data publication ensures the publishing of high-quality data through some basic procedures including data submission, peer review, data release, permanent data storage, data citation and data impact evaluation, which maximize the use of scientific data resources. The review process of data publication is very different from that of traditional academic journals. Although many data journals have proposed their evaluation criteria, peer review of data papers, especially of the data itself, is still at an initial stage.

2 Peer Review for Data Publishing

2.1 A Comparison of Peer Review Between Traditional Journals and Data Journals

Publishing has been recognized as an important process for ensuring the quality of publications. Peer Review (also known as Refereeing in some academic fields) as an academic review process has been used by publications for nearly 300 years. Back to the mid-17th century, *The Philosophical Transactions* of the Royal Society (*Phil. Trans.*) started to use peer review as a way to ensure the quality of articles to be published [5]. Based on the publishing process of traditional academic journals [6], current data publications incorporate data review into their process of data publishing.

Though drawing its main ideas from traditional journals, peer review of data has many peculiarities. (1) Focus of review criteria. Compared with traditional academic articles where arguments and conclusions are major components of the review, data review focuses on data quality and reusability, as well as consistency with its descriptor (i.e. data paper). As such, the assessment targets of data publishing include data papers, data sets, and their correlations. (2) Selection of reviewer. While traditional academic journals tend to select reviewers from the authors' field for scientific evaluations of academic content, data publications require the participation of data scientists in addition to subject area experts. (3) Mode of review. At present, many data journals use the mature review mechanism of traditional academic journals, while some data journals also exploit network resources for more efficient peer review. A typical example is *Earth System Science Data* (ESSD). ESSD uses a two-stage review mode, as shown in Fig. 2. After an initial review by the editorial department, papers are immediately published in the public discussion area (ESSDD) for public review. To ensure a rigorous, scientific and fair judgement, the paper is in the meantime sent to experts for external review. A final decision is made by the editor-in-charge based on a comprehensive consideration of public evaluation and peer review (Table 1).

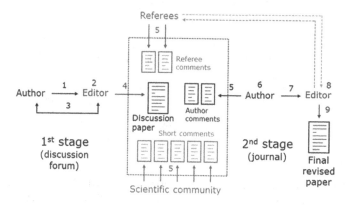

Fig. 2. Two-stage review mode of ESSD [7]

Table 1. A comparison of peer review between research journals and data journals

	Research journal	Data journal
Focus	Innovation of the research	How the data supports arguments and conclusions
Expert selection	Expert from the same field	Subject expert and data expert
Review mode	Traditional review mode	More demand for open review

2.2 Major Modes of Data Publishing and Data Quality Control

According to data dependencies and data storage modes, there are three major modes of data publishing: standalone data publishing (data published in data centers or repositories not pertaining to a specific article), data as research papers Supplementary files, and data papers (data published in correlation with a descriptor) [8] (Fig. 3 and Table 2).

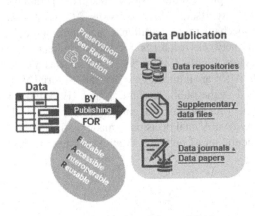

Fig. 3. Three modes of data publishing

Table 2. A comparison of data quality control among the three publishing modes

Data publishing mode	Example	Reviewer	Review criteria
Standalone data publishing	NASA [9], Figshare [10], Dryad [11], ScienceDB [12]	Database administrator	Data storage, data format and standards, data accessibility (DOI), consistency between metadata and data
Data as research papers supplementary files	Science [13], Springer Nature [14],BMC [15], PLoS [3]	Editor and subject experts	Data's support for arguments and conclusions, long-term preservation and accessibility of data
Data papers	Scientific Data [16], ESSD [17]	Editor, subject experts and data scientist	Quality of data paper, metadata and data, long-term storage and accessibility, data quality control, consistency among paper, metadata and data

This paper focuses on data papers and the peer review of data involved in data publication.

2.3   Criteria for Data Paper Review

Data paper publishing is a process during which data producers arrange the data collected and processed for a certain purpose, write papers describing the process of data collection, processing, quality control, usage, and so forth, submit data papers and data to designated places for peer review and finally have their paper and data published. In this process, both data papers and data are assigned a DOI and the two are correlated. Such a formal publishing mode enables users to acquire, understand, reanalyze, develop, and cite data in their research papers or other scientific outcomes.

According to [18, 19], almost all data journals use peer review to ensure data quality. Indeed, as an established mechanism of traditional academic publishing, peer review has great advantages in the quality control of data papers and data journals [20]. Despite this, review and evaluation criteria vary from journal to journal. We surveyed five typical data journals, including *Earth System Science Data* (ESSD) [17], *Geoscience Data Journal* [21], *Pensoft journals* [22], *Scientific Data* [16], and *Brief of Data* [23]. It can be seen that most journals perform data quality control from such aspects as data accessibility, integrity, consistency, community standards, quality control, and data production and processing methods. More details can be seen in Appendix (Table 3).

Table 3. Data quality review criteria adopted by different data journals.

	Earth system science data	Geoscience data journal	Pensoft journals	Scientific data	Brief of data
Accessibility	√	√	√	√	√
Integrity	√	√	√	√	√
Community standards	√	√	√	√	√
Accuracy	√		√		
Metadata quality	√	√	√	√	√
The methods and data-processing steps	√	√	√	√	
Quality control	√	√	√	√	
Other data files			√	√	√

Note: Some of the options here may not be reflected by the journal descriptions, so that is considered the review does not check the content.

As can be seen in our survey, most data journals have developed their own standards based on the peer review model of traditional journals. However, as review of datasets and data papers is quite different from review of traditional academic papers, data journals have integrated paper review and data review into a single process, which includes examining data papers, data sets, and their correlation or consistency. In addition, review of data sets, which are often large in size and complex in structure, has to attend to data peculiarity. As such, the peer review process of data and data papers is naturally more complicated. Up to now, no consensus has been reached regarding data evaluation standards, or standard data review procedures.

3   A Data Quality Evaluation Index for Data Journals

In 2016, Computer Network Information Center of the Chinese Academy of Sciences launched a data journal, *China Scientific Data*. Based on our work for this journal, we aim to propose a data quality evaluation index system for data publishing to promote data visibility in the peer review process, improve editing and peer review services, and promote data openness and sharing. In doing so, we aim to achieve high-efficiency data sharing so as to make data accessible, intelligible, evaluable, and reusable.

3.1   FAIR Data Principles

At the Lorentz workshop "Jointly designing a Data FAIRPORT" held in 2014, FAIR principles were first proposed. The principles determine that all experimental subjects should be discoverable, accessible, interoperable, and reusable, both for humans and for machine users. Referred to as FAIR, the principles apply not only to data in the traditional sense, but also to algorithms, processes, and tools that are used to produce data. All aspects of a research process should be made transparent, reproducible and reusable.

Based on these 4 principles, a set of 15 metrics were proposed to quantify the levels of FAIRness (Table 4).

Table 4. FAIR principle [26]

Principle	Metrics
Findability	F1. (meta)data are assigned a globally unique and eternally persistent identifier F2. data are described with rich metadata F3. (meta)data are registered or indexed in a searchable resource F4. metadata specify the data identifier
Accessibility	A1 (meta)data are retrievable by their identifier using a standardized communications protocol A1.1 the protocol is open, free, and universally implementable A1.2 the protocol allows for an authentication and authorization procedure, where necessary A2 metadata are accessible, even when the data are no longer available
Interoperability	I1. (meta)data use a formal, accessible, shared, and broadly applicable language for knowledge representation I2. (meta)data use vocabularies that follow FAIR principles I3. (meta)data include qualified references to other (meta)data
Reusability	R1. meta(data) have a plurality of accurate and relevant attributes R1.1. (meta)data are released with a clear and accessible data usage license R1.2. (meta)data are associated with their provenance R1.3. (meta)data meet domain-relevant community standards

3.2 Design of the Data Quality Evaluation Index

This study uses the FAIR principles as an important reference. Based on our previous survey and analysis, we carried out a hierarchical design of the data quality evaluation index for data publishing, where indicators of an upper level are further refined in their respective sub-levels.

Primary Indexes

For design of the primary index, we referred to principles proposed by "Science as an Open Enterprise", which include accessible, assessable, intelligible, and reusable. This is also consistent with the FAIR principles (i.e., findability, accessibility, interoperability, and reusability). The primary data quality evaluation index for data publishing is outlined as follows (Fig. 4):

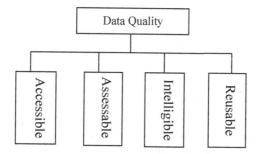

Fig. 4. Primary index for data quality evaluation

Secondary Indexes

Assume that the data is equipped with a good life-cycle quality management plan, where its production, processing and release processes are fully recorded. We then expand the primary index to obtain the secondary index as follows (Fig. 5):

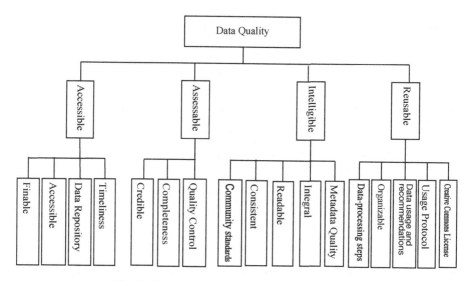

Fig. 5. Secondary index for data quality evaluation

According to the research, we designed some general indexes for data review, covering data review and paper review. We applied these evaluation indexes to *China Scientific Data* (Figs. 6 and 7).

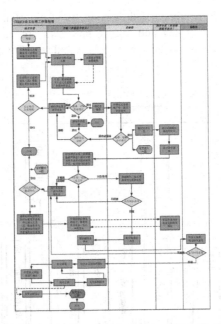

Paper review

Paper Quality

1. Title: The title is concise and reflects the article content   ☺ ☆☆☆☆☆

2. Abstract & keywords: They are concise and cover major information of the manuscript   ☺ ☆☆☆☆☆

3. Readability: The manuscript flows smoothly and includes important contents, such as data processing methods, data description and standardisation, which facilitate easy understanding of the datasets.   ☺ ☆☆☆☆☆

4. Data reuse description: (1) Data processing methods and steps are stipulated clearly enough for data reproduction; (2) Enough information is provided for data integration or other kinds of dataset reuse.   ☺ ☆☆☆☆☆

5. Description of data processing methods: The manuscript has a complete description of data processing methods and the methods are rigorously and reasonably deployed.   ☺ ☆☆☆☆☆

6. Value & significance: The manuscript needs at least one of the following criteria at different degrees in varied conditions: (1) High data reusability; (2) Rigorous and reasonable data processing methods, with novelty and reference value; (3) Ability in verifying other scientific research outcomes – the data and processing logistically consistent with relevant studies and/or support their conclusions.   ☺ ☆☆☆☆☆

Data Quality

1. Data storage: Data are saved in a reliable and suitable repository. Data files are complete and match the description of the manuscript.   ☺ ☆☆☆☆☆

2. Data accessibility: The dataset can be accessed and viewed (easily accessible by readers with a page for metadata preview). When the data are not directly accessible: (1) keywords and access conditions should be stipulated; (2) version information should be displayed by the data viewer.   ☺ ☆☆☆☆☆

3. Data quality and richness: (1) Data formats are appropriate, up to industry standard; (2) The data should have a full coverage of the author's study scope, with a proper size and full integrity, and data values should be within the expected ranges; (3) No obvious errors in details; (4) The manuscript describes credible technical validation experiments, statistical analysis, and error analysis for quality control.   ☺ ☆☆☆☆☆

4. Data consistency: The dataset does not deviate from descriptions of the manuscript.   ☺ ☆☆☆☆☆

Fig. 6. Peer review checklist of *China Scientific Data*

In the review of data papers, we assign higher weight values on the indexes about quality review of metadata, the consistency of data and papers, and data reuse. According to the appraisal application of the editorial department, it can basically meet the basic needs of manuscript review, but there are still many shortcomings that need to be corrected in future work.

Fig. 7. Peer review process of *China Scientific Data*

4 Summary

As an important part of the data publishing process, data review is still at a preliminary stage, and many complex issues await to be addressed, particularly concerning the methods for data quality judgment. Machine and manual interpretations of data have also yet to be further studied and discussed.

References

1. The Royal Society: Science as an open enterprise. http://royalsociety.org/policy/projects/science-public-enterprise/report/. Accessed 9 Mar 2014
2. Bolikowski, Ł., Houssos, N., Manghi, P., et al.: Data as "firstclass citizens". D-Lib Mag. **21**, 1–2 (2015). https://doi.org/10.1045/january2015-guest-editorial
3. PLOS: Data availability. http://journals.plos.org/plosone/s/data-availability. Accessed 11 Nov 2015
4. Springer Nature: Research Data Policy. https://www.springernature.com/gp/authors/research-data-policy/data-policy-types/12327096. Accessed 9 Aug 2018
5. Li, X.: On peer review of international S&T journals in English (Part 1), China Science Daily
6. Zhang, X., Li, X.: Critical issues on the theory and practice of data publishing. Chin. J. Sci. Tech. Period. **26**(8), 813–821 (2015)
7. ESSD: Review Process. http://www.earth-system-science-data.net/review/review_process_and_interactive_public_discussion.html. Accessed 11 May 2016
8. Zhang, X., Shen, H., Liu, F.: Scientific data and literature interoperability. In: CODATA China Committee (eds.) Research Activities in Big Data Era, pp. 149–158. Science Press, Beijing (2014)
9. NASA Administrator: Data from NASA's Missions, Research, and Activities
10. Figshare Q&A. https://figshare.com/. Accessed 1 Oct 2017
11. Dryad Policies. https://datadryad.org/pages/policies. Accessed 24 Feb 2018
12. ScienceDB. http://www.sciencedb.cn/index. Accessed 6 June 2016
13. American Association for the Advancement of Science: Science: editorial policies. http://www.sciencemag.org/authors/science-editorial-policies. Accessed 8 Mar 2018
14. Springer Nature: Availability of data, Materials and methods. http://www.nature.com/authors/policies/availability.html. Accessed 9 Aug 2018
15. BioMed Central: Open data. http://www.biomedcentral.com/about/policies/open-data. Accessed 9 Aug 2018
16. Guide for referees, Scientific Data. http://www.nature.com/sdata/policies/for-referees. Accessed 9 July 2016
17. Earth System Science Data: Review criteria. http://www.earth-system-science-data.net/peer_review/review_criteria.html. Accessed 11 May 2016
18. Candela, L., Castelli, D., Manghi, P., et al.: Data journals: a survey. J. Assoc. Inf. Sci. Technol. **66**(9), 1747–1762 (2015)
19. Ouyang, Z., Qing, X., Gu, L., et al.: Internal data journal publications: a study of cases and their characters. Chin. J. Sci. Tech. Period. **26**(5), 437–444 (2015)
20. Kenall, A.: An open future for ecological and evolutionary data? BMC Ecol. **14**(1), 280–294 (2014)
21. Guidelines for Reviewer, Geoscience Data Journal. https://rmets.onlinelibrary.wiley.com/hub/journal/20496060/features/guidelines-for-reviewers. Accessed 11 May 2016

22. Penev, L., Mietchen, D., Chavan, V., et al.: Pensoft Data Publishing Policies and Guidelines for Biodiversity Data. Pensoft Publishers. http://www.pensoft.net/J_FILES/Pensoft_Data_Publishing_Policies_and_Guidelines.pdf
23. Data in Brief. https://www.elsevier.com/journals/data-in-brief/2352-3409?generatepdf=true. Accessed 11 Mar 2017
24. Force11: fair principles. https://www.force11.org/group/fairgroup/fairprinciples. Accessed 8 July 2016

Data Quality Transaction on Different Distributed Ledger Technologies

Chao Wu[(✉)], Liyi Zhou, Chulin Xie, Yuhang Zheng, and Jiawei Yu

Zhejiang University, Hangzhou, China
`chao.wu@zju.edu.cn`

Abstract. Data quality is a bottleneck for efficient machine-to-machine communication without human intervention in Industrial Internet of Things (IIoT). Conventional centralised data quality management (DQM) approaches are not tamper-proof. They require trustworthy and highly skilled intermediation, and can only access and use data from limited data sources. This does not only impacts the integrity and availability of the IIoT data, but also makes the DQM process time and resource consuming. To address this problem, a blockchain based DQM platform is proposed in this paper, which aims to enable tamper-proof data transactions in a decentralised and trustless environment. To fit for different quality requirements, our platform supports customisable smart contracts for quality assurance. And to improve our platform's performance, we discuss and analyze different distributed ledger technologies.

Keywords: Data quality · IIoT · Blockchain · IPFS · Smart contract · DAG · IOTA

1 Introduction

Industrial Internet of Thing (IIoT) is expanding at a fast pace, which facilitates the exponential growth of sensor data. High quality IIoT data are critical to generate reliable models, make real time decisions and deliver advanced services in a wide range of application domains, such as manufacturing, logistics, transportation, health care, energy and utilities [15]. This vast amount of IIoT data, with varying complexity and reliability, are usually generated from distributed sources owned by different parties.

However, obtaining high-quality distributed sensor-based data is challenging. Firstly, the participation of human and subjective decisions is a bottleneck for efficient Data Quality Management (DQM) process. Ideally, if a reference data source exists as a standard vocabulary, the quality of other data sources can be obtained by comparing with the reference data source. However, for most cases there is no 'golden' reference data source [4]. Thus, DQM often requires active participation of domain experts to make subjective decisions, and the effectiveness of the DQM heavily depends on the expert. Secondly, since there

© Springer Nature Switzerland AG 2019
J. Li et al. (Eds.): BigSDM 2018, LNCS 11473, pp. 301–318, 2019.
https://doi.org/10.1007/978-3-030-28061-1_30

is a strong positive correlation between the data quality and price, conflict of interests exist between the data providers and consumers. Performing DQM on both sides would be a waste, and therefore a trustworthy inter-mediation is often involved to deliver quality reports at some cost. This process fails to protect the privacy of both the providers and consumers, and also makes the entire process insecure. For these reasons, enabling trustful data transactions in the decentralised and trust-less environment will help us achieve the aim of efficiency and privacy.

Some existing studies have already attempted to use blockchain based approaches to solve the efficiency and privacy issues [12,26]. By deploying smart contracts, DQM processes can be codified and carried out in a way that greatly increases reliability, immutability, transparency and security. DQM can be performed without the participation of inter-mediation. Despite the advantages of these approaches providing solutions for managing and improving data quality without centralised validation which greatly reduce the cost of DQM, there are still some challenges that need to be further addressed:

- *Data accessibility.* Data accessibility is an essential data quality factor [21]. Existing blockchain based DQM platforms store hash pointers of the content on-chain, and store the actual content off-chain. A hash pointer typically consists both the off-chain address of the data and the hash of the entire data content, which makes it useful to look up and verify the content not being tampered from the time when it was stored. However, the off-chain storage is often either a centralised database or in a third-party cloud service. If off-chain data are destroyed or lost, the pointers stored on the blockchain will become invalid, and the data will no longer be accessible.
- *DQM effectiveness and efficiency.* IIoT data quantity and complexity are increasing rapidly. The requirements for analytic skills and computational resources also increase accordingly. This is becoming a costly and time-consuming process to design and execute DQM plans.
- *Task-oriented DQM specification.* There are mainly two reasons for task-oriented DQM specification. Firstly, data consumers have different data quality needs. When purchasing data, there are a lot of constraints to consider, such as budget and data formats. Secondly, we would like to enable micropayments in IIoT. This will let consumers to only purchase the data they want. New business models are formed, such as combining the data from multiple data providers to generate a data set with the highest quality using smart contracts. For the above reason, the platform should be able to customise all subjective DQM procedures.
  Note that only subjective DQM procedures are customisable. Objective data quality measures (Quality Indicators) should be immutably stored. In fact, the platform should ensure those objective measures are never tampered.

In this paper, we propose a novel blockchain-based IIoT data DQM platform. The platform is open so that all IIoT data consumers and providers can freely join and leave the platform without affecting the entire network. To address the above challenges, peer-to-peer distributed file system is used to ensure data accessibility and platform capability. Encryption operations and smart contract mechanisms are also employed to support secure data transfer along with cryptocurrency token transfers. Attribute-based data quality validation strategy is designed and implemented to separate the subjective (Quality Attributes) and objective (Quality Indicator) parts of the DQM process. Our platform is task-oriented, with pluggable DQM plans to support customisable quality requirements, which further enables a crowd-sourcing model to utilise the community power to integrate various existing data quality algorithms. To prove the concept of the proposed platform, an experimental decentralised DQM platform is implemented, and tested with PM_{10} and $PM_{2.5}$ values from London Air Quality Network at Greenwich in 2017 [13]. The results have shown that the proposed platform is able to provide effective task-oriented data quality management efficiently in a decentralised way.

2 Related Works

Data quality is a crucial research direction with important practical values, where blockchain can be used to make the data quality management process more effective and efficient. In this section, an overview of the development of data quality researches since the 1990s will be described, followed by a review of the blockchain researches in the past few years. This section finishes with a short review of the peer-to-peer distributed file systems, which is crucial in our platform to tackle the data accessibility problem.

2.1 Data Quality

Data quality deficiencies have a significant impact on data-driven approaches in IIoT, which lead to extensive researches in this area. In this section we give a brief overview of some representative researches on data quality.

The issue about data quality has gained attention since the mid-1990s. An inspiring hierarchical framework for determining the quality characteristics of data has been put forward in 1996 [21]. As shown in Fig. 1, multiple objective quality indicators of data have been defined, such as being accurate, consistent, etc. Quality attributes, which decide the quality of data, are generated from quality indicators. Data quality is a comprehensive and multifaceted indicator requiring many data attributes. Previous studies have stated that in a fair data-market system, the most important ones among these indicators are completeness, accuracy, consistency and timeliness [14].

In 2016, a more completed data quality model was introduced to evaluate data quality using indicators such as accuracy, completeness, redundancy, readability, accessibility, consistency, usefulness, and trustworthiness [7]. In 2017, Yu and Zhang established a bi-level programming model with two kinds of cost functions to analyse the production-decision behaviour of the data platform owner and the purchasing-decision behaviour of data consumers [24].

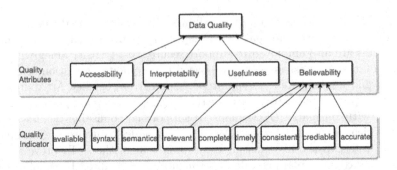

Fig. 1. Data quality hierarchy

2.2 Blockchain

The blockchain is an implementation of the distributed ledger technology (DLT). It is tamper-proof with no central governance. Blockchain revolution can be divided into 3 stages [20]. Blockchain 1.0 is the cryptocurrency for economic transactions. Bitcoin is regarded as the most famous cryptocurrency at the moment [18]. Whereas, Blockchain 2.0 utilises smart contracts technology. Smart contracts can be seen as a programme that can act as a trusted third party authority during transactions. Successful projects include Ethereum [22] for open blockchain and Hyperledger Fabric [10] for closed blockchain. Blockchain 3.0 is blockchain applications beyond currency, finance and markets. The combination of IIoT and blockchain is one of the key developments in the Blockchain 3.0 stage, which has a number of successful use cases [11]. Existing framework has employed tamper-proof log of events and management of access control to data, micropayment based purchase of assets by devices or human beings, purchase of sensors data in IIoT and many more [11]. However, to the best of authors knowledge, none of the existing researches focused on quality-oriented data collaboration on the blockchain, which is actually essential for data consumers.

There are many different consensus mechanisms for public blockchains, e.g. PoW (Proof of Work), PoS (Proof of Stake), PoST (Proof of Space-Time), PoET (Proof of Elapsed Time), and etc. The advantages and disadvantages of these consensuses have been previously discussed [6,25]. PoW provides the highest level of security. It is more robust against DDoS attacks comparing to other consensuses.

2.3 P2P Distributed File System

Data storage is a challenging problem in IIoT. IIoT data is usually large in size, sampled frequently in a distributed way.

Peer-to-peer (P2P) networks are designed to enable direct exchange of computer resources. P2P distributed file system (DFS) allows computers to have access to the same storage using a network protocol instead of block level access, in order to index, search, and obtain digital content. There have been many studies on global distributed file system since 2000s [3]. Systems were built from both the academia and industry, such as AFS [16], Napster, KaZaA, and BitTorrent.

InterPlanetary File System (IPFS) is a P2P DFS system, which could be seen as a single BitTorrent swarm, exchanging objects within one Git repository. There are several unique features of IPFS. Firstly, it uses the Merkle DAG data model, which can split large data files (>256 kB) into a list of links to file chunks that are <256 kB. This makes the system suitable for large IIoT data sets. Secondly, the hash of the root chunk can be used to both ensure the immutability of its content and retrieve the underlying data [9]. Therefore this hash can be used as our hash pointer on chain.

3   Methodology

As discussed in Sect. 1, efficient machine-to-machine communication without human intervention in IIoT is challenging. Firstly data are large in size, with varying complexity and reliability. Secondly sensors are often owned by multiple data providers, with no uptime guarantees. Lastly but not the least, data providers and consumers do not trust each other due to conflict of interests. In addition, in order to create a new data set that meets consumers' needs (budget, data format, and etc.), micropayments should be enabled to purchase IIoT data from (multiple) provider(s).

In the following paragraphs of this section, the platform design will be introduced. Firstly a problem description of DQM will be described. Then the platform overview was provided, followed by using smart contracts to manage transactions.

3.1   DQM Problem Description

Assume that there are m data sources in the system. All the data sources produce similar comparable data, with small variances. Data shards are defined as a small horizontal data chunks that partitions the entire data set produced by a data source.

Given m comparable data shards within a given time period produced by the m data sources correspondingly, denoted as $s_1, ... s_m \in S$. Let us define the set of Quality Indiactor (QI)s relevant to data shards $s_1, ... s_m \in S$ as:

$$QI(s_i) = qi_1(s_i), qi_2(s_i), \cdots \tag{1}$$

where $i \in [1, m]$, and $qi()$ are different quality indicator functions. QIs, such as *Mean, Variance, Maximum value*, etc., provide objective information about the characteristics of data. Ideally, QIs should never change once calculated.

As discussed in Sect. 1, customisable quality management must be supported. We further define Quality Attribute (QA)s as:

$$QA(s_i) = qa_1(QI(s_i)), qa_2(QI(s_i)), \cdots \tag{2}$$

where $i \in [1, m]$, and $qa()$ are different quality attribute functions. Examples of subjective data attributes include *Confidence Coefficient, Usefulness, Interpretability* and so on.

Lastly, the quality control function can be defined as:

$$f_{qc}(QA(s_1), \cdots, QA(s_m)) \in \{\mathbb{B}^m\} \tag{3}$$

The quality control function is used to accept or reject each data shard. Note that function $f_{qc}()$ is written based on data consumers' needs. For example it can output *true* for multiple shards when needed. This function should be implemented differently for each data consumer. It is not necessary to have the same quality control functions on different sets of comparable shards belonging to the same data set.

After all data shards were purchased, data consumers can post-process all data locally to produce a new data set.

3.2   Platform Overview

The architecture of the designed platform and overall execution flow is shown in Fig. 2.

1. Sharding is performed to divide data into small portions. This step is necessary due to the difficulty with scaling the DQM when there are big chunks of data in the platform. Furthermore, some data consumers might only wish to purchase a small subset of the data available. Sharding would enable micro payments in the system.
2. We use data schema as a framework to describe data, which is responsible to translate data into a consistent format. Data schema consists of two parts. Firstly the QIs are the objective quality measures. Secondly the metadata describes the data content using tags for the purposes of identification. Examples of meta-data include sensor location, sensor id, data unit and etc.
3. Data schema is stored on P2P distributed file system by the data provider.
4. The address of the data schema is stored on-chain. Users of the network should keep a local copy of all schemas on-chain.
5. The data consumer can find the hash address of data of its interest, by browsing the blockchain.
6. The data consumer then downloads all data schema of its interest from P2P DFS.
7. By using the data schemas downloaeded from the previous step, the data consumer calculates QA and decides whether or not to purchase the corresponding data shard based on the result. The data consumer then initiates a smart contract to purchase the data. Note that the consumer needs to have an asymmetric key pair and provide the public key to the smart contract in this step.
8. The data provider receives the request. It does not have to accept the trade at this stage. If there is no response from the data provider for a limited amount of time, then the trade is cancelled.

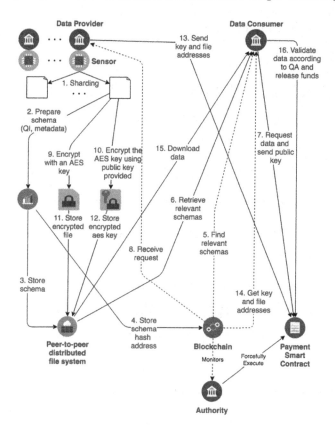

Fig. 2. Platform architecture. Solid lines can execute actively. Dashed lines require monitoring/notification. Authority can be a decentralised voting system.

9. Data provider creates an AES key, and encrypt the data content. Note that the data is not encrypted with the public key for performance reasons.
10. Data provider encrypts the AES key in the above step using the public key provided by data consumer at Step 7 and 8. This ensures that only the data consumer can decrypt the AES key, thus it is safe to transfer both the data and key using publicly accessible DFS.
11. Data provider stores encrypted data in Step 9 onto the DFS.
12. Data provider stores encrypted AES key in step 10 onto the DFS.
13. Data provider then sends the two DFS addresses in step 11 and 12 to the smart contract.
14. By monitoring the blockchain, data consumer finds that data consumer has provided the corresponding data addresses.
15. Data consumer retrieves relevant data from DFS, and decrypts the data.
16. Data consumer validates the data according to QI and QA. It releases the fund if everything is correct.

3.3   Transaction Management

A successful execution flow was demonstrated in Sect. 3.2. However, there are cases when a data transaction would fail, such as when the data provider or data consumer has disconnected from the network, or the data provider has faked the quality indicators.

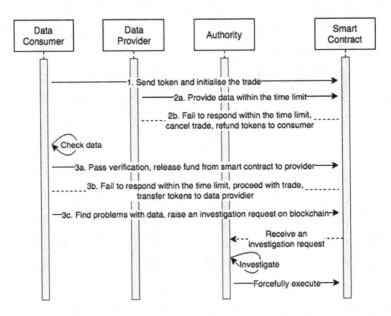

Fig. 3. Transaction sequence diagram. Dashed lines did not receive a response within the time limit

Smart contract is a program that runs on the blockchain to ensure the reliability of the transaction, and the contracts need to be verified and signed by multiple nodes in the blockchain network to ensure its correctness. Smart contracts will be used to regulate all transactions between data consumers and providers.

The transaction sequence defined in the smart contract is shown in Fig. 3. There are three major steps in the sequence:

1. Data consumer initialises the trade, and locks the tokens for the trade in the smart contract.
2. Data provider provides the data to the smart contract.
3. Data provider verifies the data and closes the trade. All tokens will be transferred to the data provider.

As shown in the sequence diagram, we have introduced a time limit in Step 2 and 3. This does not only deal with disconnections, but also prevents malicious behaviours to intentionally not respond and lock funds in the smart contracts forever.

If there are any issues with the transaction, both the data provider and consumer have the privilege to raise an investigation request. Once an investigation request is raised, all actions will be disabled for all participants, apart from the authority. Typically authority is a centralised government controlled organisation, but it could be replaced with a blockchain voting system if needed. After investigation, authority can forcefully transfer the funds in the smart contract to either the data provider or consumer. Note that authority can only forcefully transfer funds if an investigation request has been raised.

4 Platform Implementation

In this section, the details of the platform implementation will be discussed. IPFS was used to store large IIoT data efficiently off-chain. Ethereum/Solidity was used to implement DQM transaction management to target the trust issue between providers and consumers. Minimum data were stored on-chain, in order to reduce the gas cost for IIoT micropayments. Data content were symmetrically encrypted, and the symmetric keys were asymmetrically encrypted, for data privacy purposes.

4.1 Storage

In this work, IPFS was utilised as our P2P DFS for IIoT. All objects stored on the IPFS are accessible via their hash. As discussed in Sect. 2.3, large amount of data can be addressed with IPFS, which suite the need of IIoT. The hash of the merkle tree is the immutable, permanent hash pointers to data, which can be linked into blockchain transactions. This timestamps and secures the content, without having to put the data on the chain itself, which greatly improves efficiency.

The content on IPFS can be permanently available if it is 'pinned' in a cluster. This means that there is a risk to lose the data if none of the nodes in the network 'pin' the data anymore. However, participants of the network only need to permanently 'pin' the data they are interested to ensure the data are truly permanently available.

Objects are looked up using S/Kademlia DHT, which can perform queries on average contact $\left\lceil log_2(n) \right\rceil$ nodes [8].

4.2 Blockchain and Smart Contract

In this system, we use blockchain as a communication bridge between data providers and data consumers. Theoretically, we can use any implementation of the Distributed ledger technology (DLT) systems, but there are a few factors to consider. Firstly we would like to ensure that data providers and consumers will freely join and exit the platform, and equally participate in this data transaction network. Secondly as this is a public system with token trades, we would

Table 1. Gas cost for smart contract functions defined in dataTransfer.sol

Function	Gas
authority()	728
cancel()	877
dataConsumer()	816
dataProvider()	706
executeForcefulRefund()	commonly 21000
executeForcefulSettle()	commonly 21000
expirationTime()	724
initTransaction(address,bytes32,bytes32)	81112
ipfsConsumerPubKeyAddress()	500
ipfsDataAddress()	544
ipfsEncryptedAesKeyAddress()	632
ipfsSchemaAddress()	654
provideData(bytes32,bytes32)	61061
requestForcefulRefund()	20835
requestForcefulSettle()	21100
requestedForcefulRefund()	618
requestedForcefulSettle()	690
verifiedData()	commonly 21000

like to use PoW as our consensus to benefit from its highest level of security. For the above reasons, we have decided to use Ethereum.

We have forked Go Ethereum (Geth 1.8.13) and deployed a private testnet. In this version of Ethereum, it uses PoW as the consensus mechanism. Smart contracts in this study were written using Solidity Version $> 0.4.0$. On-chain data only includes three Ethereum addresses (authority, provider and consumer), four 32 bytes IPFS data addresses, a unsigned integer expiration time property, and two Boolean flags. Variable definitions and function signatures are shown in Appendix A.

Gas cost for the smart contract is shown in Table 1. Note that functions calls the 'transfer()' function could not determine the maximum cost, but 21000 is commonly seen for payable transfers that do not have additional data. The gas prices comparably low as we store the minimum amout of data on-chain.

4.3   Encryption

There are two encryption processes in this system. We have used AES-256 for efficient symmetric encryption, and encryped the AES-256 key using assymmetric encryption.

4.3.1   AES Symmetric Encryption

In this study, Advanced Encryption Standard (AES) with a key size of 256 bits was used as the symmetric encryption method. When encoding a single shard, we use block ciphers, where each block has the size of 128 bits. Therefore each shard is divided into m parts, where $m = \left\lceil \frac{\text{Number of Bits}}{128} \right\rceil$. The last undersized block is filled with PKCS5Padding.

4.3.2   ECC Asymmetric Encryption

To authorize payments, Ethereum uses the Elliptic Curve Digital Signature Algorithm (ECDSA). Elliptic-curve cryptography (ECC) secp256k1 [1] and the cryptographic hash function SHA256 [2] are the two ingredients for this algorithm. Ethereum's private key is randomly generated by the user. Then the public key can be calculated from the private key using the secp256k1 elliptic curve algorithm.

In this work we use Ethereum private and public keys as our asymmetric key pair. Note that any asymmetric key pairs could be used in our system. It does not matter how many bytes the encrypted key is, as we will be storing the actual content (encrypted AES key) in IPFS, and only the corresponding IPFS address on chain.

5   Discussion of Distributed Ledger Technologies

Various new consensus mechanisms were proposed, such as parallel blockchain extension and DAG in IOTA. The result blockchain is a trade-off between scalability and the level of decentralisation. To build a practical data transaction platform, we need to make sure its performance to meet the need of transactions, especially considering the heavy transaction tasks.

Therefore, to improve performance of our platform, we will analyze and discuss other distributed ledger technologies: parallel blockchain extension and DAG in IOTA. And the following three aspects will be analyzed as the scalability of blockchain has become the key issue when the network is used in real world application.

1. Consensus
2. Scalability: Transaction throughput (the maximum rate of processing transactions) and latency (time to confirm that a transaction has been included)
3. Support by smart contract

5.1   Parallel Blockchain Extension

The framework: parallel blockchain extension, proposed by Boyen, Carr, and Haines, is to present a solution to the two foremost challenges facing "blockchain"-based cryptocurrencies: (1) "mining pool" oligopolies and (2)

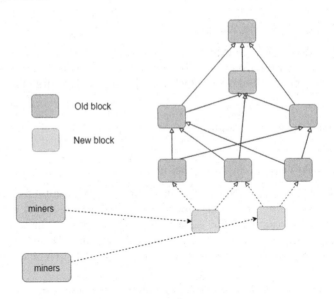

Fig. 4. Process of parallel blockchain extension

incompressibility of delays affecting validation [23]. This two challenges also limit our performance of platform. So this framework will help us to improve our platform.

5.1.1   Consensus

Traditional blockchain has a linear process of extending the blockchain: miners try to solve the puzzle, and the one that finds a solution appends the next block [5]. However, in stead of traditional structure as "chain", a new structure as "figure" is used for parallel blockchain extension. As shown in Fig. 4, with this structure, multiple miners can solve puzzles at the same time, and multiple new blocks can be appended to blockchain at the same time unlike traditional blockchain.

Parallel blockchain extension also uses proof-of-work (PoW). Though speed of generating blocks gets increasing, PoW still can prove the work and validation of a transaction because generation of blockchain still need lots of work with a specific difficulty. Once a block is appended to blockchain, it must validate two previous transactions which are his parents. Thus this new transaction will be validated by later blocks directly or indirectly. Every transaction must post a fee to offset the distributed cost of conveying and verifying the transaction [23]. And if a block validates one previous transactions successfully, it can get some reward for its validation. In addition, indirect validation will get less reword which motivates other users to create new transactions to validate previous transactions to speed up the convergence of the system.

Due to parallel blockchain extension, double-spend problem may occur. However, this problem has been taken into consideration. If this situation occurs, following validation by later blocks will detect the problem and the block with most amount of work attached to it will be accepted finally, which has been proved [23].

5.1.2   Scalability

Scalability is consisted of throughput of transactions and latency, which decides user experience in a high degree.

- *Transaction throughput.* As shown in Fig. 4, applying parallel blockchain extension can improve the performance of transaction throughput because of the concurrency. Compared with traditional blockchain, parallel blockchain extension turn the serial process to parallel process, which can increase tps (transactions per second) to N times, where N is quantity of miners. Therefore, blockchian with parallel extension perform well on throughout of transactions.
- *Latency.* Compared with traditional blockchain, parallel blockchain extension still performs well while latency of a transactions is depended on speed of generating blocks. As a result, the performance of latency can meet the need of user experience.

Therefore, both in transaction throughput and latency of parallel blockchain extension performs much better than traditional blockchain. Thus, applying scalability of parallel blockchain extension to our platform will increase the performance.

5.1.3   Support by Smart Contract

Smart contract plays an important role in our platform, which ensures equitable and reliable environments of transactions. Though smart contract is widespread used on blockchain, whether smart contract can be applied to parallel blockchain still can not be determined because of the totally different structure between tradition blockchain and parallel blockchain extension. Order of transactions has a determined influence to the result of transaction. Traditional blockchain has a chain, which guarantee the order of transactions while parallel blockchain extension only has a figure, directed acyclic graph exactly. Although timestamp of every block can guarantee the orders of transactions, the structure of "figure" may still cause some trouble while executing smart contract. Therefore, whether smart contract can be applied to parallel blockchain still can not be determined.

5.2   DAG in IOTA

IOTA is a new transaction settlement designed specifically for the Internet of Things (IoT). It is a distributed ledger that utilizes "Tangle" as its core. Tangle is a new data structure based on a Directed Acyclic Graph (DAG) to link transactions rather than grouping them into blocks like blockchain, which is illustrated

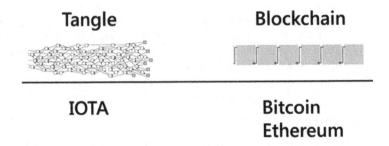

Fig. 5. IOTA's Tangle vs. Bitcoin's/ Ethereum's Blockchain

in Fig. 5. As such it has no Blocks, no Chain and also no Miners. It overcomes the inefficiencies in existing blockchain designs and creates a new approach to the decentralized Peer to Peer system consensus.

5.2.1   Consensus

Each participant that wants to make a transaction in IOTA has to actively participate in the consensus of the DAG network by approving two past transactions [17]. This is intended to be especially lightweight, as consensus does not require several peers intercommunicating or exhausting computational effort validating additions to the DAG; instead, two transactions can be validated by single peers committing a transaction themselves [19].

A confirmed transaction is accepted into the public consensus, and is very unlikely to be removed from it. To know when a transaction can be safely considered to be confirmed in the tangle, IOTA put forward two approaches to establishing consensus - the currently implemented coordinator, and the distributed approach suggested in the white paper [17]. The coordinator is an entity controlled by the IOTA Foundation. A milestone, zero-valued transaction is issued every two minutes by the coordinator and all transactions approved by it are confirmed, and the others are not. This acts as a protective mechanism because the existence of legit milestones lowers the possibility of participant connecting to malicious node.

IOTA also uses Proof of Work (Pow) as consensus protocol for spam protection but it is a short computational operation unlike the expensive PoW employed in miner-based Bitcoin. And IOTA has no miners and no transaction fees. Low PoW also reflects that IOTA is specialized for low-power cpus that are primarily used to run IoT hardware. Minimum Weight Magnitude (MWM) referring to the number of trailing zeros (in trits) in transaction hash determines the difficulty of the PoW.

5.2.2   Scalability

Take the following two metrics directly related to blockchain scalability for comparison.

- *Transaction throughput.* VISA can handle on average around 1,700 transactions per second (tps). Today the Bitcoin network is restricted to a sustained rate of 7 tps due to the bitcoin protocol restricting block sizes to 1 MB and inter-block interval to 10 min. Ethereum handles 20 tps. Iota can already handle between 500–800 tps thanks to parallelised validation of transactions.
- *Latency.* A test by Bitcoin Unlimited has shown that Bitcoin's block latency should put an end to about 500 tps. The blockchain rely on the linear transaction confirmations in a chain, but IOTA uses a whole DAG network, where all transactions are linked together. DAG eliminates blocks - and thus the latency that arises as the blocks are distributed across the network. But owing to Iota being among the most recent of emerging implementations, its latency has yet to be tested at scale.

IOTA also creates snapshots through the milestones, whereby the nodes can throw away all transactions that took place before the snapshots. This saves hard disk space and reduce the memory burden of nodes, making it easier to start a new node.

Due to the high scalability of DAG in IOTA discussed above, applying IOTA to our platform will increase our performance.

5.2.3 Support by Smart Contract

Ethereum can support smart contract because miners help to realize consensus on the order of transactions. But in DAG, every node has a different view of when transactions happened because DAG is not a linear structure like a chain. In a smart contract scenario, changing the order of the two transactions may result in different results, hence transaction sequence must be taken into account. At present, many DAG structures projects like Tangle is not smart-contract friendly. But to implement a virtual machine that can execute a smart contract is on the roadmap of IOTA. The idea is to do this with timestamp and additional transaction collation.

IOTA is extensible towards a range of IoT applications with high scalability, but our data management platform is highly relied on the built-in smart contract mechanism. We can use IOTA as our distributed ledger only when official support for IOTA smart contracts gets announced.

6   Conclusion

The fast development of IIoT facilitates the explosive growth of sensory data, which is highly demanded by data consumers to improve efficiency and deliver advanced services in a wide range of application domains. Designing a platform that can provide trustful data transactions in the decentralized and trustless environment become a crucial task. Although traditional blockchain frameworks can ensure security and privacy, they are not able to directly handle the huge amount of IIoT data due to efficiency reason. Hence, we propose a new platform using P2P DFS to store the data, and then transit the address of encrypted data through blockchain using smart contract. At the same time, symmetric keys were asymmetrically encrypted using consumer's Ethereum public key, and transmitted as well. Such a mechanism can provide efficient data accessibility as well as security. In addition, the DQM strategy is designed to provide a customised data selection for specified data consumers needs. We also discussed two consensus mechanisms to support data transaction platform: parallel blockchain extension and DAG in IOTA. The analysis of consensus, scalability, and smart contract also have shown the appropriateness of the proposed platform.

Acknowledgement. This work is supported by Cybervein-ZJU Joint Lab, Fundamental Research Funds for the Central Universities, Artificial Intelligence Research Foundation of Baidu Inc, Program of ZJU Tongdun Joint Research Lab.

A   Smart Contracts

In this section we provide the function signatures and main variables of our smart contract.

```solidity
1   pragma solidity ^0.4.25;
2   contract DataTransfer {
3       address public authority = < Address 0x... > ;
4       uint duration = < Duration here > ;
5       address public dataConsumer = msg.sender;
6       address public dataProvider;
7       bytes32 public ipfsSchemaAddress;
8       bytes32 public ipfsConsumerPubKeyAddress;
9       bytes32 public ipfsEncryptedAesKeyAddress;
10      bytes32 public ipfsDataAddress;
11      uint public expirationTime;
12      bool public requestedForcefulRefund = false;
13      bool public requestedForcefulSettle = false;
14
15      modifier onlyBy(address _account) {}
16      modifier onlyAfter(uint _time) {}
17      modifier onlyBefore(uint _time) {}
18      modifier etherProvided() {}
19      modifier dataProvidedByProvider(bool _isProvided) {}
20
21      function initTransaction(address _dataProvider, bytes32 _ipfsSchemaAddress, bytes32
        ↪ _ipfsConsumerPubKeyAddress) public payable onlyBy(dataConsumer) etherProvided {}
22      function provideData(bytes32 _ipfsEncryptedAesKeyAddress, bytes32 _ipfsDataAddress) public
        ↪ onlyBy(dataProvider) onlyBefore(expirationTime) {}
23      function cancel() public onlyBy(dataConsumer) dataProvidedByProvider(false)
        ↪ onlyAfter(expirationTime) {}
24      function verifiedData() public onlyBy(dataConsumer) dataProvidedByProvider(true) {}
25      function requestForcefulRefund() public onlyBy(dataConsumer) {}
26      function requestForcefulSettle() public onlyBy(dataProvider) dataProvidedByProvider(true) {}
27      function executeForcefulRefund() public onlyBy(authority) {}
28      function executeForcefulSettle() public onlyBy(authority) {}
29  }
```

Listing 1: Smart contract defined in dataTransfer.sol. Variable 'authority' is the Ethereum address of the Authority. Variable 'duration' is the time duration permitted to perform the trade and call the corresponding functions. Both 'authority' and 'duration' need to be pre-defiend before compilation

References

1. Recommended elliptic curve domain parameters, version 2.0. Certicom Research
2. FIPS 180–4. Secure hash standards, federal information processing standards publication 180–4. National Institute of Standards and Technology (2013)

3. Androutsellis-Theotokis, S., Spinellis, D.: A survey of peer-to-peer content distribution technologies. ACM Comput. Surv. (CSUR) **36**(4), 335–371 (2004)
4. Ballou, D.P., Tayi, G.K.: Enhancing data quality in data warehouse environments. Commun. ACM **42**(1), 73–78 (1999)
5. Bano, S., Al-Bassam, M., Danezis, G.: The road to scalable blockchain designs (2017)
6. Bano, S., et al.: Consensus in the age of blockchains. arXiv preprint arXiv:1711.03936 (2017)
7. Batini, C., Scannapieco, M.: Data and Information Quality - Dimensions, Principles and Techniques. Data-Centric Systems and Applications. Springer, Cham (2016). https://doi.org/10.1007/978-3-319-24106-7
8. Baumgart, I., Mies, S.: S/kademlia: a practicable approach towards secure key-based routing. In: 2007 International Conference on Parallel and Distributed Systems, pp. 1–8. IEEE (2007)
9. Benet, J.: IPFS-content addressed, versioned, P2P file system. arXiv preprint arXiv:1407.3561 (2014)
10. Cachin, C.: Architecture of the hyperledger blockchain fabric. In: Workshop on Distributed Cryptocurrencies and Consensus Ledgers, vol. 310 (2016)
11. Conoscenti, M., Vetro, A., De Martin, J.C.: Blockchain for the internet of things: a systematic literature review. In: 2016 IEEE/ACS 13th International Conference of Computer Systems and Applications (AICCSA), pp. 1–6. IEEE (2016)
12. Gaetani, E., Aniello, L., Baldoni, R., Lombardi, F., Margheri, A., Sassone, V.: Blockchain-based database to ensure data integrity in cloud computing environments (2017)
13. Environmental Research Group and King's College. Air quality by local authority
14. Heckman, J.R., Boehmer, E.L., Peters, E.H., Davaloo, M., Kurup, N.G.: A pricing model for data markets. In: iConference 2015 Proceedings (2015)
15. Helu, M., Libes, D., Lubell, J., Lyons, K., Morris, K.C.: Enabling smart manufacturing technologies for decision-making support. In: ASME 2016 International Design Engineering Technical Conferences and Computers and Information in Engineering Conference, p. V01BT02A035. American Society of Mechanical Engineers (2016)
16. Howard, J.H., et al.: Scale and performance in a distributed file system. ACM Trans. Comput. Syst. (TOCS) **6**(1), 51–81 (1988)
17. IOTA: Iota documents. https://docs.iota.org/introduction/what-is-iota. Accessed 16 Nov 2018
18. Nakamoto, S.: Bitcoin: a peer-to-peer electronic cash system (2008)
19. Red, V.A.: Practical comparison of distributed ledger technologies for IoT. In: Disruptive Technologies in Sensors and Sensor Systems, vol. 10206, p. 102060G. International Society for Optics and Photonics (2017)
20. Swan, M.: Blockchain: Blueprint for a New Economy. O'Reilly Media Inc., Sebastopol (2015)
21. Wang, R.Y., Reddy, M.P., Kon, H.B.: Toward quality data: an attribute-based approach. Decis. Support Syst. **13**(3–4), 349–372 (1995)
22. Wood, G.: Ethereum: a secure decentralised generalised transaction ledger. Ethereum Proj. Yellow Pap. **151**, 1–32 (2014)
23. Xavier, B., Christopher, C., Thomas, H.: Blockchain-free cryptocurrencies. A rational framework for truly decentralised fast transactions. Inf. Syst., 45–90 (2015)
24. Yu, H., Zhang, M.: Data pricing strategy based on data quality. Comput. Ind. Eng. **112**, 1–10 (2017)

25. Zheng, Z., Xie, S., Dai, H., Chen, X., Wang, H.: An overview of blockchain technology: architecture, consensus, and future trends. In: 2017 IEEE International Congress on Big Data (BigData Congress), pp. 557–564. IEEE (2017)
26. Zyskind, G., Nathan, O., et al.: Decentralizing privacy: using blockchain to protect personal data. In: 2015 IEEE of Security and Privacy Workshops (SPW), pp. 180–184. IEEE (2015)

Performance Analysis and Optimization of Alluxio with OLAP Workloads over Virtual Infrastructure

Xu Chang[1,2,3], Yongbin Liu[4], Zhanpeng Mo[4], and Li Zha[1,2,3(✉)]

[1] Key Laboratory of Network Data Science and Technology,
Chinese Academy of Sciences, Beijing, China
[2] Institute of Computing Technology, Chinese Academy of Sciences,
Beijing, China
char@ict.ac.cn
[3] University of the Chinese Academy of Sciences, Beijing, China
changxu16@mails.ucas.ac.cn
[4] G-Cloud Technology Co. Ltd., Nicosia, Cyprus
{liuyb,mozp}@g-cloud.com.cn

Abstract. With the popularity of cloud computing, decoupled compute-storage architecture has become a trend. While being able to independently scale compute and storage results in large cost savings and more flexibility, this architecture also increases the latency of data access, reducing the performance. To solve this problem, Alluxio was proposed. Alluxio achieves the goal of reducing data access latency by providing a near-compute cache when Alluxio is deployed with compute nodes. Applications and compute frameworks send requests through Alluxio and are automatically served through the cached copy. Alluxio is used in many production environments and research work, but there is no comprehensive analysis of Alluxio's acceleration effects. In this paper, we evaluate and analysis the performance of Alluxio with OLAP workloads in different application scenarios. We also summarize the shortcomings of Alluxio and optimize it. Finally, the improved performance results and conclusion are given.

Keywords: Cloud computing · Alluxio · OLAP · Virtualization

1 Introduction and Background

In recent years, with the rapid development of Internet technology, cloud computing has emerged as a new paradigm for on-demand access to a pool of computing resources, which provides resources and services to users through the network [21]. Compared with traditional approaches, cloud computing allows users to create proprietary computing clusters based on specific applications. In the cloud computing technology, the decoupled computing and storage architecture has been wildly used by the major public cloud vendors (such as AWS [3], Aliyun [1], etc).

© Springer Nature Switzerland AG 2019
J. Li et al. (Eds.): BigSDM 2018, LNCS 11473, pp. 319–330, 2019.
https://doi.org/10.1007/978-3-030-28061-1_31

The core of cloud computing is virtualization technology. It can be described that what virtualization does is to build up a few different logic views from a physical machine, each of which can be used to interact with a user simultaneously [20]. Cloud computing requirements of dynamic cutting and distribution of computing resources are owing to Virtualization. Virtualization involves the construction of an isomorphism that maps a virtual guest system to a real host. This isomorphism maps the guest state to the host state. From the perspective of resource, virtualization creates plural subsets logically from the complete set of machine.

For the past more than 10 years, computing resources and data are often tightly coupled for performance considerations. Taking the Hadoop [5] framework as an example, it usually puts the data in the HDFS of the computing node when it runs. Computing resources can easily access the data locally. However, with the prevalence of virtualization technology, the way of computing and storage coupling is obviously not applicable. From the point of view of data persistence, once the physical server fails to provide the user with storage resources or the user closes the virtual server by mistake, the data of the user would be lost. This is disastrous. From the point of view of data management and sharing, coupled computing and storage architecture is not conducive. With the development of science and technology, the growth of network bandwidth also provides conditions for the decoupling of computing and storage. In this environment, the decoupling architecture of computing and storage is proposed. Comparing with the traditional computing architecture proposed more than a decade ago, the decoupled computing and storage architecture has better elasticity and can scaling quickly according to user needs. In this architecture, only the computing resources are retained in the virtual server, and all the data can be stored in a cheaper way thanks to the object storage [18]. Object storage is a computer data storage architecture that manages data as objects, such as AWS S3, Microsoft Azure Blob Storage [9], GCS [11] and Ceph [22] storage. It can manage data uniformly. It is responsible for data sharing and disaster recovery, and it provides good data security assurance in the meantime. Decoupling computation and storage is not only beneficial to separate management of computation and storage, but also to reduce overall cost.

However, this architecture has a certain performance loss, as this can lead to network delay when data has to be accessed through the network, not locally. In order to make up for this shortcoming, Alluxio [2] was preposed. Alluxio, formerly known as Tachyon, is the world's first memory speed virtual distributed storage system. It unifies data access and bridges computation frameworks and underlying storage systems. Applications only need to connect with Alluxio to access data stored in any underlying storage systems. Additionally, Alluxio's memory-centric architecture enables data access at speeds that is orders of magnitude faster than existing solutions. Alluxio can help enterprises to meet the delay requirements of big data OALP analysis tasks well.

In past studies, researches used Alluxio as cache to improve the overall performance of the system or developed based on Alluxio. However, none of the studies evaluated and analyzed Alluxio's advantages and disadvantages in different application scenarios.

In this paper, we optimize and analyze the performance of Alluxio from the aspects of cache type, cache capacity, cache block size and file format by using Hive [7] on MapReduce [15], SparkSQL [26] and Presto [12] over virtual infrastructure. In the experiments, we find out its advantages and shortcomings in OLAP (Online Analytical Processing) systems, then improve it to apply to more application scenarios.

In the rest of this paper, we will introduce virtualization, Alluxio and related work in the Sect. 2. The experiments are described in the Sect. 3. In the Sect. 4, we show the results of the experiments and analyze the causes of these results. In the Sect. 5, we propose optimization for Alluxio deficiencies. Finally, we discuss the conclusions and future work in the Sect. 6.

2   Related Works

In the big data ecosystem, Alluxio lies between computation frameworks or jobs, such as Apache Spark, Apache MapReduce, Apache HBase [6], Apache Hive, or Apache Flink [4], and various kinds of storage systems, such as Amazon S3, Google Cloud Storage, OpenStack Swift [19], etc. Alluxio brings significant performance improvement to the ecosystem. Beyond performance, Alluxio bridges new workloads with data stored in traditional storage systems.

For user applications and computation frameworks, Alluxio is the storage underneath that usually collocates with the computation frameworks, so that Alluxio can provide fast storage, facilitating data sharing and locality between jobs, regardless of whether they are running on the same computation engine. As a result, Alluxio can serve the data at memory speed when data is local, or the computation cluster network speed when data is in Alluxio. Data is only read once from the under storage system on the first time it's accessed. Therefore, the data access can be significantly accelerated when the access to the under storage is not fast.

For under storage systems, Alluxio bridges the gap between big data applications and traditional storage systems, and expands the set of workloads available to utilize the data. Since Alluxio hides the integration of under storage systems from applications, any under storage can back all the applications and frameworks running on top of Alluxio. Also, when mounting multiple under storage systems simultaneously, Alluxio can serve as a unifying layer for any number of varied data sources.

Previous works often used alluxio as cache to improve system performance. A study [24] use Ceph Storage, Alluxio and Apache Hadoop integrated technology to collect a fully working system. It uses alluxio as memory cache to increase the performance. Another work [23] presented and designed DSA, a scalable distributed sequence alignment system and employed Alluxio as primary storage to speeds up I/O performance and reduces network traffic. Li, Zhaowei, et al. [17] analyzed the methods of In-Memory File System using HDFS Lazy Persist strategy and Alluxio to upgrade system I/O efficiency.

In addition, some researches are implemented based on Alluxio. Yu, Y., et al. [25] designed, analyzed, and developed SP-Cache, a load-balanced, redundancy-free cluster caching scheme for data-parallel clusters. They implemented SP-Cache atop Alluxio. Another study [16] proposed a packing, or bundling, layer close to the application, to transparently transform arbitrary user workloads to a write pattern more ideal for cloud storage. The layer is implemented as a distributed write-only cache, packing coalesces small files to form gigabyte sized blobs for efficient batched transfers to cloud backing stores. The packing optimization, implemented in Alluxio.

Alluxio serves as the cache layer between the compute layer and the underlying storage layer. Different configurations may affect the performance of Alluxio. Alluxio's cache type can be read cache, which temporarily stores copies of hot data in Alluxio, which speeds up the next access to this data. Alluxio can also be used as write buffer. Write temporary data into Alluxio to reduce the latency of writing data to the underlying storage and accessing this data again. However, Alluxio has a bug. When it is used as a write buffer, the bug will cause errors in the production environment. As for the optimization of this bug, we will elaborate on it later. In addition, the size of the cache block also affects the performance of Alluxio. Smaller blocks facilitate random access to data, but also mean more RPC calls to get the location of all file blocks. Because different file formats lead to different data access methods, such as sequential or partial reading, it also affects the Alluxio's performance. However, past works have not been evaluating and analyzing Alluxio's performance comprehensively. Due to the variety of production environment, Alluxio cannot be applied to all scenarios. It is necessary to study the performance and shortcoming of Alluxio under different workload. In this paper, we will discuss the performance of Alluxio under different configurations in different application scenarios.

3   Experiments and Optimization

In order to have a comprehensive evaluation and analysis of the performance of Alluxio, we designed experiments in multiple dimensions. In this section, we mainly introduce experimental environment, experimental workloads and experimental scheme.

3.1   Experimental Environment

In order to satisfy the virtualization environment, we conduct experiments on the cloud platform. All experiments are carried out on G-Cloud [10], which is the cloud computing platform of the Guoyun Technology Inc of China. We select four virtual servers (12 cores & 36G memory per server). 8 cores and 32G memory are used for the computing layer, and the remaining resources are allocated to the OS. On each virtual server, we configured a 100G SSD cloud disk and a 100G local SSD disk. The SSD cloud disk is used as data disk. The local SSD disk is used as the cache layer for Alluxio. We deploy the computing layer with the Alluxio cache layer. There is one master node and three slave nodes in all experiments. Source data is stored in underlying object storage consisting of Ceph. In the experiment, we use the classic LRU algorithm as the replacement algorithm of the cache tier.

3.2 Experimental Workloads

For testing the performance of Alluxio with OLAP workloads, so we selected ten typical queries of TPC-DS [13]. The TPC Benchmark DS (TPC-DS) is a decision support benchmark that models several generally applicable aspects of a decision support system, including queries and data maintenance. The benchmark provides a representative evaluation of performance as a general purpose decision support system. These 10 queries are carefully selected. They cover almost all data tables and query operations, and can simulate the actual production environment. In the experiment, the computing engine we selected were Hive on MapReduce, SparkSQL and Presto, which are widely used in the actual production environment.

3.3 Experimental Schemes

Memory is a scarce resource in cloud platforms, and the previous study [14] have found that using SSD as the cache is more cost-effective. Thus, using local SSD disk to construct the cache layer is more economical. In the experiments, we only use local SSD disk as the cache layer. All experiment schemes are designed from the aspects of cache type, cache capacity, cache block size and file format.

We set up three sets of experiments. In these three experiments, we used Alluxio as the read cache and adopted Hive on MapReduce, SparkSQL and Presto respectively. The size of source data is 50G. In each set of experiment, we adjusted the ratio of cache capacity to the amount of source data and set the block size of Alluxio to 4 MB or 1 GB for testing. The smaller block size is helpful for random reading and writing, while larger one is conducive to continuous reading and writing. We also set the format of files to Text or Parquet [8]. When the file format of source data is Text, the application must read full file to get the target data. When the file format of source data is Parquet, which is a column storage format, the application can only read the required data. These are the two common file formats. We executed each SQL for 4 times to eliminate the caching deviation of the first running and calculate the average running time of the latest 3 running times.

4 Results

We conduct three sets of experiments to figure out the answer to the question: What is the performance of Alluxio as a read cache in multiple workloads?

In the first set of experiments, the computing engine is Hive on MapReduce, and the result is shown in Fig. 1. The horizontal coordinate represents the ratio

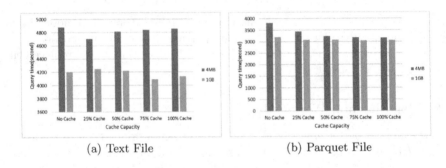

(a) Text File

(b) Parquet File

Fig. 1. Hive on MapReduce

of cache capacity to source data size. The vertical coordinate represents the total time spent for the 10 queries. The following figures and the like. It can be seen in Fig. 1(a) that when the block size is 1G, the query time is about 13% shorter than the block size of 4 MB. This is because for text, it can only read the entire file, which is sequential read operation. At the same time, intermediate data generated between different stages of Hive is written back to the object store, which is a sequential write operation. So when the block size is bigger, the performance is better. It can be seen in Fig. 1(b) that the size of the block has little effect. This is because the block size is 4 MB, which is good for random reading of Parquet. When the block size is 1G, it is beneficial to write back operation of intermediate data. So their query times are similar. On the whole, when the file format is Parquet, it has better performance than the file format is Text. This is because parquet is a column storage, and the computing engine reads only the data needed. While reading text, computing engine can only read the entire file. Compared with Text, Parquet can reduce the amount of data processed and transmitted through the network. We found that read caching does not significantly improve performance, because for Hive, query time is mainly spent on intermediate data access.

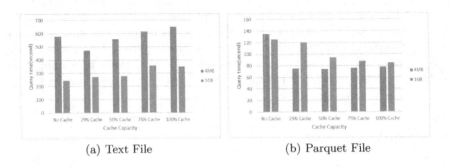

(a) Text File

(b) Parquet File

Fig. 2. SparkSQL

The computing engine of the second set of experiments is SparkSQL, and the result is shown in Fig. 2. It can be seen in Fig. 2(a) that when the block size is 1G, the query time is about 35% shorter than the block size of 4 MB. Moreover, it seems that as the cache hit ratio increases, the performance with 4 MB becomes slower. This is because when the cache hits, for Text files, applications have to read the entire file and the location policy of blocks of Alluxio is local first, which may result in access to large amounts of data from a single node. Performance is restricted by the network bandwidth of the single node. In addition, when the block size is small, a large number of RPC calls due to the addressing of file's blocks will affect the performance to a certain extent. As seen in Fig. 2(b), when the block size is 4 MB, the query time is about 15% shorter than the block size of 1 GB, because it facilitates random reading of Parquet files. Overall, the cache does not speed up text files because Text file can only be read the entire files, which can cause cache contamination and reduce cache efficiency. For Parquet files, cache has a performance improvement of up to 56%, because Parquet files allow only partial files to be read, which improves cache efficiency.

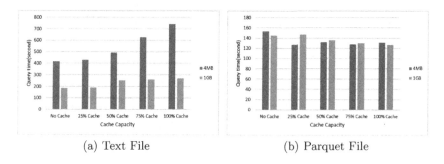

(a) Text File   (b) Parquet File

Fig. 3. Presto

The computing engine of the third set of experiments is Presto, and the result is shown in Fig. 3. It can be seen in Fig. 3(a) that when the block size is 1G, the performance is increased by about 57% compared to the block size of 4 MB. As seen in Fig. 3(b), block size has no significant effect, because Presto does not support parquet files very well. Overall, cache does not speed up performance for text files. For parquet files, cache has the highest performance improvement of 17%.

5   Optimization

Through the analysis of the experimental results, we found that Alluxio as a write buffer is also a good choice. Putting intermediate data in Alluxio avoids writing intermediate data back into object storage, which is a slow process. Moreover, while reading this part of the intermediate data, it does not need to read from the object store, but from the local read. That reduces the network latency in the query process.

5.1   Improvement

While using Alluxio as write buffer, we found a bug. If Alluxio is used as a write
buffer to handle queries with a large amount of data, intermediate data may
loss, resulting in task failure. This is because in Alluxio, cache blocks and buffer
blocks are not treated differently. But in fact, they are different. Cache block
is usually only a copy of the hot data in the underlying object store. If it is
eliminated by Alluxio, it will not cause data loss. The buffer block usually has
only one copy in Alluxio, and there is no copy in the underlying object store.
Once it is eliminated, it will cause data loss. However, Alluxio does not protect
buffer blocks. The non-persist buffer blocks may be eliminated. This process is
shown in the Fig. 4. When Alluxio space is full, the newly received buffer block
is still written to Alluxio and the old buffer block is evicted, although the buffer
block has not been persisted into object storage. This bug causes Alluxio cannot
be used as write buffer.

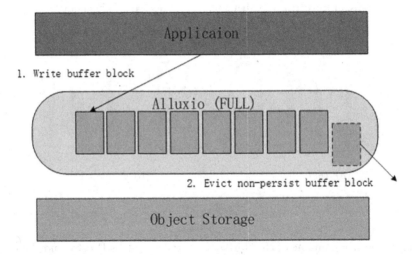

Fig. 4. Before optimization

In this paper, we optimize Alluxio and fix the bug mentioned above. In our
work, when Alluxio is used as write buffer, it protects non-persist buffer blocks
from being evicted out of Alluxio. When Alluxio has no free space, the newly
received buffer blocks are automatically persisted directly into the underlying
object storage to ensure the reliability of the data. As shown in the Fig. 5, Alluxio
checks itself for free space when it receives a new buffer block. If so, the buffer
block is written to Alluxio; if not, the buffer block is persisted directly into
the object store. The advantage is that before optimization, the applications
must ensure that the size of buffer blocks does not exceed Alluxio's capacity
in order not to lose data, but this also results in low resource utilization. After
optimization, the applications can not only make full use of the buffer resources,
but also do not have to worry about the task failure caused by the loss of data.

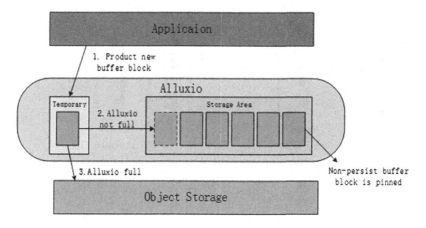

Fig. 5. After optimization

5.2   Evaluation

After optimizing Alluxio, we evaluated the performance of Alluxio write buffer on various workloads through the same three sets of experiments above. Figure 6 is an experimental result of hive on MapReduce. Write buffer has at most 49% performance improvement for the system. This is because the write buffer reduces the latency of intermediate data operations. It can be seen that with the increase of buffer capacity, the effect of acceleration is more obvious. Figures 7 and 8 are the experimental results of SparkSQL and Presto respectively. Write buffer doesn't have a noticeable acceleration effect, because the intermediate data they produce is kept in memory of the computing layer as much as possible, but not written to the object storage.

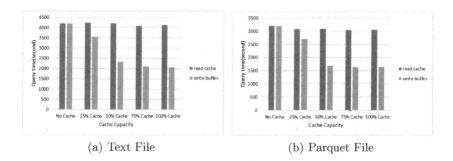

(a) Text File   (b) Parquet File

Fig. 6. Hive on MapReduce

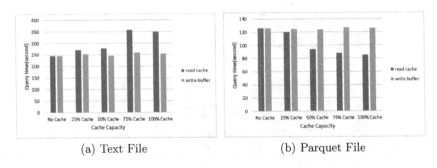

(a) Text File   (b) Parquet File

Fig. 7. SparkSQL

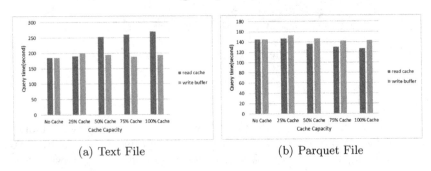

(a) Text File   (b) Parquet File

Fig. 8. Presto

6   Conclusion and Future Work

6.1   Conclusion

From the above experimental results, we can confirm that Alluxio has a good acceleration effect with OLAP workloads, which is shown as follow.

For Hive on MapReduce, if the source data file format is Text, when Alluxio's block size is 1 GB, the query time is about 13% shorter than when the block size is 4 MB. If the source data file format is Parquet, block size does not affect performance. Using Alluxio as a read cache, the performance is not significantly improved. Using Alluxio as write buffer, query time is reduced by 49% at most.

For SparkSQL, if the source data file format is Text, when Alluxio's block size is 1 GB, the query time is about 35% shorter than when the block size is 4 MB. If the source data file format is Parquet, when Alluxio's block size is 4 MB, the query time is about 15% shorter than when the block size is 1 GB. Using Alluxio as a read cache does not significantly improve the performance of Text files. For Parquet files, the maximum performance can be increased by 56%. Using Alluxio as write buffer has no obvious improvement in performance.

For Presto, if the source data file format is Text, when Alluxio's block size is 1 GB, the query time is about 57% shorter than when the block size is 4 MB. If the source data file format is Parquet, block size does not affect performance.

Using Alluxio as a read cache does not significantly improve the performance of Text files. For Parquet files, the maximum performance can be increased by 17%. Using Alluxio as write buffer has no obvious improvement in performance.

In summary, we suggest that Alluxio's block size be increased if the source data file format is Text, and Alluxio's block size be decreased if the source data file format is Parquet. For Hive on MapReduce, Alluxio should be used as write buffer. For SparkSQL and Presto, Alluxio should be used as read caching.

6.2   Future Work

In this paper, the amount of source data is not large enough to simulate the actual production environment, which can be hundreds of Gigabyte, or even a few Terabyte. At the same time, under the condition of small amount of data, the caching of the operating system will also affect the experimental results. In this paper, we only deal with three computing engines, Hive on Mapreduce, SparkSQL and Presto, and two file formats, Text and Parquet. Therefore, in future work, we need to increase the size of the source data and add more types of application scenarios and workloads to better simulate the actual production environment. In addition, for Alluxio, the cache replacement policies and cache block allocation policies are also worth our analysis in the future.

Acknowledgment. This research is partially supported by the National Key Research and Development Program of China (Grant No.2016YFB1000604).

References

1. Aliyun. https://www.aliyun.com/
2. Alluxio. https://www.alluxio.org/
3. Amazon AWS. https://amazonaws-china.com/cn/
4. Apache Flink. https://flink.apache.org/
5. Apache Hadoop. https://hadoop.apache.org/
6. Apache HBase. https://hbase.apache.org/
7. Apache Hive. https://hive.apache.org/
8. Apache Parquet. https://parquet.apache.org/
9. Azure Blob Storage. https://azure.microsoft.com/en-us/services/storage/blobs/
10. G-Cloud. http://www.g-cloud.com.cn/
11. Google Cloud Storage. https://cloud.google.com/storage/
12. Presto. https://prestodb.io/
13. TPC-DS. http://www.tpc.org/tpcds/
14. Chang, X., Zha, L.: The performance analysis of cache architecture based on Alluxio over virtualized infrastructure. In: 2018 IEEE International Parallel and Distributed Processing Symposium Workshops (IPDPSW), pp. 515–519. IEEE (2018)
15. Dean, J., Ghemawat, S.: MapReduce: simplified data processing on large clusters. Commun. ACM **51**(1), 107–113 (2008)
16. Kadekodi, S., Fan, B., Madan, A., Gibson, G.A., Ganger, G.R.: A case for packing and indexing in cloud file systems. In: 10th USENIX Workshop on Hot Topics in Cloud Computing (HotCloud 2018). USENIX (2018)

17. Li, Z., Yan, Y., Mo, J., Wen, Z., Wu, J.: Performance optimization of in-memory file system in distributed storage system. In: 2017 International Conference on Networking, Architecture, and Storage (NAS), pp. 1–2. IEEE (2017)

18. Mesnier, M., Ganger, G.R., Riedel, E.: Object-based storage. IEEE Commun. Mag. **41**(8), 84–90 (2003)

19. Sefraoui, O., Aissaoui, M., Eleuldj, M.: OpenStack: toward an open-source solution for cloud computing. Int. J. Comput. Appl. **55**(3), 38–42 (2012)

20. Vaezi, M., Zhang, Y.: Virtualization and cloud computing. In: Vaezi, M., Zhang, Y. (eds.) Cloud Mobile Networks, pp. 11–31. Springer, Cham (2017). https://doi.org/10.1007/978-3-319-54496-0_2

21. Volkov, S., Sukhoroslov, O.: Simplifying the use of clouds for scientific computing with Everest. Procedia Comput. Sci. **119**, 112–120 (2017)

22. Weil, S.A., Brandt, S.A., Miller, E.L., Long, D.D., Maltzahn, C.: Ceph: a scalable, high-performance distributed file system. In: Proceedings of the 7th Symposium on Operating Systems Design and Implementation, pp. 307–320. USENIX Association (2006)

23. Xu, B., et al.: DSA: scalable distributed sequence alignment system using SIMD instructions. In: Proceedings of the 17th IEEE/ACM International Symposium on Cluster, Cloud and Grid Computing, pp. 758–761. IEEE Press (2017)

24. Yang, C.-T., Chen, C.-J., Chen, T.-Y.: Implementation of Ceph storage with big data for performance comparison. In: Kim, K., Joukov, N. (eds.) ICISA 2017. LNEE, vol. 424, pp. 625–633. Springer, Singapore (2017). https://doi.org/10.1007/978-981-10-4154-9_72

25. Yu, Y., Huang, R., Wang, W., Zhang, J., Letaief, K.B.: SP-cache: load-balanced, redundancy-free cluster caching with selective partition (2018)

26. Zaharia, M., Chowdhury, M., Franklin, M.J., Shenker, S., Stoica, I.: Spark: cluster computing with working sets. HotCloud **10**(10–10), 95 (2010)

Author Index

Printed in the United States
By Bookmasters